Translations

of

Mathematical Monographs

Volume 51

Equations of Mixed Type

by

M. M. Smirnov

American Mathematical Society
Providence, Rhode Island
1978

УРАВНЕНИЯ СМЕШАННОГО ТИПА

М. М. СМИРНОВ

ИЗДАТЕЛЬСТВО "Наука"

Москва 1970

Translated from the Russian by
Israel Program for Scientific Translations

AMS (MOS) subject classifications (1970). Primary 35M05; Secondary 76H05.

ABSTRACT. This book is devoted to the theory of partial differential equations of mixed type. The author introduces the reader to the current state of mathematical problems closely connected with transonic gas dynamics.

Fundamental boundary value problems are considered: the Tricomi problem, the generalized Tricomi problem for Čaplygin's equation, the Frankel′ problem and the modified Tricomi problem.

Library of Congress Cataloging in Publication Data CIP

Smirnov, Modest Mikhaĭlovich.
Equations of mixed type.

(Translations of mathematical monographs ; v. 51)
Translation of Uravneniia smeshannogo tipa.
Bibliography: p.
1. Differential equations, Partial. I. Title.
II. Series.
QA377.S5713 515'.353 78-8260
ISBN 0-8218-4501-2

TABLE OF CONTENTS

Chapter I. Introduction 1
§1. Brief review of results 1
§2. Canonical forms of second-order equations of mixed type 14
Chapter II. The Tricomi problem 17
§1. Statement of the Tricomi problem 17
§2. Extremum principle and uniqueness of the solution of the Tricomi problem 19
§3. Uniqueness theorems for the Tricomi problem for the Čaplygin equation 40
§4. Existence theorem for the solution of the Tricomi problem 47
§5. Further investigation of the Čaplygin equation 83
§6. The Neumann-Tricomi problem 116
Chapter III. The generalized Tricomi problem 131
§1. Statement of the generalized Tricomi problem 131
§2. Uniqueness of the solutionsof the generalized Tricomi problem for for the Čaplygin equation 132
§3. Existence theorem for the generalized Tricomi problem 140
Chapter IV. The Frankel′ problem 183
§1. Statement of the problem 183
§2. Uniqueness of the solution of the Frankel′ problem 184
§3. Existence of the solution of the Frankel′ problem 188
Chapter V. The modified Tricomi problem for an equation of mixed type 197
§1. Statement of the problem 197
§2. Uniqueness theorem 198
§3. Some results from the theory of derivatives and integrals of fractional order for a real function of one variable 200
§4. Investigation of equation (1.1) in the hyperbolic half-plane 201
§5. Investigation of equation (1.1) in the elliptic half-plane 207
§6. Existence theorem for the Tricomi problem 210
Bibliography 222

CHAPTER I

INTRODUCTION

§1. Brief review of results

A partial differential equation is said to be an *equation of mixed type* if it is elliptic in one part of the domain and hyperbolic in the other, both parts being separated by a transition curve (or surface) upon which the equation either degenerates into a parabolic equation or is undefined. The theory of equations of mixed type has a comparatively short history.

The first really deep results were published by Tricomi in the 1920's. He studied the *fundamental boundary-value problem* (*Tricomi problem*) for the equation

$$y\frac{\partial^2 u}{\partial x^2}+\frac{\partial^2 u}{\partial y^2}=0, \tag{1.1}$$

now called the *Tricomi equation*. Let the domain D be bounded by a smooth curve Γ with endpoints A and B on the x-axis, and two characteristics l_1 and l_2 of equation (1.1), emanating from A and B and intersecting at a point C of the lower half-plane. The problem is to find a solution, regular(1) in D, satisfying the boundary conditions $u = \varphi$ on Γ and $u = \psi$ on l_1. Tricomi [69a] proved the existence and uniqueness of the solution, on the assumption that the curve Γ ends in two arbitrarily small arcs AA' and BB' of the normal curve $(x - x_0)^2 + 4y^3/9 = c^2$, lying outside the latter curve elsewhere. Under these assumptions, Tricomi reduced the problem to that of finding a function $\nu(x) = \partial u(x, 0)/\partial y$ satisfying the singular equation

(1)A function $u(x, y)$ is a *regular solution* of equation (1.1) if: 1) it is continuous in $\bar{D}$; 2) its first derivatives are continuous in $\bar{D}$, with the possible exception of the points A and B, where they may have poles of order less than 1; 3) its second partial derivatives are continuous in D, with the possible exception of points on the parabolic curve, where they may not exist; 4) it satisfies equation (1.1) at all points of $D\setminus AB$.

$$\nu(x) + \frac{1}{\pi\sqrt{3}} \int_0^1 \left(\frac{t}{x}\right)^{2/3} \left(\frac{1}{t-x} - \frac{1}{t+x-2xt}\right) \nu(t)\,dt + \int_0^1 K(x, t)\,\nu(t)\,dt = F(x), \tag{1.2}$$

where the kernel $K(x, t)$ is of Fredholm type.

In his doctoral thesis [24a], Gellerstedt solved the Tricomi problem for the equation

$$y^m \frac{\partial^2 u}{\partial x^2} + \frac{\partial^2 u}{\partial y^2} - cu = F(x, y) \tag{1.3}$$

under the same assumptions concerning Γ as adopted by Tricomi. He then [24c] studied boundary-value problems for the equation

$$y^m \frac{\partial^2 u}{\partial x^2} + \frac{\partial^2 u}{\partial y^2} = 0 \tag{1.4}$$

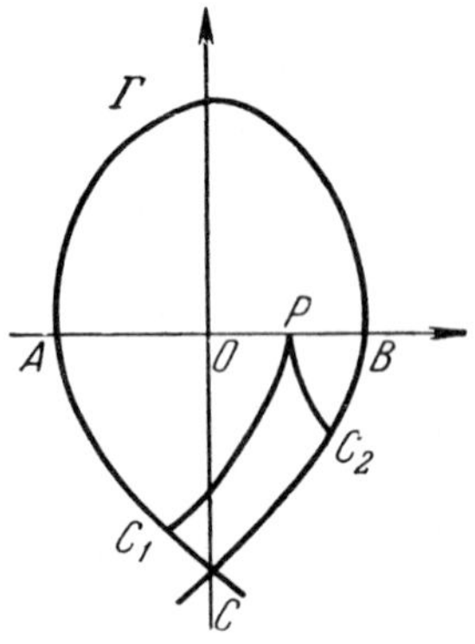

FIGURE 1.

(m is an odd natural number), where the boundary values in the hyperbolic part of the domain D are prescribed either on two characteristic arcs PC_1 and PC_2 which issue from an interior point P of the interval AB, or on characteristic arcs AC_1 and BC_2 (Figure 1). In the elliptic half-plane, the boundary values are prescribed on the curve

$$x^2 + \frac{4}{(m+2)^2} y^{(m+2)} = 1 \quad (y > 0).$$

Following Tricomi, Gellerstedt reduced these problems to a singular integral equation and subsequently, using Carleman's method [11], to a Fredhold equation.

A new stage in the theory of equations of mixed type was initiated by Frankl′ [22a]. He showed that the problem of a supersonic jet flow out of a plane-walled vessel (the velocity inside the vessel is subsonic) can be reduced to the Tricomi problem for the Čaplygin equation

$$K(y)\frac{\partial^2 u}{\partial x^2}+\frac{\partial^2 u}{\partial y^2}=0 \qquad (K(0)=0,\ K'(y)>0). \tag{1.5}$$

In another paper [221], Frankl′ elaborated an important generalization of the Tricomi problem, in which the boundary values are prescribed on the curve Γ and on a noncharacteristic arc AE lying inside the characteristic triangle and intersecting each characteristic of the second family at most once. The point E lies on a characteristic BC of equation (1.5). Henceforth we shall call this the *generalized Tricomi problem.*

Vekua [72] has pointed out the importance of equations of mixed type for the theory of infinitesimal deformations of surfaces and the zero-moment free theory of shells with curvature of variable sign.

The equations of magnetohydrodynamic flows in the region of sonic and Alfvén velocities [42], and the equations of the flow of water in an open channel at a velocity exceeding the propagation velocity of surface waves [22d], are also of mixed type.

In order to simplify the investigation of boundary-value problems for equations of mixed type, Lavrent′ev [47] suggested considering the model equation

$$\frac{\partial^2 u}{\partial x^2}+\operatorname{sgn} y\,\frac{\partial^2 u}{\partial y^2}=0. \tag{1.6}$$

A detailed investigation of the Tricomi problem and its generalizations for equation (1.6), under quite general assumptions concerning the curve bounding D in the upper half-plane, was carried out by Bicadze [10a–d, i]. He also solved the Tricomi problem for equation (1.6) in a multiply-connected domain with $\partial u/\partial n$ prescribed on Γ.

Two applications of boundary-value problems for equation (1.6) to gas dynamics are presented in Frankl′ [22b].

Using the properties of simple automorphic functions, Čibrikova [12b] obtained a solution of the Tricomi problem for equation (1.6) in explicit form, doing this without use of conformal mappings when Γ is half the boundary of one of the fundamental domains of an elementary or Fuchsian group of Möbius transformations.

The following extremum principle holds for the Tricomi problem. *A solution of the Tricomi problem vanishing on the characteristic AC achieves neither a positive maximum nor a negative minimum on an open arc AB of the type-degeneracy curve.* This principle was first stated by Bicadze [10a] for equation (1.6). Somewhat later, the principle was established for the Tricomi equation (1.1) by Germain and Bader [26b]. Babenko [4a] proved the principle for the Tricomi problem when the equation is

$$y\frac{\partial^2 u}{\partial x^2}+\frac{\partial^2 u}{\partial y^2}+a(x, y)\frac{\partial u}{\partial x}+b(x, y)\frac{\partial u}{\partial y}+c(x, y)u=0 \tag{1.7}$$

assuming that $a(x, 0) = b(x, 0) = 0$ and that the transition curve is sufficiently small.

The extremum principle is of great importance, first, because it immediately implies uniqueness of the solution, and, second, because it enables one to apply the alternating method of Schwarz to solution of the Tricomi problem under quite general assumptions on Γ.

Also worthy of mention here is an interesting maximum principle (Agmon, Nirenberg and Protter [3]) for the equation

$$K(y)u_{xx}+u_{yy}+a(x, y)u_x+b(x, y)u_y+c(x, y)u=0, \tag{1.8}$$

where $K(y) > 0$ for $y > 0$, $K(y) < 0$ for $y < 0$, and $K(0) = 0$.

Suppose that a twice continuously differentiable solution $u(x, y)$ of equation (1.8), nondecreasing as a function of y along one of the characteristics, is defined in the characteristic triangle ABC. Suppose moreover that the following conditions hold in the characteristic triangle:

$$\delta(\sqrt{-K})+a+b\sqrt{-K}\leqslant 0,$$

$$\delta\left(\frac{\delta(\sqrt{-K})}{\sqrt{-K}}+\frac{a+b\sqrt{-K}}{\sqrt{-K}}\right)+\frac{1}{2K}\left[\delta(\sqrt{-K})+a+b\sqrt{-K}\right]$$
$$\times\left[\delta(\sqrt{-K})+a-b\sqrt{-K}\right]-2c\leqslant 0, \qquad c\leqslant 0,$$

where

$$\delta\equiv\frac{\partial}{\partial y}+\sqrt{-K}\frac{\partial}{\partial x}.$$

Then, if the maximum of $u(x, y)$ is positive, it is attained on the interval AB of the parabolic curve. The uniqueness theorem for the Tricomi problem for equation (1.8) follows easily from this maximum principle. Pul'kin [64c] proved that the generalized solutions(2) of the Čaplygin equation satisfy this maximum principle. A further generalization of this result for equation (1.7) is due to Volkodavov [73a].

The first proof of a uniqueness theorem for solutions of the Tricomi problem for the Čaplygin equation (1.5) is due to Frankl′ [22a]. He assumed that the function

$$F(y)=2\left(\frac{K}{K'}\right)'+1 \tag{1.9}$$

---

(2) For a definition of generalized solutions, see p. 33.

is nonnegative for $y < 0$. Note that this condition holds everywhere in the case of the Tricomi equation (1.1).

Protter [63b] generalized the uniqueness theorem, allowing $F(y)$ to be negative for $y < 0$. He also proved [63c] a uniqueness theorem on the assumption that $K(y)$ has a continuous third derivative such that $K''' < 0$ for $y < 0$. In his uniqueness proof, Protter used the so-called a, b, c-method, based on an idea of Friedrichs. The essence of the a, b, c-method is as follows. Let $u(x, y)$ be a quasi-regular solution(3) of equation (1.5) defined in a domain D, and satisfying homogeneous boundary-value conditions. Consider the integral

$$\iint_D (au + bu_x + cu_y)(Ku_{xx} + u_{yy})\,dx\,dy,$$

where a, b and c are sufficiently smooth functions of (x, y). By virtue of (1.5), this integral vanishes. The functions a, b and c are chosen in such a way that, after transformation of the integral by Green's formula, one obtains a positive definite expression which vanishes only if $u(x, y) \equiv 0$. The a, b, c-method was also used by Wu and Ding [79] and Wang [76] in order to prove a uniqueness theorem for the Tricomi problem for the Čaplygin equation.

Babenko [4a] carried out a detailed investigation of the Tricomi problem for the Čaplygin equation (1.5). He proved the existence and uniqueness of a generalized solution of the Tricomi problem for this equation, on the assumption that $|dx/ds| < Cy^2(s)$ in a neighborhood of the points A and B on Γ, where C is a constant and $x = x(s)$, $y = y(s)$ are parametric equations of Γ. For the Tricomi equation (1.1), Babenko was able to drop the restrictions on the behavior of Γ near the axis $y = 0$.

Pul'kin [64b] investigated the Tricomi problem for the equation

$$\frac{\partial^2 u}{\partial x^2} + \operatorname{sgn} y \frac{\partial^2 u}{\partial y^2} + a(x, y)\frac{\partial u}{\partial x} + b(x, y)\frac{\partial u}{\partial y} + c(x, y)\,u = 0. \tag{1.10}$$

He proved the existence and uniqueness of a generalized solution to the Tricomi problem, assuming that the tangent to the smooth curve Γ at A and B is parallel to the y-axis, and imposing certain restrictions on the coefficients a, b, c. Vostrova [75b] proved the existence and uniqueness of the solution of the boundary-value problem for equation (1.10), the values of $\partial u/\partial n$ being prescribed on Γ and those of the unknown function on a segment of the characteristic $x - y = 0$. Krikunov [44] considered the problem for equation (1.10) with derivatives in the boundary condition on Γ, assuming the coefficients analytic in the upper half-plane. Smirnov [66e] investigated the boundary-value problem for

(3) For the definition of quasi-regular solutions, see p. 41.

equation (1.4), with the quantity

$$y^m \frac{dy}{ds}\frac{\partial u}{\partial x} - \frac{dx}{ds}\frac{\partial u}{\partial y}$$

prescribed on Γ and the values of the unknown function on the characteristic l_1.

Karatoprakliev [39] grneralized the Tricomi problem for equation (1.1) to the case that the solution $u(x, y)$ and its derivative $\partial u/\partial n$ may have a discontinuity of the first kind on the parabolic curve and satisfy there the following "matching" conditions:

$$u(x, -0) = \alpha(x)\, u(x, +0) + \gamma(x), \qquad -1 \leqslant x \leqslant 1,$$
$$\frac{\partial u(x, -0)}{\partial y} = \beta(x)\frac{\partial u(x, +0)}{\partial y} + \delta(x), \qquad -1 < x < 1.$$

He proved the uniqueness of the solution to this problem, also demonstrating its existence when Γ is the normal curve $x^2 + 4y^{3/9} = 1, y > 0$.

Karol′ [40a] proved existence and uniqueness theorems for the Tricomi problem for the equation

$$\frac{\partial^2 u}{\partial x^2} + \operatorname{sgn} y\, |y|^m \frac{\partial^2 u}{\partial y^2} = 0 \qquad (0 < m < 1) \tag{1.11}$$

on the assumption that Γ is the normal curve

$$y^{2-m} = \left(\frac{2-m}{2}\right)^2 x(1-x) \qquad (0 \leqslant x \leqslant 1).$$

A new approach to the Tricomi problem was presented by Germain and Bader [26c, d] for the case of the Tricomi equation. They made use of certain fundamental solutions $H_k(P, M)$, which depend on a real parameter k. Here $M(x, y)$ is a variable point and $P(x_0, y_0)$ a fixed point. These solutions are symmetric with respect to M and P. The most important property of the Germain-Bader fundamental solutions is that they are defined on the whole plane. If P is in the elliptic half-plane, then $H_k(P, M)$ is regular throughout the plane except at the point B, where it has a logarithmic singularity. If P lies on the parabolic curve $y = 0$, the solution has the form $\gamma r_p^{-1/3}$, where the constant γ has different values in the domains $r_p^2 > 0$ and $r_p^2 < 0$, with $r_p^2 = (x - x_0)^2 + 4y^3/9$. If P is in the hyperbolic half-plane, then $H_k(P, M)$ is analytic in the whole plane excluding the characteristics passing through P and their reflections. The singularity along these curves is the sum of a logarithmic singularity with the same coefficient on both sides and a discontinuity of the first kind. Using solutions with the same singularities as $H_k(P, M)$, but vanishing on the normal curve, Germain and Bader solved the Tricomi problem in explicit form, assuming that Γ coincides with the normal curve; the solution is expressed in terms of the problem data and the values of the unknown function $u(x, y)$ along a characteristic on which no data are prescribed. These values are determined from a certain uniquely solvable singular integral equation. Germain and Bader also constructed the Green's function of the Tricomi problem.

Another method for solution of the Tricomi problem, also based on the use of integral equations, was developed by Agmon [2b]. Using the fundamental solutions $H_k(P, M)$ of equation (1.1), he sought a solution of the Tricomi problem as a sum of potentials of simple and double layer distributions. He derived a system of integral equations for the unknown densities from the boundary conditions and proved its solvability. The existence of a strong solution is thus established, provided the curve Γ is orthogonal to the x-axis at the points $A\ (-1, 0)$ and $B\ (1, 0)$, the boundary values $u|_\Gamma = \varphi$ and $u|_{l_2} = \psi(x)$ are continuous, and $\psi''(x)$ exists and satisfies a Hölder condition on $[0, 1]$. This method was also applied in the case when the conormal derivative is given on Γ. Agmon also proved that the solution of the Tricomi problem for the equation

$$y \frac{\partial^2 u}{\partial x^2} + \frac{\partial^2 u}{\partial y^2} + c\,(x, y)\,u = F\,(x, y)$$

may be reduced to solution of the Riesz-Schauder equation.

Frankl′ [22c] proved existence and uniqueness theorems for the generalized Tricomi problem for equation (1.1), on the assumption that the noncharacteristic curve γ coincides with the characteristic l_1 in some arbitrarily small neighborhood of the point A and is close to l_1, while Γ is a "normal" curve. Bicadze [10e] proved existence and uniqueness theorems for the generalized Tricomi problem for equation (1.6), under certain assumptions on the curves Γ and γ. He also studied the generalized Tricomi problem in the case that both characteristics are replaced by two noncharacteristic curves lying inside the characteristic triangle. Using Gellerstedt's method, Protter [63d] established the existence of a solution to the generalized Tricomi problem for the Čaplygin equation (1.5), retaining Tricomi's assumptions for Γ and assuming that γ coincides with the characteristic in a small neighborhood of A. Babenko [4a] proved the existence of a generalized solution to the generalized Tricomi problem for equation (1.5), assuming that on the noncharacteristic curve γ

$$-(-K\,(y))^{-1/2} < \frac{dy}{dx} < 0,$$

without requiring γ to coincide with the characteristic in a neighborhood of A.

Mihaĭlov [54] proved the existence of a unique generalized solution of the generalized Tricomi problem with trivial boundary conditions, for the equation

$$\begin{aligned} K\,(y)\,u_{xx} + u_{yy} + a\,(x, y)\,u_x \qquad\qquad \\ + b\,(x, y)\,u_y + c\,(x, y)\,u - \lambda u = f\,(x, y), \end{aligned} \tag{1.12}$$

for any function $f(x, y) \in \overset{0}{W}{}_2^{-1}$, where $a(x, y)$ and $b(x, y)$ are continuously differentiable, $c(x, y)$ is continuous in $\bar{D}_\gamma$ and λ is a real parameter, provided $\lambda \geqslant \lambda_0$ for some real λ_0. He also proved that this is a Fredholm problem for any λ.

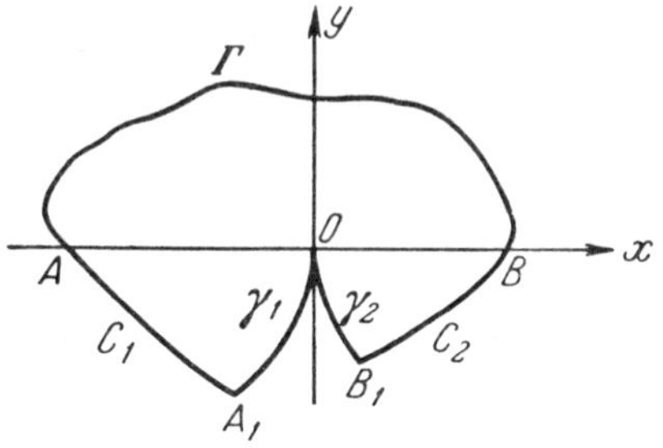

FIGURE 2.

The a, b, c-method has been used successfully to prove a uniqueness theorem for the generalized Tricomi problem (Morawetz [56a]; Dong [19]). In the case of equation (1.5), Morawetz proved a uniqueness theorem for the domain D shown in Figure 2, assuming that the inequalities

$$0 \geqslant \frac{dy}{dx} \geqslant -\frac{1}{\sqrt{-K(y)}} \text{ (on } C_1\text{)}, \quad 0 \leqslant \frac{dy}{dx} \leqslant \frac{1}{\sqrt{-K(y)}} \text{ (on } C_2\text{)} \tag{1.13}$$

hold along the curves C_1 and C_2 and that Γ is star-shaped with respect to the point O; in analytical terms, this means that, on going counterclockwise around Γ, we have $xdy - ydx \geqslant 0$.

In another approach to the proof of uniqueness theorems (Morawetz [56b]), one associates with each solution of (1.5) a function

$$\Psi(x, y) = \int_{(x_0,\, y_0)}^{(x,\, y)} (Ku_x^2 - u_y^2)\, dy - 2u_x u_y\, dx.$$

This line integral does not depend on the integration path. In addition, one assumes that the derivatives of $u(x, y)$ satisfy a condition ensuring that $\Psi(x, y)$ is continuous in the closed domain $\bar{D}$ (Figure 2). In the half-plane $y > 0$, the function $\Psi(x, y)$ satisfies an elliptic equation

$$K\Psi_{xx} + \Psi_{yy} - \alpha\Psi_x - \beta\Psi_y = 0,$$

where

$$\alpha = \frac{K'\Psi_x}{2\sqrt{K\Psi_x^2 + \Psi_y^2}}, \qquad \beta = \frac{K'}{2K}\left[1 + \frac{\Psi_y}{\sqrt{K\Psi_x^2 + \Psi_y^2}}\right].$$

This means that $\Psi(x, y) \not\equiv \text{const}$ satisfies the maximum principle in the elliptic part of D. Moreover, along the characteristics

$$\frac{dy}{dx} = \pm \frac{1}{\sqrt{-K(y)}}$$

we have

$$\frac{d\Psi}{dy} = -\left(\frac{du}{dy}\right)^2 < 0.$$

Consequently, $\Psi(x, y)$ attains its maximum on $\Gamma + C_1 + C_2$. Using the method of auxiliary functions, Morawetz proved a uniqueness theorem for the generalized Tricomi problem for equation (1.5), assuming, besides (1.13), that

$$0 \leqslant \alpha \leqslant 2\pi \quad \text{on} \quad \Gamma, \tag{1.14}$$

where α is the angle between the tangent vector to Γ and the positive x-axis in the sense of increasing arc-length on going counterclockwise around Γ.

Under certain conditions, the method of auxiliary functions makes it possible to prove a uniqueness theorem for the analog of the Neumann problem for equation (1.5). The conditions relate to the curve Γ and may be formulated as follows [56c]. Let

$$\cos\tilde{\alpha} = \frac{dx}{d\tilde{s}}, \qquad \sin\tilde{\alpha} = \frac{d\tilde{y}}{d\tilde{s}},$$

where $\tilde{s}$ is the arc-length along $\tilde{\Gamma}$ on the plane

$$\left(x,\ \tilde{y} = \int_0^y \sqrt{K(y)}\,dy\right).$$

Suppose that $\tilde{\alpha}$ satisfies the conditions:

$$\tilde{\alpha} < \pi \quad \text{at} \quad B, \qquad \tilde{\alpha} > \pi \quad \text{at} \quad A, \qquad -\pi < \tilde{\alpha} < 3\pi \quad \text{on} \quad \tilde{\Gamma}. \tag{1.15}$$

The geometric condition on Γ is essential; Morawetz gave an example showing that conditions (1.15) may be violated at A and B, while the Neumann problem with trivial data has a nontrivial solution.

Frankl′ [22a] considered the problem of sufficiently wide supersonic jet flow impinging on a wedge, when a zone of subsonic velocities is formed ahead of the wedge; he reduced this to the boundary-value problem for the Čaplygin equation (1.5), with $u = 0$ on part of Γ near A and on the characteristic AC, and on the other part of Γ a homogeneous equation

$$P\frac{\partial u}{\partial x} + Q\frac{\partial u}{\partial y} = 0, \tag{1.16}$$

where P and Q are given functions. This problem is homogeneous and therefore determines the solution up to a constant factor. Frankl′ proved a uniqueness theorem for this problem, assuming condition (1.9). Melent′ev [53a] proved the uniqueness of the solution to the Frankl′ problem for equations (1.1) and (1.6), and also for a more general problem, in which (1.16) is replaced by

$$P\frac{\partial u}{\partial x}+Q\frac{\partial u}{\partial y}+Ru=0$$

on the assumption that

$$[P\cos(nx)+Q\cos(ny)]\,R<0.$$

He proved an existence theorem for the Frankl′ problem for equation (1.6) in the domain D, when the curve Γ has a special form. He also studied [53e] the Hilbert-Poincaré problem for the Tricomi equation (1.1).

Frankl′ [22e] formulated a new boundary-value problem for the Čaplygin equation (1.5) (the *Frankl′ problem*). This problem may be stated as follows. *Determine a regular solution of equation* (1.5) *in the domain D bounded by an interval $A'A$ on the y-axis, $-a \leqslant y \leqslant a$, a characteristic $A'C$ of equation* (1.5), *$C = C(a_1, 0)$, $a_1 > 0$, an interval CB on the x-axis, $a_1 \leqslant x \leqslant b$, and a curve Γ in the upper half-plane with endpoints A and B. The solution is required to satisfy the boundary conditions*

$$u\,|_{\Gamma}=\varphi_1,\quad u\,|_{CB}=\varphi_2,\quad \left.\frac{\partial u}{\partial x}\right|_{A'A}=0,$$

$$u(0,\,y)-u(0,\,-y)=f(y),\qquad 0\leqslant y\leqslant a.$$

Bicadze [10e, g] proved the uniqueness of solutions of the Frankl′ problem for equations (1.5) and (1.6), under the additional assumption that $dy/ds \geqslant 0$ along Γ, where $x = x(s)$, $y = y(s)$ are parametric equations of Γ and s is arc-length on Γ, reckoned from the point $B\ (b,\ 0)$. He also proved an existence theorem for this problem in the case of equation (1.6). Both the method of auxiliary functions and the a, b, c-method have been used successfully to prove uniqueness theorems for the Frankl′ problem in the case of the Čaplygin equation (Devingtal′ [16b], Lin [51]). Devingtal′ proved an existence theorem for the Frankl′ problem with $K(y) = \operatorname{sgn} y\,|y|^m$, $m > 1$.

Ovsjannikov [61] investigated generalized solutions of (1.1). He studied functional-analytic properties of certain operators which play an important role in the theory of the Tricomi problem. On this basis, he suggested a scheme for application of the least-squares method to approximate solution of the Tricomi problem.

Functional analysis was first used in this context by Morawetz [56e] in order to prove the existence of weak solutions of boundary-value problems for equations of mixed type. She considered a first-order system equivalent to the Čaplygin equation. Friedrichs had studied more general systems. Morawetz proved the existence of a weak solution of the generalized Tricomi problem, under certain additional assumptions on the boundary of the domain. Friedrichs [23] and Lax and Phillips [48] have shown that Morawetz' weak solution is also a strong solution (in the sense of Friedrichs) of the generalized Tricomi problem. Morawetz' technique for obtaining energy inequalities was applied to other equations and boundary conditions by Frankl′ [22g], Bicadze [10j] and Lin [51]. A functional-analytic approach to existence proofs for weak solutions of the (ordinary or generalized) Tricomi problem was developed by Berezanskiĭ [7a–c]. He established energy inequalities which provided estimates in other norms and for equations more general than Čaplygin's. He also extended these results to domains with unbounded elliptic regions. Sorokina [68] proved the uniqueness of the weak solution of the Tricomi problem and showed that it coincides with the strong solution.

Karol′ [40b, c, e] investigated the equation of mixed type

$$u_{xx} + yu_{yy} + \alpha u_y = 0 \qquad (\alpha = \text{const}) \tag{1.17}$$

in a domain D bounded above by a curve (or several curves) Γ in the half-plane $y > 0$ with endpoints on the x-axis, and bounded below by two (or more) characteristic arcs. Assuming $\alpha < 0$ and imposing certain restrictions on the smoothness of the solution u and its derivatives on the parabolic curve, he proved the existence and uniqueness of the solution with the boundary condition prescribed on the entire boundary of the domain. The boundary conditions on Γ may be of different types. Karol′ generalized this boundary-value problem, replacing the characteristics by a smooth curve lying wholly in the characteristic triangle. He solved the boundary-value problems for $\alpha > 0$ [40d] with boundary conditions prescribed only on Γ. He also studied the Tricomi problem for equation (1.17). Since the derivative $\partial u/\partial y$ of the solution to this problem may, generally speaking, go to infinity on the parabolic curve, even when the boundary conditions are smooth, the following condition is introduced, "matching" the derivatives in the domains D^+ and D^- on the x-axis:

$$\lim_{y\to+0} y^{\alpha}\frac{\partial u}{\partial y} = \lim_{y\to-0} (-y)^{\alpha}\frac{\partial u}{\partial y}.$$

Frankl′ [22g] reduced the problem of flow inside a plane-parallel Laval nozzle of given shape (the direct problem of the theory of the Laval nozzle) to a new boundary-value problem for equation (1.11) with $m = ½$. He showed that it is no longer sufficient to specify $u(x, y)$ on Γ and on one of the characteristics in order to ensure the existence and uniqueness of solutions of equation (1.11) with $0 < m < 1$. Instead of the usual continuity condition on the parabolic curve, i.e. $u_y(x, +0) = u_y(x, -0)$, one requires the Frankl′ discontinuity condition:

$$\frac{\partial u(x, +0)}{\partial y} = -\frac{\partial u(x, -0)}{\partial y}.$$

Using finite differences, Karmanov [39a, b] proved the existence of a solution to the Tricomi problem for the Lavrent′ev-Bicadze equation (1.6). under very weak restrictions on Γ. Assuming the existence of a solution to the Tricomi problem for equation (1.6), Volkov [74], Ladyženskaja [45] and Halilov [29] presented difference schemes converging to the exact solution; for the Tricomi equation (1.1) this was done by Filippov [20], and later by Ogawa [60] for the general Tricomi equation (1.7). Kovalenko [43a, b] sharpened and generalized the estimates obtained in [20] and [60]. Using finite differences, he proved the uniqueness of the generalized solution of the Tricomi problem for the equation

$$K(y)u_{xx} + u_{yy} + a(x, y)u_x + b(x, y)u_y + c(x, y)u = f(x, y),$$

where

$$K(y) = \operatorname{sgn} y |y|^m q(y), \quad m > 0, \quad q(y) > 0,$$

under additional restrictions on Γ in the neighborhoods of A and B and on the coefficients a, b, c and K.

Smirnov [66a–c] investigated the fourth-order equation

$$\frac{\partial^4 u}{\partial x^4} + 2 \operatorname{sgn} y \frac{\partial^4 u}{\partial x^2 \partial y^2} + \frac{\partial^4 u}{\partial y^4} = 0 \tag{1.18}$$

in a domain D bounded by a smooth curve Γ in the upper half-plane with endpoints A (0, 0) and B (1, 0), and by the characteristics AC: $y = -x$ and BC: $y = x - 1$ of equation (1.18). He proved the existence and uniqueness of solutions of (1.18) satisfying the boundary conditions

$$u|_\Gamma = \varphi_1(s), \qquad \left.\frac{\partial u}{\partial n}\right|_\Gamma = \varphi_2(s), \tag{1.19}$$

$$\left.\frac{\partial u}{\partial n}\right|_{y=-x} = \psi_1(x) \qquad \left(0 \leqslant x \leqslant \frac{1}{2}\right),$$

$$\left.\frac{\partial u}{\partial n}\right|_{y=x-1} = \psi_2(x) \qquad \left(\frac{1}{2} \leqslant x \leqslant 1\right). \tag{1.20}$$

He also proved uniqueness theorems for boundary-value problems in the case that, besides (1.19), the unknown function $u(x, y)$ is prescribed on both characteristics, or the case that the unknown function is prescribed on one of the characteristics and the normal derivative $\partial u/\partial n$ on the other.

Žegalov [80a–c] considered certain boundary-value problems for the equation

$$\Lambda^n u \equiv \left(\frac{\partial^2}{\partial x^2} + \operatorname{sgn} y \frac{\partial^2}{\partial y^2}\right)^n u = 0. \tag{1.21}$$

He proved the uniqueness and existence of solutions of equations (1.21) in a domain D, satisfying the boundary conditions

$$\Lambda^k u = \begin{cases} \varphi_k(s) & \text{on} \quad \Gamma, \\ \psi_k(x) & \text{on} \quad AC \end{cases} \tag{1.22}$$
$$(k = 0,\ 1,\ \ldots,\ n-1).$$

Žegalov also studied boundary-value problems for equation (1.21) when the values of $\Lambda^k u$ on one of the characteristics are replaced by linear conbinations of its values on both characteristics with variable coefficients. The unknown function and its derivative may have a discontinuity of the first kind on the type-degeneracy curve, and satisfy "matching" conditions:

$$\alpha_{0k}(x)\,\Lambda_+^k u^+ + \beta_{0k}(x)\,\Lambda_-^k u^- = \gamma_{0k}(x),$$
$$\alpha_{1k}\frac{\partial \Lambda_+^k u^+}{\partial y} + \beta_{1k}\frac{\partial \Lambda_-^k u^-}{\partial y} = \gamma_{1k}(x)$$
$$(k = 0,\ 1,\ \ldots,\ n-1),$$

where $\alpha_{0k}(x)$ and $\beta_{0k}(x)$ are given functions and α_{1k} and β_{1k} are constants.

Bicadze [10f] considered a boundary-value problem for the equation

$$\sum_{k=1}^{n} \frac{\partial^2 u}{\partial x_k^2} + \operatorname{sgn} z \frac{\partial^2 u}{\partial z^2} = 0$$

in a domain D bounded by the hemisphere σ: $r^2 + z^2 = 1$, $z \geqslant 0$, and by the characteristic cones

$$K_1: \quad z = -1 + r, \qquad -1 \leqslant z \leqslant 0,$$

and

$$K_2: \quad z + r = 0, \quad -\tfrac{1}{2} \leqslant z \leqslant 0, \quad r = \sqrt{x_1^2 + \ldots + x_n^2},$$

where u assumes given values on σ and K_1. If the apex of the cone K_2 is displaced from the origin into the domain $r \leqslant 1$, $z = 0$, the problem may be overdetermined. Uniqueness for this problem follows from the extremum principle. Bicadze proved an existence theorem for the particular case in which the conditions on K_1 depend only on r, while $u = 0$ on σ.

Equations of mixed type were also studied by Abutaliev [1a, b], Bakievič [5a–e], Fleischer [21], Kapilevič [36a–e], Lemešinskaja [50], Nahušev [57], Petruško [62], Šabat [65] and others.

§2. Canonical forms of second-order equations of mixed type

Consider the second-order quasi-linear equation

$$A\frac{\partial^2 u}{\partial x^2}+2B\frac{\partial^2 u}{\partial x\,\partial y}+C\frac{\partial^2 u}{\partial y^2}+F\left(x,\ y,\ u,\ \frac{\partial u}{\partial x},\ \frac{\partial u}{\partial y}\right)=0, \tag{2.1}$$

where A, B and C are real functions of x and y in some domain D. We shall assume that A, B and C do not vanish simultaneously anywhere in D.

Equation (2.1) is of elliptic, hyperbolic or parabolic type in D, according as $B^2-AC<0$, $B^2-AC>0$ or $B^2-AC=0$, respectively. However, these three types do not exhaust the whole class of second-order equations in two independent variables, since the expression B^2-AC need not be of constant sign in D. Thus, if $\Delta=B^2-AC$ changes sign in D, we shall say that equation (2.1) is of *mixed type*. The curve γ defined by the equation $\Delta=B^2-AC=0$ will be called the *parabolic curve* (or *line*) of equation (2.1) or the *type-degeneracy curve* of the equation. We confine the discussion to two possible cases as regards the position of the parabolic curve γ on the boundary of the field of characteristic directions of equation (2.1):

1) The parabolic curve γ is nowhere tangent to the characteristic directions of equation (2.1), i.e.

$$A\,dy^2-2B\,dx\,dy+C\,dx^2\neq 0$$

everywhere on γ.

2) At every point of γ, the tangent coincides with a characteristic direction of equation (2.1). This means that along γ

$$A\,dy^2-2B\,dx\,dy+C\,dx^2=0,$$

i.e. γ is also a characteristic of equation (2.1) (actually, the envelope of the family of characteristics).

In the first case, (2.1) is called an equation of mixed type of the *first kind*; in the second, an equation of mixed type of the *second kind*. The simplest examples are

$$yu_{xx}+u_{yy}=0,\qquad u_{xx}+yu_{yy}=0, \tag{2.2}$$

for which $y=0$ is the parabolic curve; for the second equation, $y=0$ is also a characteristic. For $y>0$ equations (2.2) are elliptic; for $y<0$ they are hyperbolic.

Consider the simplest and, at the same time, very important case of power degeneracy: in a neighborhood σ of γ, the function $\Delta(x, y)$ may be expressed as

$$\Delta(x,\ y)=H^n(x,\ y)\,M(x,\ y), \tag{2.3}$$

where $M(x, y)$ is a function nonvanishing and finite-valued in σ; $H(x, y) = 0$ is the equation of γ, H_x and H_y do not vanish simultaneously, and n is an odd positive integer.(4) Since by assumption the function $\Delta(x, y)$ changes sign across γ, it follows that for all odd n equation (2.1) is elliptic on one side of γ and hyperbolic on the other side, i.e. it is of mixed type. Tricomi [69a] and Cinquini-Cibrario [13a] showed that, under certain restrictions on the smoothness of A, B and C, equation (2.1) may be reduced in a neighborhood of γ by a nonsingular real transformation of the independent variables (we retain the previous notation for the new independent variables) to the canonical form

$$y^n u_{xx} + u_{yy} = \varphi_1(x, y, u, u_x, u_y), \tag{2.4}$$

if (2.1) is an equation of mixed type of the first kind, and to the form

$$u_{xx} + y^n u_{yy} = \varphi_2(x, y, u, u_x, u_y), \tag{2.5}$$

if it is of the second kind. Of special interest are the linear equations

$$\begin{aligned} y^n u_{xx} + u_{yy} + a_1(x, y) u_x & \\ + b_1(x, y) u_y + c_1(x, y) u &= f_1(x, y), \\ u_{xx} + y^n u_{yy} + a_2(x, y) u_x & \\ + b_2(x, y) u_y + c_2(x, y) u &= f_2(x, y). \end{aligned} \tag{2.6}$$

In addition to the above examples, equations (2.1) with a power degeneracy of arbitrary order $\alpha > 0$ are of great interest. For these equations we have

$$\Delta(x, y) = \operatorname{sgn} H(x, y) |H(x, y)|^\alpha M(x, y).$$

The simplest examples are

$$\begin{aligned} \operatorname{sgn} y |y|^\alpha u_{xx} + u_{yy} &= 0, \\ u_{xx} + \operatorname{sgn} y |y|^\alpha u_{yy} &= 0. \end{aligned} \tag{2.7}$$

In particular, if $\alpha = 0$ this gives the even simpler equation

$$u_{xx} + \operatorname{sgn} y u_{yy} = 0. \tag{2.8}$$

We make special mention of the Čaplygin equation, which is of paramount importance in gas dynamics of transonic flows. This equation for the stream function has the form

$$K(\sigma)\psi_{\theta\theta} + \psi_{\sigma\sigma} = 0, \tag{2.9}$$

where $K(\sigma)$ is a monotone increasing function of σ and $K(0) = 0$.

---

(4) When $n = 2m$ ($m = 1, 2, 3, \ldots$) we get degenerate equations of elliptic-parabolic or hyperbolic-parabolic type.

The substitution

$$x = \theta, \qquad y = \left[\frac{3}{2}\int_0^\sigma \sqrt{K(t)}\, dt\right]^{2/3}$$

reduces equation (2.9) to canonical form:

$$y\psi_{xx} + \psi_{yy} + b(y)\psi_y = 0, \tag{2.10}$$

where

$$b(y) = \frac{d^2y}{d\sigma^2} : \left(\frac{dy}{d\sigma}\right)^2.$$

Setting $\psi = u \exp(-\frac{1}{2}\int_0^y b(y)\,dy)$, we bring (2.10) to the form

$$yu_{xx} + u_{yy} + c(y)u = 0, \tag{2.11}$$

where $c(y) = -\frac{1}{4}[b^2(y) + 2b'(y)]$.

CHAPTER II

THE TRICOMI PROBLEM

§1. Statement of the Tricomi problem

Consider the equation

$$K(y)\frac{\partial^2 u}{\partial x^2}+\frac{\partial^2 u}{\partial y^2}+a(x, y)\frac{\partial u}{\partial x}+b(x, y)\frac{\partial u}{\partial y} + c(x, y)u=f(x, y), \tag{1.1}$$

where $K(y)$ is a continuous monotone increasing function of y and $K(0) = 0$. Equation (1.1) is elliptic for $y > 0$ and hyperbolic for $y < 0$; $y = 0$ is its parabolic curve.

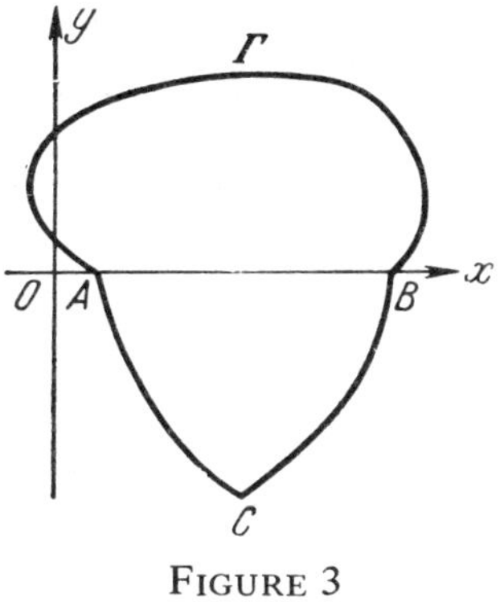

FIGURE 3

Let Γ be a rectifiable Jordan curve lying in the upper half-plane with endpoints at A $(a, 0)$ and B $(b, 0)$. For $y < 0$, (1.1) has two families of real characteristics, defined by the equations

$$\frac{dy}{dx}=-\frac{1}{\sqrt{-K(y)}}, \tag{1.2a}$$

$$\frac{dy}{dx}=\frac{1}{\sqrt{-K(y)}}. \tag{1.2b}$$

Let D denote the domain bounded by Γ and by characteristics AC and BC belonging to families (1.2a) and (1.2b), respectively (Figure 3). The region of D in the half-plane $y > 0$ $(y < 0)$ is denoted by D^+ (D^-).

TRICOMI PROBLEM. *Find a solution in D of equation* (1.1), *continuous in* $\overline{D}$, *which assumes on the curve* Γ *and on one of the characteristics, say AC, prescribed continuous values* $u|_\Gamma = \varphi(s)$ *and* $u|_{AC} = \psi(x)$, *with* $\varphi(l) = \psi(0)$, *where l is the length of* Γ.

On the parabolic curve $y = 0$ of equation (1.1), we have matching conditions

$$\lim_{y\to+0} u(x, y) = \lim_{y\to-0} u(x, y) \qquad (a \leqslant x \leqslant b),$$

$$\lim_{y\to+0} \frac{\partial u}{\partial y} = \lim_{y\to-0} \frac{\partial u}{\partial y} \qquad (a < x < b).$$

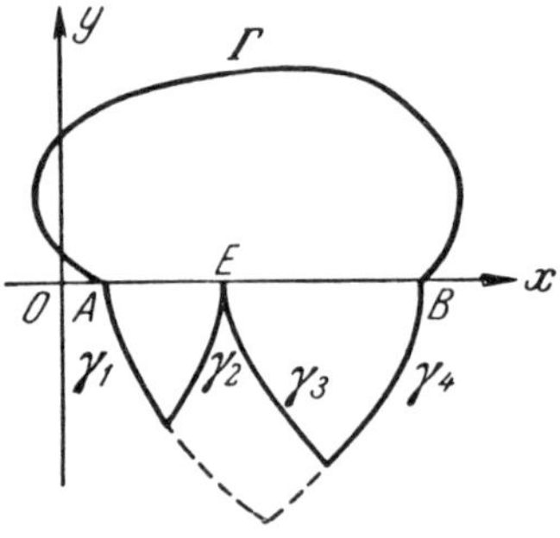

FIGURE 4

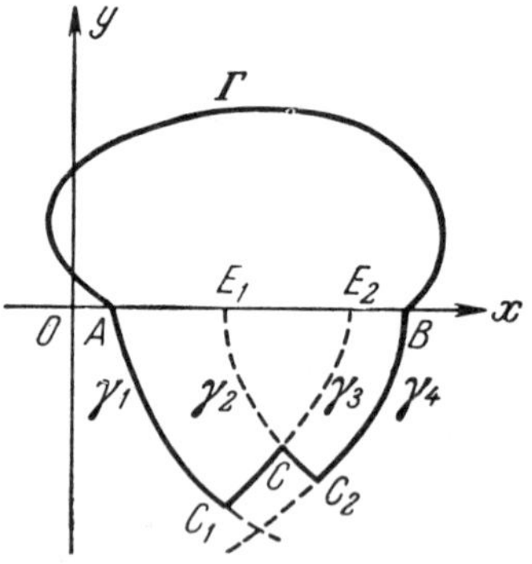

FIGURE 5

The Tricomi problem may be studied in a more complicated domain. Take a point E on the x-axis between A and B. Let γ_1 and γ_3 be the characteristics of family (1.2a) emanating from A and E, respectively, and γ_2 and γ_4 the characteristics of family (1.2a) emanating from E and B, respectively. Denote by D_1 the domain bounded by the curves Γ and $\gamma_1, \gamma_2, \gamma_3, \gamma_4$ (Figure 4).

Problem T_1 is to find a solution of equation (1.1) in D_1, assuming prescribed values either on Γ and on the characteristics γ_1 and γ_3, or on Γ and on γ_2 and γ_4.

The Tricomi problem can be further generalized. For example, let E_1 and E_2 be two points on the x-axis between A and B; let γ_1 and γ_2 be the characteristics of family (1.2a) emanating from A and E_1 respectively, and γ_3 and γ_4 the characteristics of family (1.2b) emanating from E_2 and B, respectively. The characteristics γ_2 and γ_3 intersect at a point C, γ_1 and γ_3 at a point C_1, and γ_2 and γ_4 at a point C_2 (Figure 5). Denote by D_2 the domain bounded by Γ and the arcs AC_1, C_1C, CC_2 and C_2B.

Problem T_2 is to find a solution in D_2 of equation (1.1), assuming prescribed values on Γ and on the characteristic arcs AC_1 and CC_2.

§2. Extremum principle and uniqueness of the solution of the Tricomi problem

1. *Auxiliary propositions.* Consider the equation

$$L(u) \equiv yu_{xx} + u_{yy} + a(x, y)u_x + b(x, y)u_y + c(x, y)u = 0, \quad (2.1)$$

where $a(x, y)$, $b(x, y)$ and $c(x, y)$ are twice continuously differentiable functions, with $c(x, y) \leqslant 0$ for $y > 0$. We first consider the Cauchy problem for equation (1.2) in the half-plane $y < 0$, i.e. the problem of *determining a function* $u(x, y) \in C^{(2)}$ $(y < 0)$ *which satisfies equation* (2.1) *in the half-plane* $y < 0$ *and the initial data*

$$u|_{y=0} = \tau(x), \qquad \left.\frac{\partial u}{\partial y}\right|_{y=0} = \nu(x). \quad (2.2)$$

In characteristic coordinates

$$\xi = x - \frac{2}{3}(-y)^{3/2}, \qquad \eta = x + \frac{2}{3}(-y)^{3/2}$$

equation (2.1) takes the form

$$\frac{\partial^2 u}{\partial \xi\, \partial \eta} + A(\xi, \eta)\frac{\partial u}{\partial \xi} + B(\xi, \eta)\frac{\partial u}{\partial \eta} + C(\xi, \eta)u = 0, \quad (2.3)$$

where

$$A(\xi, \eta) = \frac{1}{6(\eta - \xi)} - \frac{1}{4}\left(\frac{4}{3}\right)^{2/3}\frac{a}{(\eta - \xi)^{2/3}} - \frac{1}{4}\left(\frac{4}{3}\right)^{1/3}\frac{b}{(\eta - \xi)^{1/3}} = \frac{1}{6(\eta - \xi)} + A_1(\xi, \eta),$$

$$B(\xi, \eta) = -\frac{1}{6(\eta - \xi)} - \frac{1}{4}\left(\frac{4}{3}\right)^{2/3}\frac{a}{(\eta - \xi)^{2/3}} + \frac{1}{4}\left(\frac{4}{3}\right)^{1/3}\frac{b}{(\eta - \xi)^{1/3}} = -\frac{1}{6(\eta - \xi)} + B_1(\xi, \eta),$$

and

$$C(\xi, \eta) = -\frac{1}{4}\left(\frac{4}{3}\right)^{2/3}\frac{c}{(\eta - \xi)^{2/3}}.$$

The initial data (2.2) become

$$\lim_{\eta - \xi \to 0} u(\xi, \eta) = \tau(\xi),$$
$$\lim_{\eta - \xi \to 0}\left(\frac{3}{4}\right)^{1/3}(\eta - \xi)^{1/3}\left(\frac{\partial u}{\partial \xi} - \frac{\partial u}{\partial \eta}\right) = \nu(\xi). \quad (2.4)$$

A solution of (2.3) satisfying conditions (2.4) is given (see [4a]) by

$$u(\xi, \eta) = \int_{\xi}^{\eta} g(\xi, \eta; t)\, \tau(t)\, dt - \int_{\xi}^{\eta} h(\xi, \eta; t)\, \nu(t)\, dt. \tag{2.5}$$

Here

$$\begin{aligned} h(\xi, \eta; t) = \frac{1}{2}\left(\frac{4}{3}\right)^{1/3} \frac{\Gamma(2/3)}{\Gamma^2(5/6)} \Big\{ &(t-\xi)^{-1/6}(\eta - t)^{-1/6} \\ &+ \int_{\xi}^{t} (t-\xi')^{-1/6}(\eta - t)^{-1/6} B_1(\xi', \eta)\, \widetilde{B}_1(\xi'; \xi, \eta)\, d\xi' \\ &+ \int_{\eta}^{t} (\eta' - t)^{-1/6}(t-\xi)^{-1/6} A_1(\xi, \eta')\, \widetilde{A}_1(\eta'; \xi, \eta)\, d\eta' \\ &+ \int_{\xi}^{t}\int_{\eta}^{t} (\eta' - t)^{-1/6}(t-\xi')^{-1/6} D[v(\xi', \eta'; \xi, \eta)]\, d\xi'\, d\eta' \Big\}, \end{aligned} \tag{2.6}$$

$$\begin{aligned} g(\xi, \eta; t) = -\, b(t, 0)\, h(\xi, \eta; t) + \frac{\Gamma(1/3)}{\Gamma^2(1/6)} \Big\{ &\frac{(\eta-\xi)^{2/3}}{(t-\xi)^{5/6}(\eta-t)^{5/6}} \\ &+ \int_{\xi}^{t} \frac{(\eta-\xi')^{2/3}}{(t-\xi')^{5/6}(\eta-t)^{5/6}} B_1(\xi', \eta)\, \widetilde{B}_1(\xi'; \xi, \eta)\, d\xi' \\ &+ \int_{\eta}^{t} \frac{(\eta'-\xi)^{2/3}}{(t-\xi)^{5/6}(\eta'-t)^{5/6}} A_1(\xi, \eta')\, \widetilde{A}_1(\eta'; \xi, \eta)\, d\eta' \\ &+ \int_{\xi}^{t}\int_{\eta}^{t} \frac{(\eta'-\xi')^{2/3}}{(t-\xi')^{5/6}(\eta'-t)^{5/6}} D[v(\xi', \eta'; \xi, \eta)]\, d\xi'\, d\eta' \Big\}, \end{aligned} \tag{2.7}$$

where we have set

$$\widetilde{A}_1(\eta'; \xi, \eta) = \left(\frac{\eta'-\xi}{\eta-\xi}\right)^{1/6} e^{\int_{\eta}^{\eta'} A_1(\xi, t)\, dt},$$

$$\widetilde{B}_1(\xi'; \xi, \eta) = \left(\frac{\eta-\xi'}{\eta-\xi}\right)^{1/6} e^{\int_{\xi}^{\xi'} B_1(t, \eta)\, dt},$$

$$D[v(\xi', \eta'; \xi, \eta)] = \frac{\partial(A_1 v)}{\partial \xi'} + \frac{\partial(B_1 v)}{\partial \eta'} - Cv,$$

and $v(\xi', \eta'; \xi, \eta)$ is the Riemann function of equation (2.3). As a function of (ξ', η'), it satisfies the adjoint equation

$$\frac{\partial^2 v}{\partial \xi'\, \partial \eta'} - \frac{\partial(Av)}{\partial \xi'} - \frac{\partial(Bv)}{\partial \eta'} + Cv = 0 \tag{2.8}$$

and the conditions

$$\begin{aligned} v(\xi', \eta'; \xi, \eta)|_{\xi'=\xi} &= \tilde{A}_1(\eta'; \xi, \eta), \\ v(\xi', \eta'; \xi, \eta)|_{\eta'=\eta} &= \tilde{B}_1(\xi'; \xi, \eta). \end{aligned} \tag{2.9}$$

The Riemann function of the equation

$$\frac{\partial^2 u}{\partial \xi\, \partial \eta} - \frac{1}{6(\eta - \xi)}\left(\frac{\partial u}{\partial \eta} - \frac{\partial u}{\partial \xi}\right) = 0$$

is known to be

$$v_0(\xi', \eta'; \xi, \eta) = \frac{(\eta' - \xi')^{1/3}}{(\eta - \xi')^{1/6}(\eta' - \xi)^{1/6}} F\left(\frac{1}{6}, \frac{1}{6}, 1; \sigma\right), \tag{2.10}$$

where $F(1/6, 1/6, 1; \sigma)$ is the hypergeometric function and

$$\sigma = \frac{(\xi' - \xi)(\eta' - \eta)}{(\xi' - \eta)(\eta' - \xi)}.$$

The Riemann function of equation (2.8) may be expressed as a series:

$$v(\xi', \eta'; \xi, \eta) = \sum_{n=0}^{\infty} v_n(\xi', \eta'; \xi, \eta), \tag{2.11}$$

where

$$\begin{aligned} v_1(\xi', \eta'; \xi, \eta) = & \int_{\xi}^{\xi'}\int_{\eta}^{\eta'} v_0(\xi', \eta'; \xi_1, \eta_1) D[v_0(\xi_1, \eta_1; \xi, \eta)]\, d\xi_1\, d\eta_1 \\ & + \int_{\xi}^{\xi'} v_0(\xi', \eta'; t, \eta) B(t, \eta) \tilde{B}(\xi, \eta, t)\, dt \\ & + \int_{\eta}^{\eta'} v_0(\xi', \eta'; \xi, t) A(\xi, t) \tilde{A}(\xi, \eta, t)\, dt, \end{aligned}$$

$$v_{n+1}(\xi', \eta'; \xi, \eta) = \int_{\xi}^{\xi'}\int_{\eta}^{\eta'} v_0(\xi', \eta'; \xi_1, \eta_1) D[v_n(\xi_1, \eta_1; \xi, \eta)]\, d\xi_1\, d\eta_1$$
$$(n = 1, 2, 3, \ldots).$$

LEMMA 1. *The series* (2.11) *is absolutely and uniformly convergent for* $\xi < \xi' \leqslant \eta' < \eta$ *when* $0 < \eta - \xi < \infty$. *This series can be differentiated term by term, and its sum satisfies* (2.8) *and* (2.9). *Moreover,*

$$|v(\xi', \eta'; \xi, \eta)| \leqslant \frac{(\eta' - \xi')^{1/3}}{(\eta' - \xi)^{1/6}(\eta - \xi)^{1/6}} C(\xi', \eta'; \xi, \eta), \tag{2.12}$$

where $C(\xi', \eta'; \xi, \eta)$ *is a bounded function.*

Frankl′ [22k] proved Lemma 1 for $0 < \eta - \xi < \epsilon$, where $\epsilon > 0$ is sufficiently small. Babenko [4a] showed that the lemma is valid without any restrictions on $\eta - \xi$.

It can be shown [22k] that if $\tau(t)$ and $\nu(t)$ are twice continuously differentiable for $x_1 < t < x_2$, then the function $u(\xi, \eta)$ defined by (2.5) has continuous derivatives up to second order with respect to x and y in the domain bounded by the characteristics $\xi = x_1$, $y = x_2$ and the parabolic curve $\eta = \xi$. In this case we have the classical solution of the Cauchy problem (2.1)–(2.2).

The expression (2.5) will be called a *generalized solution* of equation (2.3) if $\tau(t)$ and $\nu(t)$ are continuous for $x_1 < t < x_2$.

A necessary condition for a generalized solution to possess some smoothness property is that the functions $\tau(t)$ and $\nu(t)$ be smooth in a suitable sense. Henceforth, we consider the generalized solution in the triangle ABC bounded by the segment AC of the characteristic $\xi = 0$, the segment CB of the characteristic $\eta = l$ and an arc of the parabolic curve $\eta = \xi$ ($\Delta ABC = \{\xi, \eta \in [0, l], \eta > \xi\}$).

Consider the following class of generalized solutions of the Cauchy problem introduced by Babenko [4a].

CLASS R_1. *A generalized solution* (2.5) *of equation* (2.3) *belongs to class* R_1 *if* $\tau(t)$ *satisfies a Hölder condition with exponent* $\alpha_1 > 5/6$ *for* $0 \leqslant t < l$ *and* $\nu(t)$ *satisfies a Hölder condition with exponent* $\alpha_2 > 1/6$ *for* $0 \leqslant t < l$.

Below we shall often use expressions of the type

$$f_\alpha(x) = \frac{1}{\Gamma(\alpha)} \int_a^x (x-t)^{\alpha-1} f(t)\, dt \qquad (a \leqslant x \leqslant b),$$

where $f(t)$ is integrable on (a, b); such integrals are known as *integrals of fractional order* α.

LEMMA 2 [4a]. *If a generalized solution* u *belongs to the class* R_1, *then its derivatives* u_x *and* u_y *are continuous in* ΔABC, u_y *is continuous right up to the parabolic curve, and*

$$\lim_{y \to 0} \frac{\partial u}{\partial y} = \nu(x) \qquad (0 < x < l).$$

PROOF. Set

$$\gamma_1 = \frac{\Gamma\left(\frac{1}{3}\right)}{\Gamma^2\left(\frac{1}{6}\right)}, \quad \gamma_2 = \frac{1}{2}\left(\frac{4}{3}\right)^{1/3} \frac{\Gamma\left(\frac{2}{3}\right)}{\Gamma^2\left(\frac{5}{6}\right)}$$

and consider the function

$$u_0(\xi, \eta) = \gamma_1 \int_\xi^\eta \frac{(\eta-\xi)^{2/3}\, \tau(t)\, dt}{(\eta-t)^{5/6}(t-\xi)^{5/6}} - \gamma_2 \int_\xi^\eta \frac{\nu(t)\, dt}{(\eta-t)^{1/6}(t-\xi)^{1/6}}. \tag{2.13}$$

Note that $u_0(\xi, \eta)$ is a generalized solution in class R_1 of equation (2.3) with $a = b = c = 0$.

It is known [30] that if $f(x)$ satisfies a Hölder condition on (0, 1) with exponent α, it may be expressed as

$$f(x) = f(0) + \int_0^x (x-t)^{\beta-1} g(t)\, dt,$$

where $g(t)$ satisfies a Holder condition with exponent $\alpha - \beta > 0$. Therefore, $\tau(t)$ and $\nu(t)$ may be expressed as

$$\begin{aligned} \tau(t) &= \tau(0) + \int_0^t (t-s)^{-1/6+\varepsilon} \varphi(s)\, ds, \\ \nu(t) &= \nu(0) + \int_0^t (t-s)^{-5/6+\varepsilon} \psi(s)\, ds, \end{aligned} \tag{2.14}$$

where $\epsilon > 0$ is sufficiently small and $\psi(s)$ and $\varphi(s)$ are continuous functions for $0 \leqslant s < 1$.

Substituting (2.14) into (2.13), changing the order of integration and using the integral representation of the hypergeometric function,[1] we get

$$\begin{aligned} u_0(\xi, \eta) = &\int_0^{\xi} \varphi_1(s)(\eta - s)^{-1/6+\varepsilon} F\left(\frac{1}{6}, \frac{1}{6} - \varepsilon, \frac{1}{3}; \frac{\eta-\xi}{\eta-s}\right) ds \\ &+ \int_{\xi}^{\eta} \varphi_2(s)(\eta-\xi)^{-1/6}(\eta-s)^{\varepsilon} F\left(\frac{1}{6}, \frac{5}{6}, 1+\varepsilon; \frac{\eta-s}{\eta-\xi}\right) ds \\ &+ \int_0^{\xi} \varphi_3(s)(\eta-s)^{-1/6}(\xi-s)^{\varepsilon} F\left(\frac{1}{6}+\varepsilon, \frac{1}{6}, 1+\varepsilon; \frac{\xi-s}{\eta-s}\right) ds \\ &+ \tau(0) - \left(\frac{3}{4}\right)^{2/3} (\eta-\xi)^{2/3} \nu(0), \end{aligned} \tag{2.15}$$

where

---

(1) $$\int_0^1 t^{a-1}(1-t)^{c-a-1}(1-zt)^{-b}\, dt = \frac{\Gamma(a)\Gamma(c-a)}{\Gamma(c)} F(a, b, c; z)$$

$(0 < \operatorname{Re} a < \operatorname{Re} c)$.

$$\varphi_1(s) = \varphi(s) - \sqrt{3}\,\frac{\Gamma\left(\frac{2}{3}\right)\Gamma\left(\frac{1}{6}+\varepsilon\right)}{\Gamma\left(\frac{5}{6}+\varepsilon\right)}\,\gamma_2\psi(s),$$

$$\varphi_2(s) = \frac{\Gamma\left(\frac{1}{6}\right)\Gamma\left(\frac{5}{6}+\varepsilon\right)}{\Gamma(1+\varepsilon)}\,\gamma_1\varphi(s) - \frac{\Gamma\left(\frac{5}{6}\right)\Gamma\left(\frac{1}{6}+\varepsilon\right)}{\Gamma(1+\varepsilon)}\,\gamma_2\psi(s), \tag{2.16}$$

$$\varphi_3(s) = \sqrt{3}\,\frac{\Gamma\left(\frac{5}{6}\right)\Gamma\left(\frac{1}{6}+\varepsilon\right)}{\Gamma(1+\varepsilon)}\,\gamma_2\psi(s).$$

(2.15) implies the existence of $\partial u_0/\partial\xi$ and $\partial u_0/\partial\eta$ and their continuity in ΔABC.

Let

$$M_1(a) = \max_{0\leqslant s\leqslant a} |\varphi(s)|, \qquad M_2(a) = \max_{0\leqslant s\leqslant a} |\psi(s)|.$$

It is easily seen that

$$\left.\begin{array}{l}\left|\dfrac{\partial u_0}{\partial\xi}\right| \\ \left|\dfrac{\partial u_0}{\partial\eta}\right|\end{array}\right\} \leqslant C\,(M_1(\eta) + M_2(\eta))\,(\eta-\xi)^{-1/3}, \tag{2.17}$$

where C is a constant.

The first derivatives of the first two integrals in (2.15) are bounded by $C_1(\eta-\xi)^{-1/6+\epsilon}$. Thus, we have

$$\begin{aligned}&\left(\frac{3}{4}\right)^{1/3}(\eta-\xi)^{1/3}\left(\frac{\partial u_0}{\partial\xi} - \frac{\partial u_0}{\partial\eta}\right)\\ &= \nu(0) + \left(\frac{3}{4}\right)^{2/3}\frac{\frac{1}{6}\left(\frac{1}{6}+\varepsilon\right)}{1+\varepsilon}\int_0^{\varepsilon}\varphi_3(s)(\eta-s)^{-5/6}(\xi-s)^{\varepsilon}\left(\frac{\xi-s}{\eta-s}+1\right)\\ &\qquad\times F\left(\frac{5}{6}, \frac{5}{6}+\varepsilon, 2+\varepsilon; \frac{\xi-s}{\eta-s}\right)ds + O\left((\eta-\xi)^{1/6+\varepsilon}\right).\end{aligned}$$

Consequently, by (2.16) and (2.14),

$$\begin{aligned}\lim_{y\to 0}\frac{\partial u_0}{\partial y} &= \lim_{\eta-\xi\to 0}\left(\frac{3}{4}\right)^{1/3}(\eta-\xi)^{1/3}\left(\frac{\partial u_0}{\partial\xi} - \frac{\partial u_0}{\partial\eta}\right)\\ &= \nu(0) + \int_0^{\xi}(\xi-s)^{-5/6+\varepsilon}\,\psi(s)\,ds = \nu(\xi)\end{aligned} \tag{2.18}$$

uniformly for $0 \leqslant \xi \leqslant \theta < 1$.

We now turn to the generalized solution defined by (2.5). Using $u_0(\xi,\eta)$, we can write (2.5) as

$$u(\xi,\eta)=u_0(\xi,\eta)-\int_\xi^\eta \tau(t)\,b(t,0)\,h(\xi,\eta,t)\,dt$$
$$+\int_\xi^\eta u_0(\xi',\eta)\,\tilde{B}_1(\xi',\eta)\,\tilde{B}_1(\xi';\xi,\eta)\,d\xi'$$
$$-\int_\xi^\eta u_0(\xi,\eta')\,A_1(\xi,\eta')\,\tilde{A}_1(\eta';\xi,\eta)\,d\eta' \tag{2.19}$$
$$-\int_\xi^\eta d\xi'\int_{\xi'}^\eta u_0(\xi',\eta')\,D\,[v(\xi',\eta';\xi,\eta)]\,d\eta'.$$

Integrating the double integral in (2.19) by parts, we get

$$\int_\xi^\eta d\xi'\int_{\xi'}^\eta u_0(\xi',\eta')\,D\,[v(\xi',\eta';\xi,\eta)]\,d\eta'$$
$$=-\int_\xi^\eta \tau(t)\,b(t,0)\,h(\xi,\eta,t)\,dt$$
$$+\int_\xi^\eta u_0(\xi',\eta)\,B_1(\xi',\eta)\,\tilde{B}_1(\xi';\xi,\eta)\,d\xi' \tag{2.20}$$
$$-\int_\xi^\eta u_0(\xi,\eta')\,A_1(\xi,\eta')\,\tilde{A}_1(\eta';\xi,\eta)\,d\eta'$$
$$-\int_\xi^\eta d\xi'\int_{\xi'}^\eta \left(A_1\frac{\partial u_0}{\partial \xi'}+B_1\frac{\partial u_0}{\partial \eta'}+Cu_0\right)v(\xi',\eta';\xi,\eta)\,d\eta'.$$

We have here used the fact that

$$\lim_{\eta'-\xi'\to 0} A_1(\xi',\eta')\,v(\xi',\eta';\xi,\eta)$$
$$=-\frac{1}{2}\,b(\xi',0)\lim_{\eta'-\xi'\to 0}\frac{1}{2}\left(\frac{4}{3}\right)^{1/3}\frac{v(\xi',\eta';\xi,\eta)}{(\eta'-\xi')^{1/3}}=-\frac{1}{2}\,b(\xi',0)\,h(\xi,\eta;\xi'),$$

and $a(x,0)=0$.

Substituting (2.20) into (2.19), we get

$$u(\xi,\eta)=u_0(\xi,\eta)$$
$$+\int_\xi^\eta d\xi'\int_{\xi'}^\eta \left(A_1\frac{\partial u_0}{\partial \xi'}+B_1\frac{\partial u_0}{\partial \eta'}+Cu_0\right)v(\xi',\eta';\xi,\eta)\,d\eta'. \tag{2.21}$$

The derivatives $\partial v/\partial\xi$ and $\partial v/\partial\eta$ satisfy the following inequalities, which may be proved by rather tedious calculations [4a]:

$$\left|\frac{\partial v}{\partial \xi}\right| \leqslant \frac{(\eta' - \xi')^{1/3}}{(\eta' - \xi)^{7/6}(\eta - \xi')^{1/6}} C_1(\xi, \eta; \xi', \eta'),$$
$$\left|\frac{\partial v}{\partial \eta}\right| \leqslant \frac{(\eta' - \xi')^{1/3}}{(\eta' - \xi)^{1/6}(\eta - \xi')^{7/6}} C_2(\xi, \eta; \xi', \eta'), \tag{2.22}$$

where C_1 and C_2 are bounded functions for $\xi \leqslant \xi' \leqslant \eta' \leqslant \eta$, $0 < \eta - \xi < \infty$.

By virtue of (2.17) and (2.22), we may differentiate inside the double integral in (2.21); denoting the latter integral by I, we get

$$\frac{\partial I}{\partial \xi} = O\left((\eta - \xi)^{1/3}\right), \qquad \frac{\partial I}{\partial \eta} = O\left((\eta - \xi)^{1/3}\right). \tag{2.23}$$

In view of (2.17) and (2.23), we now conclude that $\partial u/\partial \xi$ and $\partial v/\partial \eta$ exist and are continuous in ΔABC. It is easily seen that

$$\left.\begin{array}{l}\left|\dfrac{\partial u}{\partial \xi}\right| \\ \left|\dfrac{\partial u}{\partial \eta}\right|\end{array}\right\} \leqslant C\left(M_1(\eta) + M_2(\eta)\right)(\eta - \xi)^{-1/3}, \tag{2.24}$$

and therefore, by (2.18) and (2.23),

$$\lim_{y \to 0} \frac{\partial u}{\partial y} = \lim_{\eta - \xi \to 0} \left(\frac{3}{4}\right)^{1/3} (\eta - \xi)^{1/3} \left(\frac{\partial u}{\partial \xi} - \frac{\partial u}{\partial \eta}\right) = \nu(\xi)$$

uniformly for $0 \leqslant \xi \leqslant \theta < 1$.

LEMMA 3 [4a]. *For any generalized solution $u \in R_1$, there exists a sequence $\{u_n\}$ of twice continuously differentiable solutions of equation* (2.3), *such that in any closed subtriangle $\Delta A'B'C$ of ΔABC*

$$\lim_{n \to \infty} u_n(\xi, \eta) = u(\xi, \eta) \tag{2.25}$$

and for any $\epsilon > 0$, for $n \geqslant N(\epsilon)$,

$$\left|\frac{\partial u_n}{\partial \xi} - \frac{\partial u}{\partial \xi}\right| < \varepsilon(\eta - \xi)^{-1/3}, \quad \left|\frac{\partial u_n}{\partial \eta} - \frac{\partial u}{\partial \eta}\right| < \varepsilon(\eta - \xi)^{-1/3}. \tag{2.26}$$

PROOF. Since $\tau(x)$ and $\nu(x)$ may be expressed in the form (2.14), where $\varphi(s)$ and $\psi(s)$ are continuous for $0 \leqslant s < 1$, there exist sequences $\{\varphi_n(s)\}$ and $\{\psi_n(s)\}$ of twice continuously differentiable functions on $0 \leqslant s < 1$ such that $\lim_{n\to\infty}\varphi_n(s) = \varphi(s)$ and $\lim_{n\to\infty}\psi_n(s) = \psi(s)$ uniformly on any interval $0 \leqslant s \leqslant \theta < 1$. The corresponding functions $\tau_n(x)$ and $\nu_n(x)$ are twice continuously differentiable on $0 \leqslant x < 1$, and so the solution $u_n(\xi, \eta)$ is twice continuously differentiable in ΔABC. Inequalities (2.26) now follow from (2.24), and the truth of (2.25) is obvious. Q.E.D.

We now consider the *Goursat problem* for equation (2.3) in ΔABC: *to determine a twice continuously differentiable solution* $u(\xi, \eta)$ *of equation* (2.3), *continuous in the closed triangle* ΔABC *and satisfying the conditions*

$$u\,|_{\eta=\xi} = \tau(\xi), \qquad u\,|_{\xi=0} = 0. \tag{2.27}$$

In order to solve the Goursat problem, we shall use the Riemann-Hadamard function $v(\xi, \eta; \xi_0, \eta_0)$, which is defined by the following conditions:

1) as a function of (ξ_0, η_0), it satisfies the equation

$$E(u) \equiv \frac{\partial^2 u}{\partial\xi\,\partial\eta} - \frac{1}{6(\eta-\xi)}\left(\frac{\partial u}{\partial\eta} - \frac{\partial u}{\partial\xi}\right) = 0;$$

2) as a function of (ξ, η), it satisfies the adjoint equation $E^*(u) = 0$;

3) $v(\xi_0, \eta_0; \xi_0, \eta_0) = 1$;

4) $\partial[v]/\partial\xi + [v]/6(\eta - \xi) = 0$ for $\eta = \xi_0$, where

$$[v] = \lim_{\varepsilon\to 0}\,[v(\xi, \xi_0+\varepsilon; \xi_0, \eta_0) - v(\xi, \xi_0-\varepsilon; \xi_0, \eta_0)].$$

The Riemann-Hadamard function $v(\xi, \eta; \xi_0, \eta_0)$ was constructed by Gellerstedt [24a]. It has the form

$$v(\xi, \eta; \xi_0, \eta_0) = \begin{cases} v_1(\xi, \eta; \xi_0, \eta_0) = \left(\dfrac{\eta-\xi}{\eta_0-\xi_0}\right)^{1/6} F\left(\dfrac{1}{6}, \dfrac{5}{6}, 1; s\right) & \text{for } \eta > \xi_0, \\[2ex] v_2(\xi, \eta; \xi_0, \eta_0) = k\,\dfrac{(\eta-\xi)(\eta_0-\xi_0)^{2/3}}{(\eta_0-\eta)^{5/6}(\xi_0-\xi)^{5/6}}\, F\left(\dfrac{5}{6}, \dfrac{5}{6}, \dfrac{5}{3}; \dfrac{1}{s}\right) & \text{for } \eta < \xi_0, \end{cases} \tag{2.28}$$

where

$$s = \frac{(\xi_0-\xi)(\eta_0-\eta)}{(\eta-\xi)(\eta_0-\xi_0)}, \qquad k = \frac{\Gamma\left(\frac{5}{6}\right)}{\Gamma\left(\frac{1}{6}\right)\Gamma\left(\frac{5}{3}\right)}.$$

Let $u(\xi, \eta)$ be a twice differentiable solution of equation (2.3) in ΔABC, satisfying the boundary conditions (2.27).

Equation (2.3) may be written

$$E(u) = -A_1(\xi, \eta)\frac{\partial u}{\partial\xi} - B_1(\xi, \eta)\frac{\partial u}{\partial\eta} - C(\xi, \eta)\,u.$$

Integrate the identity

$$vE(u) - uE^*(v) = -\left(A_1\frac{\partial u}{\partial\xi} + B_1\frac{\partial u}{\partial\eta} + Cu\right)v$$

over a triangle $\Delta A'B'C'$ bounded by a segment $A'C'$ of the characteristic $\xi = 0$, a segment $C'B'$ of the characteristic $\eta = \xi_0 - \epsilon$ and a segment $A'B'$ of the

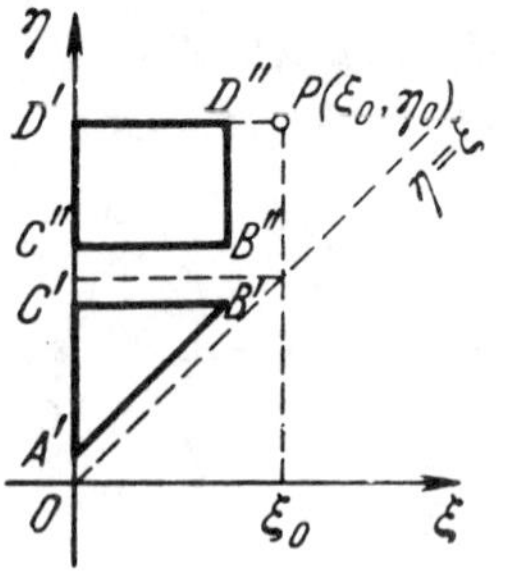

FIGURE 6

straight line $\eta - \xi = \epsilon$; then do the same over a rectangle $C''B''D''D'$ bounded by segments $C''B''$ and $D'D''$ of the characteristics $\eta = \xi_0 + \epsilon$ and $\eta = \eta_0$, respectively, and segments $B''D''$ and $C''D'$ of the characteristics $\xi = \xi_0 - 2\epsilon$ and $\xi = 0$ (Figure 6). Finally, letting $\epsilon \longrightarrow 0$, we obtain

$$u(\xi, \eta) = \frac{\Gamma\left(\frac{5}{6}\right)}{\Gamma\left(\frac{1}{6}\right)\Gamma\left(\frac{2}{3}\right)} \int_0^{\xi} \frac{(\eta - \xi)^{2/3}}{(\eta - t)^{5/6}(\xi - t)^{5/6}} \tau(t)\, dt$$
$$- \int_0^{\xi} d\xi' \int_{\xi'}^{\eta} \left(A_1 \frac{\partial u}{\partial \xi'} + B_1 \frac{\partial u}{\partial \eta'} + Cu\right) v(\xi', \eta'; \xi, \eta)\, d\eta'. \tag{2.29}$$

Here we have used conditions 1)–4) and the boundary conditions (2.27).

Let $u(\xi, \eta) \in R_1$ be a generalized solution of (2.3) satisfying the boundary conditions (2.27). By Lemma 3, there exists a sequence $\{u_n(\xi, \eta)\}$ of twice continuously differentiable solutions of (2.3) such that

$$\lim_{n \to \infty} u_n(\xi, \eta) = u(\xi, \eta)$$

uniformly in any closed triangle $\Delta A'B'C'$, and such that inequalities (2.26) are valid. Therefore, we may let $n \longrightarrow \infty$ in (2.29) (replacing u by u_n). Thus any generalized solution of (2.3) which belongs to class R_1 and vanishes on the characteristic $\xi = 0$ may be expressed in the form (2.29).

It is readily verified that

$$\left|\frac{\partial v}{\partial \xi'}\right| < \frac{(\eta - \xi)^{2/3}}{(\eta - \xi')^{5/6} |\xi - \eta'|^{5/6}}. \tag{2.30}$$

Therefore, we may integrate by parts with respect to ξ' in the double integral in (2.29) containing the term

$$A_1 \frac{\partial u}{\partial \xi'} v(\xi', \eta'; \xi, \eta).$$

Setting

$$K_0(\xi, \eta, t) = k' \frac{(\eta-\xi)^{2/3}}{(\eta-t)^{5/6}(\xi-t)^{5/6}}, \qquad k' = \frac{\Gamma\left(\frac{5}{6}\right)}{\Gamma\left(\frac{1}{6}\right)\Gamma\left(\frac{2}{3}\right)}$$

and integrating as described, we express (2.29) in the form

$$u(\xi, \eta) = \int_0^\xi K_0(\xi, \eta, t)\,\tau(t)\,dt - \int_\xi^\eta \left(\frac{\eta'-\xi}{\eta-\xi}\right)^{1/6} A_1(\xi, \eta')\,u(\xi, \eta')\,d\eta' - \int_0^\xi d\xi' \int_{\xi'}^\eta \left(B_1 v \frac{\partial u}{\partial \eta'} + uw\right) d\eta', \tag{2.31}$$

where

$$w(\xi', \eta'; \xi, \eta) = Cv - \frac{\partial (A_1 v)}{\partial \xi'}.$$

We shall seek a solution of (2.31) in the form

$$u(\xi, \eta) = \int_0^\xi K(\xi, \eta, t)\,\tau(t)\,dt, \tag{2.32}$$

where $K(\xi, \eta, t)$ is an unknown function. Substituting (2.32) into (2.31), we get

$$K(\xi, \eta, t) = K_0(\xi, \eta, t) - \int_\xi^\eta \left(\frac{\eta'-\xi}{\eta-\xi}\right)^{1/6} A_1(\xi, \eta')\,K(\xi, \eta', t)\,d\eta' - \int_t^\xi d\xi' \int_{\xi'}^\eta \left(B_1 v \frac{\partial K}{\partial \eta'} + wK\right) d\eta'. \tag{2.33}$$

Let us assume that the function $K(\xi, \eta, t)$ is a series

$$K(\xi, \eta, t) = \sum_{n=0}^{\infty} (-1)^n K_n(\xi, \eta, t). \tag{2.34}$$

Then, by (2.33),

$$K_{n+1}(\xi, \eta, t) = \int_\xi^\eta \left(\frac{\eta'-\xi}{\eta-\xi}\right)^{1/6} A_1(\xi, \eta')\,K_n(\xi, \eta', t)\,d\eta' + \int_t^\xi d\xi' \int_{\xi'}^\eta \left(B_1 v \frac{\partial K_n}{\partial \eta'} + wK_n\right) d\eta' \tag{2.35}$$

$$(n = 0, 1, 2, \ldots).$$

We must show that the series (2.34) is convergent. It is easy to see that

$$v(\xi', \eta'; \xi, \eta) < \frac{(\eta'-\xi')(\eta-\xi)^{2/3}}{(\eta-\xi')^{5/6}\,|\xi-\eta'|^{5/6}}. \tag{2.36}$$

Now, since

$$w = -\left[\frac{1}{4}\left(\frac{4}{3}\right)^{2/3}\frac{c(\xi', \eta')}{(\eta'-\xi')^{2/3}} + \frac{\partial A_1}{\partial \xi'}\right]v - A_1\frac{\partial v}{\partial \xi'}, \tag{2.37}$$

we have

$$|w| < M_3\frac{v}{\eta'-\xi'} + \left|\frac{\partial v}{\partial \xi'}\right| M_1$$

or, by (2.30) and (2.36),

$$|w| < (M_1 + M_3)\frac{(\eta-\xi)^{2/3}}{(\eta-\xi')^{5/6}|\xi-\eta'|^{5/6}}, \tag{2.38}$$

where

$$M_1(l) = \max_{0\leqslant\xi\leqslant\eta\leqslant l}|A_1(\xi,\eta)|,$$

$$M_3(l) = \max_{0\leqslant\xi\leqslant\eta\leqslant l}\left|\frac{1}{4}\left(\frac{4}{3}\right)^{2/3}c(\xi,\eta)(\eta-\xi)^{1/3} + (\eta-\xi)\frac{\partial A_1}{\partial \xi}\right|.$$

In view of (2.36) and (2.38), we infer from (2.35) that

$$\begin{aligned}|K_{n+1}| \leqslant \mathfrak{M}\Bigg[&\int_\xi^\eta\left(\frac{\eta'-\xi}{\eta-\xi}\right)^{1/6}|K_n(\xi,\eta',t)|\,d\eta' \\ &+\int_t^\xi d\xi'\int_{\xi'}^\eta\frac{(\eta-\xi)^{2/3}}{(\eta-\xi')^{5/6}|\xi-\eta'|^{5/6}}\left\{(\eta'-\xi')\left|\frac{\partial K_n}{\partial \eta'}\right| + |K_n|\right\}d\eta'\Bigg],\end{aligned} \tag{2.39}$$

where

$$\mathfrak{M}(l) = \max(M_2, M_1 + M_3), \qquad M_2(l) = \max_{0\leqslant\xi\leqslant\eta\leqslant l}|B_1(\xi,\eta)|.$$

It is readily checked that

$$\begin{aligned}\left|\frac{\partial v}{\partial \eta}\right| &< (\eta'-\xi')(\eta-\xi)^{-1/3}(\eta-\xi')^{-5/6}|\xi-\eta'|^{-5/6}, \\ \left|\frac{\partial^2 v}{\partial \xi'\,\partial \eta}\right| &< 4(\eta-\xi)^{-1/3}(\eta-\xi')^{-5/6}|\xi-\eta'|^{-5/6}.\end{aligned} \tag{2.40}$$

It then follows from (2.37) that

$$\left|\frac{\partial w}{\partial \eta}\right| < (M_3 + 4M_1)(\eta-\xi)^{-1/3}(\eta-\xi')^{-5/6}|\xi-\eta'|^{-5/6}. \tag{2.41}$$

Differentiating (2.35) with respect to η and using (2.40) and (2.41), we obtain

$$\left|\frac{\partial K_{n+1}}{\partial \eta}\right| \leqslant \mathfrak{M}\left\{|K_n(\xi, \eta, t)| + \frac{1}{6}\int_{\xi}^{\eta} \frac{(\eta'-\xi)^{1/6}}{(\eta-\xi)^{7/6}}|K_n(\xi, \eta', t)|\,d\eta'\right.$$

$$+\int_t^{\xi}(\eta-\xi')^{-5/6}(\eta-\xi)^{-1/6}\left[(\eta-\xi')\left|\frac{\partial K_n(\xi', \eta, t)}{\partial \eta}\right| + |K_n(\xi', \eta, t)|\right]d\xi'$$

$$\left.+4\int_t^{\xi} d\xi'\int_{\xi'}^{\eta}(\eta-\xi)^{-1/3}(\eta-\xi')^{-5/6}|\xi-\eta'|^{-5/6}\left[(\eta'-\xi')\left|\frac{\partial K_n}{\partial \eta'}\right| + |K_n|\right]d\eta'\right\}. \tag{2.42}$$

Let us denote the right-hand side of (2.39) by $L_{n+1}(\xi, \eta, t)$ and set

$$N_n(\xi, \eta, t) = (\eta-\xi)\left|\frac{\partial K_n}{\partial \eta}\right| + |K_n|.$$

Then, by (2.39) and (2.42),

$$\left.\begin{aligned} N_n(\xi, \eta, t) < \mathfrak{M}\left\{|K_n| + \int_t^{\xi}(\eta-\xi')^{-5/6}(\eta-\xi)^{-1/6}N_n(\xi', \eta, t)\,d\xi'\right\} \\ \times(\eta-\xi) + 5L_{n+1}(\xi, \eta, t) \qquad (n = 0, 1, 2, \ldots). \end{aligned}\right\} \tag{2.43}$$

Using (2.39) and (2.43), we get

$$\left.\begin{aligned} |K_1(\xi, \eta, t)| &< \pi^2\mathfrak{M}\left[\left(\frac{\eta-\xi}{\xi-t}\right)^{2/3} + \left(\frac{\eta-\xi}{\xi-t}\right)^{5/6}\right] \\ &< 2\pi^2\mathfrak{M}\,\frac{(\eta-\xi)^{2/3}(\eta-t)^{1/6}}{(\xi-t)^{5/6}}, \\ |N_1(\xi, \eta, t)| &< 6\pi^2\mathfrak{M}\left[\left(\frac{\eta-\xi}{\xi-t}\right)^{2/3} + \left(\frac{\eta-\xi}{\xi-t}\right)^{5/6}\right] \\ &< 12\pi^2\mathfrak{M}\,\frac{(\eta-\xi)^{2/3}(\eta-t)^{1/6}}{(\xi-t)^{5/6}}. \end{aligned}\right\} \tag{2.44}$$

Hence, by induction, we obtain

$$\left.\begin{aligned} |K_n(\xi, \eta, t)| &< \mathfrak{M}^n C^n\,\frac{(\eta-\xi)^{2/3}(\eta-t)^{1/6+n-1}}{(\xi-t)^{5/6}}, \\ |N_n(\xi, \eta, t)| &< \mathfrak{M}^n C^n\,\frac{(\eta-\xi)^{2/3}(\eta-t)^{1/6+n-1}}{(\xi-t)^{5/6}}, \end{aligned}\right\} \tag{2.45}$$

where C is an absolute constant; it is readily seen that $C < 124\pi$.

Estimates (2.45) imply that the series (2.34) is absolutely convergent for $t < \xi$ and $\eta < 1/\mathfrak{M}C = l_0$. It is obvious that (2.32) is the unique solution of (2.3) satisfying the boundary conditions (2.27) in the class of functions such that

$$\frac{\partial u}{\partial \eta} = O\left((\eta - \xi)^{-1/3}\right).$$

We have thus proved

LEMMA 4. *If $y < 0$, any solution of equation* (2.1) *in class R_1 which vanishes on the characteristic $\xi = 0$ may be expressed as*

$$u(\xi, \eta) = \int_0^\xi K(\xi, \eta, t)\, \tau(t)\, dt, \tag{2.46}$$

for $\eta < l_0 = 1/\mathfrak{M}(l_0)C$.

REMARK. In the case of the Tricomi equation, i.e. $a \equiv b \equiv c \equiv 0$, the solution (2.46) takes the simpler form

$$u(\xi, \eta) = \frac{\Gamma\left(\frac{5}{6}\right)}{\Gamma\left(\frac{1}{6}\right)\Gamma\left(\frac{2}{3}\right)} \int_0^\xi \frac{(\eta - \xi)^{2/3}}{(\eta - t)^{5/6}(\xi - t)^{5/6}}\, \tau(t)\, dt.$$

LEMMA 5 ([4a]; see also [15]). *Let the function $u(x, y) \in C^2(D^+) \cap C(\bar{D}^+)$ satisfy the inequality $L(u) \geqslant 0$ ($L(u) \leqslant 0$) in the domain D^+, and let this function attain its maximum positive (minimum negative) value at a point x_0, $0 < x_0 < l$, in the interval $[0, l]$. If the values of $u(x, y)$ on Γ are less (greater) than $u(x_0, 0)$, then*

$$\lim_{y \to 0} \frac{\partial u(x_0, y)}{\partial y} < 0 \qquad (> 0)$$

if the limit exists.

PROOF. Clearly, we cannot have

$$\lim_{y \to 0} \frac{\partial u(x_0, y)}{\partial y} > 0.$$

Assume that

$$\lim_{y \to 0} \frac{\partial u(x_0, y)}{\partial y} = 0.$$

Let $\beta = \max_D |b(x, y)|$, and let d denote the diameter of D. We can assume without loss of generality that $u(x_0, 0) = 1$. In view of the conditions of the lemma we have $\max_\Gamma u(x, y) \leqslant 1 - \epsilon$, where $0 < \epsilon < 1$.

Consider the function

$$v(x, y) = \frac{\varepsilon u(x, y)}{e^{\beta d} - \varepsilon e^{\beta y}}.$$

On Γ we have

$$v(x, y) \leqslant \frac{\varepsilon(1 - \varepsilon)}{e^{\beta d} - \varepsilon e^{\beta d}} = \varepsilon e^{-\beta d} < \frac{\varepsilon}{e^{\beta d} - \varepsilon},$$

and on $[0, l]$

$$v(x, 0) < \frac{\varepsilon}{e^{\beta d} - \varepsilon}, \qquad v(x_0, 0) = \frac{\varepsilon}{e^{\beta d} - \varepsilon}.$$

Since $L(u) \geqslant 0$, the function $v(x, y)$ satisfies the inequality

$$\begin{gathered} y v_{xx} + v_{yy} + a(x, y) v_x + \bar{b}(x, y) v_y + \bar{c}(x, y) v \geqslant 0 \\ (\bar{c}(x, y) < 0), \end{gathered} \tag{$*$}$$

and also, according to our assumption

$$\lim_{y \to 0} \frac{\partial v(x_0, y)}{\partial y} = \frac{\varepsilon\beta}{(e^{\beta d} - \varepsilon)^2} > 0.$$

Therefore, $v(x, y)$ should attain its maximum value inside the domain D, but this is impossible in view of ($*$). The case of the minimum can be treated in the same fashion. Q.E.D.

2. *Extremum principle. Uniqueness theorem.* The equation

$$y u_{xx} + u_{yy} + a(x, y) u_x + b(x, y) u_y + c(x, y) u = 0 \tag{2.1}$$

will be considered in a domain D which is the union of a simple Jordan curve Γ in the half-plane $y > 0$, subtended by the interval $[0, l]$ on the x-axis, and of two characteristics

$$AC\colon\ \xi = x - \frac{2}{3}(-y)^{3/2} = 0 \ \text{ and } \ BC\colon\ \eta = x + \frac{2}{3}(-y)^{3/2} = 1$$

of equation (2.1). Let D^+ and D^- denote the regions of D located in the half-planes $y > 0$ and $y < 0$, respectively.

DEFINITION. A function $u(x, y)$ is said to be a *generalized solution in class* $\mathfrak{R}_1$ *of equation* (2.1) if it satisfies the following conditions:

1) $u(x, y)$ is continuous in the closed domain $\bar{D}$;

2) $u(x, y)$ is twice continuously differentiable in D^+ and satisfies (2.1);

3) for any x, $0 < x < l$, the limit $\lim_{y\to 0} \partial u(x, y)/\partial y = \nu(x)$ exists;

4) $u(x, y) \in R_1$ in D^-.

THEOREM 2.1 [4a]. *Assume that the following conditions are satisfied*:

1) *The coefficients a, b, c of equation* (2.1) *are twice continuously differentiable in the domain D, $c \leqslant 0$, $y \geqslant 0$.*

2) $a(x, 0) = b(x, 0) = 0$.

3) $l\mathfrak{M}(l) < \vartheta$, *where ϑ is an absolute constant.*

Then a solution $u(x, y) \in \mathfrak{R}_1$ of the Tricomi problem for equation (2.1), *which vanishes on the characteristic $\xi = 0$ (or $\eta = l$), attains its positive maximum and negative minimum in the closed domain $\bar{D}^+$ on the curve Γ.*

PROOF. Let $u(x, y)$ be a solution of equation (2.1) in class $\mathfrak{R}_1$ which vanishes on the characteristic $\xi = 0$. Let $l < l_0$. The function $u(x, y)$ obviously cannot attain an extremum in the interior of D^+. Let us assume that the positive maximum in the closed domain $\overline{D}^+$ is achieved at an interior point $P(x_0, 0)$ of the interval $(0, l)$. Let $u(x_0, 0) = M'$. By Lemma 4,

$$u(\xi, \eta) = \int_0^\xi K(\xi, \eta, t)\tau(t)\,dt$$

$$= \int_0^\xi K_0(\xi, \eta, t)\tau(t)\,dt + \int_0^\xi \tau(t)\sum_{n=1}^\infty (-1)^n K_n(\xi, \eta, t)\,dt.$$

Using the estimates (2.45), we have

$$u(\xi, \eta) < M' \int_0^\xi K_0(\xi, \eta, t)\,dt + M'(\eta - \xi)^{2/3} \int_0^\xi \sum_{n=1}^\infty C^n \mathfrak{M}^n \frac{(\eta - t)^{1/6+n-1}}{(\xi - t)^{5/6}}\,dt$$

$$= M'\left\{1 - k' \int_{\frac{\xi}{\eta-\xi}}^\infty \zeta^{5/6}(1+\zeta)^{-5/6}\,d\zeta + (\eta - \xi)^{2/3} \int_0^\xi \frac{C\mathfrak{M}(\eta - t)^{1/6}\,dt}{(\xi - t)^{5/6}\left[1 - C\mathfrak{M}(\eta - t)\right]}\right\}.$$

Hence

$$u(\xi, \eta) < M'\left\{1 - k' \int_{\frac{\xi}{\eta-\xi}}^\infty (1+\zeta)^{-5/3}\,d\zeta + \frac{C\mathfrak{M}(\eta - \xi)^{2/3}}{1 - C\mathfrak{M}\eta} \int_0^\xi \frac{(\eta - t)^{1/6}}{(\xi - t)^{5/6}}\,dt\right\}$$

or

$$u(\xi, \eta) < M'\left\{1 - (\eta - \xi)^{2/3}\left[\frac{3k'}{2}\eta^{-2/3} - \frac{C\mathfrak{M}}{1 - C\mathfrak{M}\eta} \int_0^\xi \frac{(\eta - t)^{1/6}}{(\xi - t)^{5/6}}\,dt\right]\right\}.$$

For sufficiently small $\eta - \xi$ we have

$$\frac{3k'}{2}\eta^{-2/3} - \frac{C\mathfrak{M}}{1 - C\mathfrak{M}\eta} \int_0^\xi \frac{(\eta - t)^{1/6}}{(\xi - t)^{5/6}}\,dt \geqslant 0,$$

since if $\eta = \xi$ the left-hand side of the last inequality equals

$$\xi^{-2/3}\left(\frac{3}{2}k' - \frac{3C\mathfrak{M}\xi}{1 - C\mathfrak{M}\xi}\right) > 0,$$

provided that

$$\xi < \frac{k'}{C\mathfrak{M}(k' + 2)} = \frac{\vartheta}{\mathfrak{M}}.$$

Therefore, if

$$l\mathfrak{M}(l) < \vartheta, \tag{2.47}$$

then

$$u(\xi, \eta) < M' \qquad (\text{in } D^-). \tag{2.48}$$

On the other hand, applying Lemma 5 to $u(x, y)$ in D^+, we get

$$\lim_{y \to +0} \frac{\partial u(x_0, y)}{\partial y} < 0$$

and thus, for sufficiently small $|y|$, $y < 0$,

$$u(x_0, y) > u(x_0, 0) = M' \qquad (\text{in } D^-). \tag{2.49}$$

Inequalities (2.48) and (2.49) are contradictory. Q.E.D.

As a corollary we have

THEOREM 2.2. *If* $a(x, 0) = b(x, 0) = 0$ *and* (2.47) *holds, then problem* T *for equation* (2.1) *has a unique solution in class* $\mathfrak{R}_1$.

The proof is obvious.

The strongest restriction imposed on the coefficients a and b of equation (2.1) is that they vanish at $y = 0$. This restriction can be slightly weakened. It is sufficient to demand that $b(x, 0) = \text{const} < 0$. By a simple substitution this case can be reduced to the one considered above.

Note that for the Tricomi equation $yu_{xx} + u_{yy} = 0$ the extremum principle and the uniqueness theorem are valid without any restrictions on the length of the parabolic curve.

In this subsection we have basically followed Babenko [4a].

3. *Extremum principle for hyperbolic equations.*(2) Consider the equation

$$L_1(u) \equiv \frac{\partial^2 u}{\partial \xi \, \partial \eta} + \alpha(\xi, \eta) \frac{\partial u}{\partial \xi} + \beta(\xi, \eta) \frac{\partial u}{\partial \eta} + \gamma(\xi, \eta) u = 0 \tag{2.50}$$

in a domain D_1 bounded by segments AC and CB of characteristics $\xi = 0$ and $\eta = 1$ of equation (2.50) and by an interval AB of the straight line $\eta = \xi$.

Assume that 1) $\alpha(\xi, \eta)$ has a continuous first derivative with respect to ξ in $D_1 + AC$, 2) the coefficients α, β and γ are continuous in the closed domain $\bar{D}_1$, and 3) they satisfy the conditions

$$\text{(I)} \quad \begin{cases} \alpha \geqslant 0, & (2.51) \\ \alpha_\xi + \alpha\beta - \gamma \geqslant 0, & (2.52) \\ \gamma \geqslant 0 & (2.53) \end{cases}$$

in D_1.

(2) See [3].

Conditions (I) can be written in a slightly different form. The differential operator obtained by multiplying $L_1(u)$ by $\beta_1 = e^{\int \beta d\xi}$ can be written as

$$\beta_1 L_1(u) = \frac{\partial}{\partial \xi}\left(\beta_1 \frac{\partial u}{\partial \eta}\right) + \frac{\partial}{\partial \xi}(\alpha_1 u) + \gamma_1 u,$$

where $\alpha_1 = \beta_1 \alpha$ and $\gamma_1 = \beta_1 \gamma - \alpha_{1\xi}$. Then conditions (I) can be rewritten as

$$\begin{cases} \alpha_1 \geqslant 0, & (2.51') \\ \gamma_1 \leqslant 0. & (2.52') \end{cases}$$

We shall assume that $u(\xi, \eta)$ is twice continuously differentiable in D_1 and continuous together with its first derivatives in $\overline{D}_1$.

THEOREM 2.3 [3] (MAXIMUM PRINCIPLE). *Assume that* 1) $u_\eta(0, \eta) \leqslant 0$; 2) *the coefficients of the operator* $L_1(u)$ *satisfy conditions* (I); *and* 3) $L_1(u) \leqslant 0$ *in* $D_1 + AC + CB$. *If the maximum of* u *in* $\overline{D}_1$ *is positive, then it is attained on* $\overline{AB}$.

Note that if $\gamma \equiv 0$, Theorem 2.3 holds without assuming that $\max u > 0$ in $\overline{D}_1$.

We first prove a weaker version of the maximum principle. Let us say that the coefficients of equation (2.50) satisfy conditions (I′) if 1) they satisfy conditions (I) and 2) either (2.51) is a strict inequality or on every interval $\eta = \text{const}$ the set of points at which (2.52) and (2.53) are equalities has measure zero.

THEOREM 2.3′. *Let* $u(\xi, \eta)$ *satisfy the conditions of Theorem* 2.3 *and let conditions* (I′) *hold. Then the conclusion of Theorem* 2.3 *holds.*

Theorem 2.3′ follows from

LEMMA 6. *Let* $[P, Q]$ *be a characteristic segment, parallel to the* ξ*-axis and located entirely in* $D_1 + AC + CB$. *Let the coefficients of the operator* $L_1(u)$ *satisfy conditions* (I′), *and* $L_1(u) \leqslant 0$ *on* $[P, Q]$. *If* $\max u$ *on* $[P, Q]$ *is nonnegative and is attained at* Q, *then*

$$(\beta_1 u_\eta)_Q \leqslant (\beta_1 u_\eta)_P.\text{(}^3\text{)}$$

If moreover $u(Q) > 0$ *and* $u(Q) > u(P)$, *then*

$$(\beta_1 u_\eta)_Q < (\beta_1 u_\eta)_P.$$

PROOF. According to our assumption, $L_1(u) \leqslant 0$ on $[P, Q]$; hence

$$\beta_1 L_1(u) = (\beta_1 u_\eta)_\xi + (\alpha_1 u)_\xi + \gamma_1 u \leqslant 0 \quad \text{on} \quad [P, Q].$$

---

(3) $(\beta_1 u_\eta)_Q$ denotes the value of $\beta_1 u_\eta$ at the point Q.

Integration along $[P, Q]$ gives

$$\int_P^Q [(\beta_1 u_\eta)_\xi + (\alpha_1 u)_\xi + \gamma_1 u]\, d\xi \leqslant 0,$$

and thus

$$(\beta_1 u_\eta)_Q - (\beta_1 u_\eta)_P \leqslant -\int_P^Q \gamma_1 u\, d\xi + (\alpha_1 u)_P - (\alpha_1 u)_Q. \tag{2.54}$$

In the right-hand side of (2.54) we add and subtract

$$u(Q) \int_P^Q \gamma_1 d\xi = u(Q) \int_P^Q \beta_1 \gamma\, d\xi - u(Q)\,[\alpha_1(Q) - \alpha_1(P)],$$

so that we have

$$\begin{aligned}(\beta_1 u_\eta)_Q - (\beta_1 u_\eta)_P \leqslant & \int_P^Q [u(Q) - u]\, \gamma_1\, d\xi \\ & - u(Q) \int_P^Q \beta_1 \gamma\, d\xi - [u(Q) - u(P)]\, \alpha_1(P).\end{aligned} \tag{2.55}$$

Using conditions (I) and the fact that $u(Q) \geqslant u(P)$, we have

$$(\beta_1 u_\eta)_Q \leqslant (\beta_1 u_\eta)_P.$$

Therefore, if $u(Q) > 0$ and $u(Q) > u(P)$, then at least one term on the right of (2.55) is negative, by conditions (I′). Consequently,

$$(\beta_1 u_\eta)_Q < (\beta_1 u_\eta)_P.$$

Q.E.D.

The proof of Theorem 2.3′ follows easily from Lemma 6.

Indeed, suppose that the maximum of $u(\xi, \eta)$ in $\overline{D}_1$ is not attained on $\overline{AB}$. It cannot be attained on AC ($\xi = 0$) either, since max u on AC is attained at the point A (as $u_\eta(0, \eta) \leqslant 0$), which belongs to $\overline{AB}$. Thus the maximum point Q lies in $D + CB$. Let the straight line $\eta =$ const through the point Q cut AC at a point P. The maximum of u on $[P, Q]$ is attained at the point Q; and $u(Q) > 0$ and $u(Q) > u(P)$. By Lemma 6,

$$(\beta_1 u_\eta)_Q < (\beta_1 u_\eta)_P,$$

and since $u_\eta(P) \leqslant 0$, we get $u_\eta(Q) < 0$. But then Q cannot be a maximum point of u in $\overline{D}_1$. Q.E.D.

Now, using Theorem 2.3′, we shall prove Theorem 2.3.

Let $g(\xi, \eta)$ be a function satisfying the conditions

$$M_1(g) \equiv g_{\xi\eta} + \alpha g_\xi + \beta g_\eta > 0 \quad \text{in} \quad \overline{D}_1, \tag{2.56}$$

$$g_\eta \geqslant 0 \text{ in } D_1, \quad g_\eta = 0 \quad \text{for} \quad \xi = 0. \tag{2.57}$$

Such a function is easily constructed (for example, $g(\xi, \eta) = \sinh k\xi \cosh k\eta$ for sufficiently large k). If u is a function satisfying the conditions of Theorem 2.3, consider the set of functions $u_\epsilon = e^{-\epsilon g}u$, $\epsilon > 0$. It easily is seen that

$$L_{1\varepsilon}(u_\varepsilon) = \frac{\partial^2 u_\varepsilon}{\partial\xi\,\partial\eta} + \alpha_\varepsilon \frac{\partial u_\varepsilon}{\partial\xi} + \beta_\varepsilon \frac{\partial u_\varepsilon}{\partial\eta} + \gamma_\varepsilon u_\varepsilon \leqslant 0 \quad \text{in } \bar{D}_1,$$

where

$$\alpha_\varepsilon = \alpha + \varepsilon g_\eta, \quad \beta_\varepsilon = \beta + \varepsilon g_\xi, \quad \gamma_\varepsilon = \gamma + \varepsilon[\varepsilon g_\xi g_\eta + M_1(g)]. \tag{2.58}$$

We claim that for sufficiently small $\epsilon > 0$ the function u_ϵ and the coefficients of the operator $L_{1\epsilon}(u)$ satisfy the conditions of Theorem 2.3. First note that, by the assumptions made about u, and the fact that $e^{-\epsilon g} > 0$, the function u_ϵ has a positive maximum in $\bar{D}_1$. By (2.57), $u_{\epsilon\eta} \leqslant 0$ for $\xi = 0$. By (2.51) and (2.57), we have $\alpha_\epsilon = \alpha + \epsilon g_\eta \geqslant 0$. A direct check shows that

$$\frac{\partial \alpha_\varepsilon}{\partial\xi} + \alpha_\varepsilon\beta_\varepsilon - \gamma_\varepsilon = \frac{\partial\alpha}{\partial\xi} + \alpha\beta - \gamma \geqslant 0.$$

Next, using (2.58) and (2.53), we have

$$\gamma_\varepsilon \geqslant \varepsilon[\varepsilon g_\xi g_\eta + M_1(g)].$$

By (2.56), $\gamma_\epsilon > 0$ in $\bar{D}_1$ for sufficiently small $\epsilon > 0$. Thus conditions (I$'$) hold. According to Theorem 2.3$'$, if $\epsilon > 0$ is sufficiently small, the function u_ϵ attains a positive maximum on $\overline{AB}$. Since $u_\epsilon \to u$ as $\epsilon \to 0$, it follows that u also attains a positive maximum on $\overline{AB}$. Q.E.D.

4. *Maximum principle for degenerate hyperbolic equations. Uniqueness theorem.* Consider the equation

$$K(y)u_{xx} + u_{yy} + a(x, y)u_x + b(x, y)u_y + c(x, y)u = 0, \tag{2.59}$$

where $K(y)$ is a monotone increasing function with a continuous second derivative, and $K(0) = 0$.

Let D_1^- denote the domain bounded by the characteristics

$$AC\colon x + \int_0^y \sqrt{-K(t)}\,dt = 0 \text{ and } BC\colon x - \int_0^y \sqrt{-K(t)}\,dt = 1$$

and by the interval AB on the x-axis. Let us assume that the coefficients of (2.59) are continuous in the closed domain $\bar{D}_1^-$ and that $a(x, y)$ and $b(x, y)$ have continuous first derivatives in $D_1^- + AC + BC$.

The results of subsection 2 can be extended to equation (2.59). In order to obtain similar conditions on the coefficients of (2.59), we introduce characteristic coordinates:

$$\xi = x + \int_0^k \sqrt{-K(t)}\,dt, \qquad \eta = x - \int_0^y \sqrt{-K(t)}\,dt.$$

In the characteristic coordinates, (2.59) is of the form (2.50). The domain D_1^- ($y < 0$) becomes Δ: $0 < \xi < \eta < 1$. Conditions (I) for equation (2.59) in D_1^- take the form

$$(\mathrm{I}'') \quad \begin{cases} a + b\sqrt{-K} - \dfrac{K'}{2\sqrt{-K}} \leqslant 0, \\ \delta\left(\dfrac{a + b\sqrt{-K}}{\sqrt{-K}} + \dfrac{K'}{2K}\right) \\ \qquad + \dfrac{1}{2K}\left(a + b\sqrt{-K} - \dfrac{K'}{2\sqrt{-K}}\right) \\ \qquad \times \left(a - b\sqrt{-K} - \dfrac{K'}{2\sqrt{-K}}\right) - 2c \leqslant 0, \\ c(x, y) \leqslant 0, \end{cases}$$

where $\delta \equiv \partial/\partial y + \sqrt{-K}\partial/\partial x$.

THEOREM 2.4. *Assume that*: 1) $u(\xi, \eta)$ *is a generalized solution in class* R_1 *of equation* (2.50), *continuous in* $\bar{\Delta}^-$ *and vanishing on the characteristic* $\xi = 0$; 2) *the coefficients* α, β *and* γ *satisfy conditions* (I′). *Then, if the maximum of* $u(\xi, \eta)$ *in* $\bar{\Delta}^-$ *is positive, it is attained on* $\overline{AB}$.

PROOF. Suppose that $\max u = M > 0$ in $\bar{\Delta}^-$ and that it is not attained on AB. It cannot be attained on AC either, since $u = 0$ for $\xi = 0$. Then there clearly exists a closed set F disjoint from $\overline{AB}$, at every point of which $u(\xi, \eta) = M$. Since F and $\overline{AB}$ are closed and bounded sets, the distance between them is positive. Thus, there exists $\epsilon > 0$ such that the segment $A'B'$ of the straight line $\eta = \xi + \epsilon$ has no points in common with F and such that F is in the closed triangle $\Delta A'B'C$. Using Lemma 3 of subsection 1, a generalized solution $u(\xi, \eta) \in R_1$ can be represented as the limit of a sequence $\{u_n(\xi, \eta)\}$ of twice continuously differentiable solutions of (2.50) in $\Delta A'B'C$ such that $u_n(0, \eta) = 0$. All the conditions of Theorem 2.3 hold in the closed triangle $\Delta A'B'C$, and so $u_n(\xi, \eta)$ has a positive maximum on $A'B'$. Let M_1 denote the maximum of $u(\xi, \eta)$ on $A'B'$. Evidently, $M_1 < M$. Set $\epsilon = (M - M_1)/2$. Since the sequence $\{u_n(\xi, \eta)\}$ is uniformly convergent in the closed triangle $\Delta A'B'C$, it follows that for the above $\epsilon_1 > 0$ there exists an N such that for any $n > N$

$$u - \varepsilon_1 < u_n(\xi, \eta) < u + \varepsilon_1.$$

On F, we have

$$u_n(\xi, \eta) > \frac{M + M_1}{2},$$

whereas on $A'B'$

$$u_n(\xi, \eta) < \frac{M + M_1}{2} \quad \text{for} \quad n > N.$$

Hence $u_n(\xi, \eta)$ does not attain its maximum on $A'B'$, and this contradicts Theorem 2.3. Thus $u(\xi, \eta)$ assumes the value M on $\overline{AB}$. Q.E.D.

THEOREM 2.5. *If the coefficients of equation* (2.59) *satisfy conditions* (I″), *problem T has a unique solution in class* $\mathfrak{R}_1$.

PROOF. Let $v(x, y)$ be a solution of (2.59) in $\mathfrak{R}_1$ which satisfies the conditions $v|_\Gamma = 0$ and $v|_{\xi=0} = 0$. We claim that $v(x, y) \equiv 0$ in $\overline{D}$. Suppose that $v(x, y) \neq 0$ in $\overline{D}$, and let $v(x, y)$ have a positive maximum M in $\overline{D}$. According to Theorem 2.4, max $v(x, y)$ in $\overline{D}$ is attained on AB in D^+ or on Γ. Inside D^+, $v(x, y)$ cannot attain an extremum, in view of the maximum principle for elliptic equations. Since $v = 0$ on Γ, it follows that max $v(x, y)$ is attained at an interior point of AB. Let $v(x_0, 0) = M$, $0 < x_0 < 1$. According to Lemma 5,

$$\lim_{y \to 0} \frac{\partial v(x_0, y)}{\partial y} < 0 \qquad (y > 0),$$

whereas, by Theorem 2.4,

$$\lim_{y \to 0} \frac{\partial v(x_0, y)}{\partial y} \geqslant 0 \qquad (y < 0),$$

but this is impossible since $\partial v / \partial y$ is continuous on the parabolic curve. Thus $v \equiv 0$ in $\overline{D}$. Q.E.D.

REMARK. For the Čaplygin equation

$$K(y) u_{xx} + u_{yy} = 0, \qquad K(0) = 0, \qquad K'(y) > 0,$$

condition (I″) takes the form

$$\frac{d^2}{dy^2} |K|^{-1/4} \geqslant 0 \quad \text{for} \quad y < 0.$$

§3. Uniqueness theorems for the Tricomi problem for the Čaplygin equation

Consider the Čaplygin equation

$$K(y) u_{xx} + u_{yy} = 0, \tag{3.1}$$

where $K(y)$ is a monotone increasing function with continuous second derivative, and $K(0) = 0$.

Let D be the domain bounded by a simple Jordan curve Γ with endpoints $O(0, 0)$ and $A(a, 0)$, located in the upper half-plane $y > 0$, and by the characteristic arcs

$$\gamma_1\colon\ x = -\int_0^y \sqrt{-K(t)}\,dt, \qquad \gamma_2\colon\ x - a = \int_0^y \sqrt{-K(t)}\,dt \tag{3.2}$$

of equation (3.1).

The Tricomi problem consists in finding a solution of (3.1) in the domain D which has prescribed values on Γ and on the characteristic γ_1.

Frankl′ [22a] proved a uniqueness theorem for the Tricomi problem for equation (3.1), assuming that the function

$$F(y) = 2\left(\frac{K}{K'}\right)' + 1 \tag{3.3}$$

is nonnegative for $y < 0$. Protter [63b] generalized this theorem, including cases in which $F(y)$ may be negative for $y < 0$.

We shall prove uniqueness theorems for the Tricomi problem for two types of domains D. In the first case the curve Γ lying in the elliptic half-plane is arbitrary, but the part of the domain located in the hyperbolic half-plane is restricted by the condition $F(y) > d$, where $d < 0$. In the second case the hyperbolic part of the domain is arbitrary and $F(y)$ may have arbitrary values for $y < 0$, but Γ cannot extend too far in the direction of the y-axis.

DEFINITION. A function $u(x, y)$ is said to be a *quasi-regular solution of equation* (3.1) *in the domain* D if it belongs to class $C^{(2)}(D) \cap C(\overline{D})$, satisfies (3.1) in D, the integrals

$$\int_0^a u(x, 0)\, u_y(x, 0)\, dx, \qquad \iint_{D^+} (Ku_x^2 + u_y^2)\, dx\, dy$$

exist, and Green's formula is applicable to the integrals

$$\iint_D uL(u)\, dx\, dy, \quad \iint_D u_xL(u)\, dx\, dy, \quad \iint_D u_yL(u)\, dx\, dy.$$

The contour integral in Green's formula is the limit of integrals along curves approaching the boundary of the domain D from its interior.

THEOREM 2.6 [63b]. *Assume that*: 1) *$K(y)$ is a monotone increasing function with continuous second derivative, such that $K(0) = 0$, $K'(y) \neq 0$ for $y < 0$, $F(0) > 0$*; 2) *there exists a constant $d_1 < 0$ such that $F(y) \geqslant d_1$ in D*; *and*

3) *$u(x, y)$ is a quasi-regular solution of equation* (3.1), *vanishing on* Γ *and* γ_1. *Then* $u(x, y) \equiv 0$ *in* D.

THEOREM 2.7 [63b]. *Assume that*: 1) $K(y)$ *satisfies condition* 1) *of Theorem* 2.6; 2) *there exists a constant* $d_2 > 0$ *such that* $y_m < d_2$, *where* y_m *is the maximum of the ordinates of the points of* Γ, *and* 3) $u(x, y)$ *is a quasi-regular solution of equation* (3.1), *vanishing on* Γ *and* γ_1. *Then* $u(x, y) \equiv 0$ *on* D.

Theorems 2.6 and 2.7 will be proved simultaneously.

Consider the integral

$$\iint\limits_D (au + bu_x + cu_y)(Ku_{xx} + u_{yy})\, dx\, dy, \tag{3.4}$$

where $a(x, y)$, $b(x, y)$ and $c(x, y)$ are sufficiently smooth functions and $u(x, y)$ is a quasi-regular solution of (3.1).

Applying Green's formula to the integral (3.4), we get

$$\begin{aligned}0 = &\iint\limits_D \Big[\frac{1}{2}(Ka_{xx} + a_{yy})u^2 - a(Ku_x^2 + u_y^2) - \frac{1}{2}b_x(Ku_x^2 - u_y^2)\\ &- b_y u_x u_y - \frac{1}{2}(cK)_y u_x^2 - c_x K u_x u_y - \frac{1}{2}c_y u_y^2\Big]\, dx\, dy\\ &+ \int\limits_{\Gamma+\gamma_1+\gamma_2} \Big[- auu_y + \frac{1}{2}a_y u^2 - bu_x u_y + \frac{1}{2}c(Ku_x^2 - u_y^2)\Big]\, dx\\ &+ \Big[aKuu_x - \frac{1}{2}a_x K u^2 + \frac{1}{2}b(Ku_x^2 - u_y^2) + cKu_x u_y\Big]\, dy = J_1 + J_2.\end{aligned}$$

Let the solution $u(x, y)$ vanish on Γ and γ_1. Take $b = c \equiv 0$ in D^+. Then J_2 can be written as

$$\begin{aligned}J_2 = &\int\limits_{\gamma_1} \Big[- bu_x u_y + \frac{1}{2}c(Ku_x^2 - u_y^2)\Big]\, dx\\ &+ \Big[\frac{1}{2}b(Ku_x^2 - u_y^2) + cKu_x u_y\Big]\, dy\\ &+ \int\limits_{\gamma_2} \Big[- auu_y + \frac{1}{2}a_y u^2 - bu_x u_y + \frac{1}{2}c(Ku_x^2 - u_y^2)\Big]\, dx\\ &\qquad + \Big[aKuu_x - \frac{1}{2}a_x K u^2 + \frac{1}{2}b(Ku_x^2 - u_y^2) + cKu_x u_y\Big]\, dy\end{aligned}$$

or, using the equalities

$$dx = -\sqrt{-K}\, dy \quad (\text{on } \gamma_1), \qquad dx = \sqrt{-K}\, dy \quad (\text{on } \gamma_2),$$

as

$$
\begin{aligned}
J_2 = {} & \frac{1}{2}\int_{\gamma_1} (b - c\sqrt{-K})(\sqrt{-K}\,u_x - u_y)(u_x\,dx + u_y\,dy) \\
& - \frac{1}{2}\int_{\gamma_2} (b + c\sqrt{-K})\left(\sqrt{-K}\,u_x^2 + 2u_xu_y + \frac{1}{\sqrt{-K}}\,u_y^2\right) dx \qquad (3.5) \\
& - \int_{\gamma_2} a\sqrt{-K}\,u\,du - \frac{1}{2}\sqrt{-K}\,u^2(a_x\,dx + a_y\,dy) = I_1 + I_2 + I_3.
\end{aligned}
$$

Since $u \equiv 0$ on γ_1, it follows that $u_x dx + u_y dy = 0$ on γ_1, and hence $I_1 = 0$.

The integral J_1 can be divided into three integrals in the following way:

$$
\begin{aligned}
J_1 = {} & -\iint_{D^+} a(Ku_x^2 + u_y^2)\,dx\,dy \\
& - \frac{1}{2}\iint_{D^-} \{[2aK + Kb_x - (cK)_y]\,u_x^2 + 2(Kc_x + b_y)\,u_xu_y \qquad (3.6) \\
& + (2a - b_x + c_y)\,u_y^2\}\,dx\,dy + \iint_D (Ka_{xx} + a_{yy})\,u^2\,dx\,dy \\
& = I_4 + I_5 + I_6.
\end{aligned}
$$

We claim that it is possible to choose the functions a, b and c so that all the integrals $I_2, \ldots, I_6$ are nonpositive. As their sum is equal to zero, it will follow that each of them must vanish. Then $I_4 = 0$ (for $a > 0$) will imply that $u(x, y) \equiv c$ in D^+, and since $u \equiv 0$ on Γ, it will follow that $u(x, y) \equiv 0$ in D^+. In particular, $u(x, 0) = 0$ and $\partial u(x, 0)/\partial y = 0$, so that $u(x, y) \equiv 0$ in D^-, because of the uniqueness of the solution of the Cauchy problem for equation (3.1). Thus, $u(x, y) \equiv 0$ everywhere in D.

For $y < 0$ we set

$$
c = \frac{4aK}{K'}, \qquad b = -c\sqrt{-K}. \tag{3.7}
$$

Obviously, with this choice of $b(x, y)$ we have $I_2 = 0$. Integration of I_3 by parts gives

$$
I_3 = \int_{\gamma_2} \left(\sqrt{-K}\,a_x + a_y + \frac{aK'}{4K}\right) u^2\,dx.
$$

This integral is nonpositive if

$$
\sqrt{-K}\,a_x + a_y + \frac{aK'}{4K} \leqslant 0 \qquad \text{for} \quad y < 0
$$

or if

$$
a(x, y) \geqslant 0 \qquad \text{for} \quad y \leqslant 0, \tag{3.8}
$$

$$\frac{\sqrt{-K}\, a_x + a_y}{a} \leqslant 0 \qquad \text{for} \quad y \leqslant 0. \tag{3.9}$$

The integral I_5 is nonpositive if

$$(Kc_x + b_y)^2 \leqslant (2a - b_x + c_y)(2aK + Kb_x - (cK)_y) \quad \text{for} \quad y \leqslant 0 \tag{3.10}$$

and

$$2a - b_x + c_y \geqslant 0 \qquad \text{for} \quad y \leqslant 0. \tag{3.11}$$

By (3.7), inequality (3.10) holds for any function $a(x, y)$, and (3.11) becomes

$$\sqrt{-K}\, a_x + a_y + a \frac{K'}{2K} F(y) \leqslant 0. \tag{3.12}$$

We now distinguish two different cases.

Case 1. $F(y) > 0$ for all $y < 0$. Define $a(x, y)$ to be any positive constant for all (x, y) in D. Inequalities (3.8) and (3.9) hold, and $I_3 \leqslant 0$. It is clear that $I_4 \leqslant 0$, $I_5 \leqslant 0$ and $I_6 = 0$. Since $I_1 + \cdots + I_6 = 0$, this implies that each integral equals zero, and so $u(x, y) \equiv 0$ in D.

Case 2. The function $F(y)$ may be negative for $y < 0$. Set

$$a(x, y) = \begin{cases} e^{-\beta x}, & y \leqslant 0, \\ e^{-\beta x} \cos \gamma y, & y \geqslant 0, \end{cases} \tag{3.13}$$

where β and γ are constants.

In the proof of Theorem 2.6 we let $\gamma = \pi/2y_m$ and $\beta = \gamma/\sqrt{K(y_m)}$. Inequality (3.12) will hold if

$$F(y) \geqslant \frac{2K\sqrt{-K}\,\beta}{K'}. \tag{3.14}$$

Let

$$d_1 = \min_{y \in D^-} \frac{2K\sqrt{-K}\,\beta}{K'};$$

then, if $F(y) > d_1$, it follows from (3.14) that $I_5 \leqslant 0$. It is easy to see that all the other integrals are nonpositive. Hence, as before, it follows that $u(x, y) \equiv 0$ in D.

To prove Theorem 2.7, we set

$$\beta = \max_{y<0} \frac{K'}{2K\sqrt{-K}} F(y).$$

Then, if $y_m < \pi/2\beta\sqrt{K(y_m)}\,(= d_2)$, we let $\gamma = \beta\sqrt{K(y_m)}$, and simple calculations show that all the integrals are positive; hence $u(x, y) \equiv 0$ in D.

We now consider yet another theorem of Protter [63b].

THEOREM 2.8. *Assume that* 1) *$K(y)$ has a continuous third order derivative such that $K'''(y) \leqslant 0$ whenever $F(y) < 0$ for $y < 0$;* 2) *$u(x, y)$ is a quasi-regular solution of equation* (3.1), *vanishing on Γ and γ_1. Then $u(x, y) \equiv 0$ in D.*

PROOF. Using the same techniques as in the proof of Theorems 2.6 and 2.7, we again obtain (3.5) and (3.6). We claim that the functions a, b and c can be so chosen that all the integrals $I_2, \ldots, I_6$ are nonpositive. As in the preceding proof, it will then follow easily that $u(x, y) \equiv 0$ in D.

For $y < 0$ we again set

$$c = \frac{4aK}{K'}, \qquad b = -c\sqrt{-K}. \tag{3.7}$$

Recall that for $y \geqslant 0$ we set $b = c \equiv 0$. Obviously, with this choice of $b(x, y)$ we have $I_2 = 0$.

For the sake of simplicity, let us assume that $F(y)$ is nonnegative for $y_0 \leqslant y \leqslant 0$ and negative for $y < y_0$.

By virtue of the choice of b and c, I_5 can be written as

$$I_5 = -\iint_{D^-} \left[aF(y) + \frac{2K}{K'}\left(a_y + \sqrt{-K}\,a_x\right)\right]\left(\sqrt{-K}\,u_x + u_y\right)^2 dx\,dy.$$

This integral is positive if

$$aF(y) + \frac{2K}{K'}\left(a_y + \sqrt{-K}\,a_x\right) \geqslant 0 \quad \text{for} \quad y < 0.$$

Let x_1 be the maximum abscissa of the points of Γ. Define

$$a(x, y) = \begin{cases} [-K(y)]^{-3/2} K'(y)\left[(x_1 - x) + \int_0^y \sqrt{-K(t)}\,dt\right], & y \leqslant y_0, \\ [-K(y_0)]^{-3/2} K'(y_0)\left[(x_1 - x) + \int_0^y \sqrt{-K(t)}\,dt\right], & y_0 \leqslant y \leqslant 0, \\ [-K(y_0)]^{-3/2} K'(y_0)(x_1 - x), & y \geqslant 0. \end{cases} \tag{3.15}$$

It is easy to see that $a(x, y)$ has continuous first order derivatives and piecewise-continuous second order derivatives in D. In fact, for $y = 0$ we have $a_y(x, 0+) = 0$ and

$$a_y(x, 0-) = [-K(y_0)]^{-3/2} K'(y_0)\sqrt{-K(0)} = 0.$$

On the line $y = y_0$ we have

$$a_y(x, y_0 + 0) = [-K(y_0)]^{-3/2} K'(y_0) \sqrt{-K(y_0)}$$

and

$$a_y(x, y_0 - 0) = [-K(y_0)]^{-3/2} K'(y_0) \sqrt{-K(y_0)} + \frac{1}{2}[-K(y_0)]^{-5/2}[3K'^2(y_0) - 2K(y_0)K''(y_0)] \times \left[(x_1 - x) + \int_0^{y_0} \sqrt{-K(t)}\,dt\right].$$

Hence, since $F(y_0) = 0$,

$$a_y(x, y_0 + 0) = a_y(x, y_0 - 0).$$

Obviously, $a(x, y) \geqslant 0$ in D. Indeed, if $y \geqslant 0$ this follows from the choice of x_1. For negative y, we observe that the quantity $-x + \int_0^y \sqrt{-K(t)}\,dt$ is constant on one of the two families of characteristics of equation (3.1). The quantity $x_1 - x + \int_0^y \sqrt{-K(t)}\,dt$ is positive everywhere in D^-, since it is positive at $y = 0$.

By (3.15) and (3.3),

$$aF(y) + \frac{2K}{K'}(a_y + \sqrt{-K}\,a_x) = 0 \qquad \text{for} \quad y \leqslant y_0,$$

$$aF(y) + \frac{2K}{K'}(a_y + \sqrt{-K}\,a_x) = aF(y) \quad \text{for} \quad y_0 \leqslant y \leqslant 0.$$

Hence it follows that $I_5 \leqslant 0$, since $F(y) \geqslant 0$ for $y_0 \leqslant y \leqslant 0$. Integrating I_3 by parts and dividing it into two integrals gives

$$I_3 = \int_{\gamma_2'} \left(\sqrt{-K}\,a_x + a_y + \frac{aK'}{4K}\right) u^2\,dx + \int_{\gamma_2''} \left(\sqrt{-K}\,a_x + a_y + \frac{aK'}{4K}\right) u^2\,dx,$$

where γ_2' is the part of the characteristic γ_2 located above $y = y_0$ and γ_2'' the part located below $y = y_0$. Substituting $a(x, y)$ from (3.15), we get

$$I_3 = \int_{\gamma_2'} [-K(y_0)]^{-3/2} K'(y_0) \left[(x_1 - x) + \int_0^y \sqrt{-K(t)}\,dt\right] \frac{K'}{4K} u^2\,dx + \int_{\gamma_2''} \left[-\frac{K'(y)}{2K(y)} aF(y) + a\frac{K'}{4K}\right] u^2\,dx.$$

The first integral is obviously nonpositive. The second integral is also nonpositive, since $F(y) \leqslant 0$ for $y \leqslant y_0 < 0$. Thus $I_3 \leqslant 0$. It is clear that $I_4 \leqslant 0$. Consider the last integral, I_6, which can be written in the following form thanks to (3.15):

$$I_6 = \frac{1}{2} \iint\limits_{D^-} a_{yy} u^2 \, dx \, dy.$$

Let D_1^- denote the part of D^- located above the line $y = y_0$, and D_2^- the part below $y = y_0$. Then I_6 can be divided into two integrals:

$$I_6 = -\frac{1}{2} \iint\limits_{D_1^-} [-K(y_0)]^{-3/2} K'(y_0) \frac{K'(y)}{\sqrt{-K(y)}} u^2 \, dx \, dy + \frac{1}{2} \iint\limits_{D_2^-} a_{yy} u^2 \, dx \, dy. \tag{3.16}$$

For $y \leqslant y_0$,

$$a_{yy} = \frac{1}{2} [-K(y)]^{-2} [5K'^2(y) - 4K(y) K''(y)] + \left[(x_1 - x) + \int_0^y \sqrt{-K(t)} \, dt \right] [-K(y)]^{-7/2} \times \left[K^2 K''' + \frac{3}{4} K' (5K'^2 - 6KK'') \right]. \tag{3.17}$$

By assumption, $3K'^2 \leqslant 2KK''$ for $y \leqslant y_0$. Therefore, $5K'^2 \leqslant 4KK''$, and the first term on the right of (3.17) is negative. It is clear that $5K'^2 \leqslant 6KK''$ for $y \leqslant y_0$, and the second term in (3.17) is negative if $K''' < 0$ for $y \leqslant y_0$. In fact, a sharper sufficient condition can be given:

$$K''' \leqslant \frac{3K'}{4K^2} (6KK'' - 5K'^2), \qquad y \leqslant y_0.$$

Thus it follows from (3.16) that $I_6 \leqslant 0$. Q.E.D.

§4. Existence theorem for the solution of the Tricomi problem[4]

Consider the equation

$$y u_{xx} + u_{yy} + c(x, y) u = 0. \tag{4.1}$$

Let D be a domain bounded by a simple Jordan curve Γ in the upper half-plane with endpoints A (0, 0), B (1, 0), and by characteristics

$$AC: \ \xi = x - \frac{2}{3}(-y)^{3/2} = 0 \quad \text{and} \quad BC: \ \eta = x + \frac{2}{3}(-y)^{3/2} = 1$$

of equation (4.1). Let D^+ and D^- denote the regions of D in the half-planes $y > 0$ and $y < 0$, respectively.

Let $x = x(s)$, $y = y(s)$ be parametric equations of the curve Γ, where s is

---

(4) This section and the next present results of Babenko.

arc-length reckoned from the point B. Concerning the curve Γ, we make the following assumptions:

1) The functions $x(s)$ and $y(s)$ have continuous derivatives $x'(s)$ and $y'(s)$ which do not vanish simultaneously on $[0, l]$, and the second derivatives $x''(s)$ and $y''(s)$ satisfy a Hölder condition on $[0, l]$, where l is the length of Γ.

2) In neighborhoods of the points A and B on Γ,

$$\left|\frac{dx}{ds}\right| \leqslant C^2 y^2(s),$$

where C is a constant.

Tricomi Problem. *Find a solution of equation* (4.1) *in the domain D assuming prescribed values on Γ and on the characteristic AC*:

$$u|_\Gamma = \varphi(s), \qquad u|_{\xi=0} = \psi(\eta),$$

where $\varphi(s)$ satisfies a Hölder condition with exponent α and $\psi(\eta)$ has a bounded first derivative satisfying a Hölder condition with exponent β.

We shall look for a solution $u(x, y)$ of the problem T in class $\mathfrak{R}_1$.

The fundamental idea (due to Tricomi himself [69a]) of the existence proof for solutions of the Tricomi problem for equation (4.1) is as follows. One first solves problem N (see below) in the domain D^+ and the Cauchy-Goursat problem in D^-, treating $\nu(x) = \partial u(x, 0)/\partial y$ as a known function. The two solutions obtained are then matched, together with their first derivatives, on the interval AB of the parabolic curve. Thus the existence proof for a solution of the Tricomi problem is reduced to solution of a singular integral equation for $\nu(x)$, whose solvability follows from the uniqueness theorem.

1. Problem N. *Find a regular solution of equation* (4.1) *in the domain D^+, continuous in the closed domain $\overline{D}^+$, having continuous derivatives u_x and u_y in $D + \Gamma$ (except that at A and B the derivatives may go to infinity of order less than $2/3$*(5)*), and satisfying the boundary conditions*

$$u|_\Gamma = \varphi(s), \qquad \lim_{y\to+0} \frac{\partial u}{\partial y} = \nu(x), \tag{4.2}$$

where $\nu(x)$ is continuous for $0 < x < 1$ and may go to infinity of order less than $2/3$ as $x \longrightarrow 0$ and $x \longrightarrow 1$ (as a function of x and $(1 - x)$).

The uniqueness of the solution of problem N follows from Lemma 5 of §2.1.

The solution of problem N for the equation

$$yu_{xx} + u_{yy} = 0 \tag{*}$$

(5) $u_x = O(r^{-2/3+\epsilon})$, $u_y = O(r^{-2/3+\epsilon})$, where r is the distance from A or B.

is given (see [4a] or [66g]) by

$$u_0(x, y) = -\int_0^1 \nu(t)\, G(t, 0;\ x, y)\, dt - \int_0^l \varphi(s)\, \rho(s;\ x, y)\, ds. \tag{4.3}$$

Here

$$G(x, y;\ x_0, y_0) = q(x, y;\ x_0, y_0) + p(x, y;\ x_0, y_0)$$

is the Green's function of problem N for equation (*),

$$q(x, y;\ x_0, y_0) = k r_1^{-1/3} F\left(\frac{1}{6}, \frac{1}{6}, \frac{1}{3};\ 1 - \sigma\right) \tag{4.4}$$

is a fundamental solution of (*), where

$$\left.\begin{matrix} r^2 \\ r_1^2 \end{matrix}\right\} = (x - x_0)^2 + \frac{4}{9}\left(y^{3/2} \mp y_0^{3/2}\right)^2, \quad \sigma = \frac{r^2}{r_1^2}, \quad k = \left(\frac{4}{3}\right)^{1/3} \frac{\Gamma^2\left(\frac{1}{6}\right)}{4\pi\Gamma\left(\frac{1}{3}\right)};$$

$p(x, y; x_0, y_0)$ is a regular solution of (*) everywhere in the interior of D^+, and $\rho(s; x, y)$ is a solution of the integral equation

$$\rho(s;\ x, y) - 2\int_0^l K(t, s)\, \rho(t;\ x, y)\, dt = 2A(s;\ x, y), \tag{4.5}$$

where

$$A(s;\ x;\ y) \equiv \eta \frac{d\eta}{ds} \frac{\partial q(\xi, \eta;\ x, y)}{\partial \xi} - \frac{d\xi}{ds} \frac{dq(\xi, \eta;\ x, y)}{\partial \eta}, \quad (\xi, \eta) \in \Gamma,$$

$$K(t, s) = A(s;\ x(t), y(t)), \qquad (x(t), y(t)) \in \Gamma. \tag{4.6}$$

In the case of a normal domain D_0 bounded by the interval [0, 1] of the x-axis and by the normal curve

$$C_{1/2}: \left(x - \frac{1}{2}\right)^2 + \frac{4}{9} y^3 = \frac{1}{4}, \quad y \geqslant 0,$$

the Green's function of problem N for equation (*) may be written out explicitly:

$$G_0(x, y;\ x_0, y_0) = q(x, y;\ x_0, y_0) - (2r_0)^{-1/3}\, q\left(x - \frac{1}{2}, y;\ \bar{x}_0, \bar{y}_0\right), \tag{4.7}$$

where

$$r_0^2 = \left(x_0 - \frac{1}{2}\right)^2 + \frac{4}{9} y_0^3, \quad \bar{x}_0 = \frac{x_0 - \frac{1}{2}}{4r_0^2}, \quad \bar{y}_0^{3/2} = \frac{y_0^{3/2}}{4r_0^2}.$$

Then

$$\begin{aligned} G(x, y;\ x_0, y_0) &- G_0(x, y;\ x_0, y_0) \\ &= \int_0^l \rho(s;\ x, y)\, G_0(\xi, \eta;\ x_0, y_0)\, ds. \end{aligned} \tag{4.8}$$

Setting $y = 0$ in (4.3) and using (4.8), we get

$$u_0(x, 0) = -\int_0^1 \nu(t)\, G_0(t, 0;\ x, 0)\, dt - \int_0^1 \nu(t)\, H(t, x)\, dt - \int_0^l \varphi(s)\, \rho(s;\ x, 0)\, ds$$

or, by (4.4) and (4.7),

$$\tau(x) = -k \int_0^1 \left[\frac{1}{|x-t|^{1/3}} - \frac{1}{(x+t-2xt)^{1/3}} \right] \nu(t)\, dt - \int_0^1 H(t, x)\, \nu(t)\, dt - \int_0^l \varphi(s)\, \rho(s;\ x, 0)\, ds, \tag{4.9}$$

where

$$H(t, x) = G(t, 0;\ x, 0) - G_0(t, 0;\ x, 0) = \int_0^l \rho(s;\ t, 0)\, G_0(\xi, \eta;\ x, 0)\, ds. \tag{4.10}$$

LEMMA 1 [4a]. *For* $0 < x < 1$ *and* $0 < t < 1$,

$$\left| \frac{\partial H(t, x)}{\partial x} \right| \leqslant \frac{C}{(x+t-2xt)^{1/3}}, \tag{4.11}$$

where C *is a constant depending only on the domain* D^+.

PROOF. Setting $(\xi - 1/2)^2 + 4\eta^3/9 = R^2$, we infer from (4.7) and (4.4) that

$$G_0(\xi, \eta;\ x, 0) = k\left[R^2 - 2\left(\xi - \frac{1}{2}\right)\left(x - \frac{1}{2}\right) + \left(x - \frac{1}{2}\right)^2\right]^{-1/6} - k\left[4\left(x - \frac{1}{2}\right)^2 R^2 - 2\left(\xi - \frac{1}{2}\right)\left(x - \frac{1}{2}\right) + \frac{1}{4}\right]^{-1/6}$$

or $G_0(\xi, \eta; x, 0) = k(R_1^{-1/3} - R_2^{-1/3})$, where we have put

$$R_1^2 = R^2 - 2\left(\xi - \frac{1}{2}\right)\left(x - \frac{1}{2}\right) + \left(x - \frac{1}{2}\right)^2,$$
$$R_2^2 = 4\left(x - \frac{1}{2}\right)^2 R^2 - 2\left(\xi - \frac{1}{2}\right)\left(x - \frac{1}{2}\right) + \frac{1}{4}.$$

By the conditions imposed on the curve Γ, we have $R^2 = 1/4 + O(\eta^3)$ for small

η. Then

$$\frac{3}{k}\frac{\partial G_0}{\partial x} = (x-\xi)\left(R_2^{-7/3} - R_1^{-7/3}\right) + \left(x - \frac{1}{2}\right) R_2^{-7/3} O\left(\eta^3\right).$$

Note that

$$\left| R_2^{-7/3} - R_1^{-7/3} \right| \leqslant \left| R_2^2 - R_1^2 \right| \left(\frac{1}{R_1^2 R_2^{7/3}} + \frac{1}{R_2^2 R_1^{7/3}} \right)$$
$$\leqslant C x (1-x) \left(\frac{1}{R_1^2 R_2^{7/3}} + \frac{1}{R_2^2 R_1^{7/3}} \right) \eta^3.$$

For $0 \leqslant x \leqslant 1/4$ we have

$$\frac{1}{R_1^2} \leqslant \frac{C_2}{x^2 + \frac{4}{9}\eta^3}, \qquad \frac{1}{R_2^2} \leqslant \frac{C_3}{x^2 + \frac{4}{9}\eta^3},$$

where the constants C_2 and C_3 do not depend on x, ξ or η. Hence,

$$\left| \frac{\partial G_0}{\partial x} \right| \leqslant C_4 \frac{|x-\xi|\, x(1-x)\, \eta^3}{\left(x^2 + \frac{4}{9}\eta^3\right)^{13/6}} + C_5 \frac{\eta^3}{\left(x^2 + \frac{4}{9}\eta^3\right)^{7/6}} < \frac{C_6}{\left(x^2 + \frac{4}{9}\eta^3\right)^{1/6}}.$$

Similarly, for $3/4 \leqslant x \leqslant 1$,

$$\left| \frac{\partial G_0}{\partial x} \right| \leqslant \frac{C_7}{\left((x-1)^2 + \frac{4}{9}\eta^3\right)^{1/6}}.$$

Thus, for $0 \leqslant x \leqslant 1/4$, it follows from (4.10) and (4.5) that

$$\left| \frac{\partial H(t,x)}{\partial x} \right| \leqslant \int_0^l |\rho(s;t,0)| \left| \frac{\partial G_0(\xi,\eta;x,0)}{\partial x} \right| ds$$
$$\leqslant 2C_6 \int_0^l \frac{|A(s;t,0)|\, ds}{\left(x^2 + \frac{4}{9}\eta^3\right)^{1/6}} + 4C_6 \int_0^l |A(t_1;t,0)|\, dt_1 \int_0^l \frac{|R(s,t_1,2)|\, ds}{\left(x^2 + \frac{4}{9}\eta^3\right)^{1/6}}$$
$$\leqslant 2C_6 \int_{l-\varepsilon}^l \frac{|A(s;t,0)|\, ds}{\left(x^2 + \frac{4}{9}\eta^3\right)^{1/6}} + C_7.$$

Then

$$A(s;t,0) = \frac{k}{3} \frac{(t-\xi)\eta \frac{d\eta}{ds} - \frac{2}{3}\eta^2 \frac{d\xi}{ds}}{\left[(t-\xi)^2 + \frac{4}{9}\eta^3\right]^{7/6}},$$

and so

$$|A(s;\, t,\, 0)| \leqslant C_8 \frac{\eta^{1/2}\left|\dfrac{d\eta}{ds}\right|}{\left[(t-\xi)^2+\dfrac{4}{9}\eta^3\right]^{1/2}} + C_9.$$

Hence,

$$\left|\frac{\partial H(t,\, x)}{\partial x}\right| \leqslant C_{10} \int\limits_{l-\varepsilon}^{l} \frac{\eta^{1/2}\left|\dfrac{d\eta}{ds}\right| ds}{\left[(t-\xi)^2+\dfrac{4}{9}\eta^3\right]^{1/2}\left(x^2+\dfrac{4}{9}\eta^2\right)^{1/6}} + C_{11}$$

$$\leqslant C_{12} \int\limits_{l-\varepsilon}^{l} \frac{\eta^{1/2}\left|\dfrac{d\eta}{ds}\right| ds}{\left(t^2+\dfrac{4}{9}\eta^3\right)^{1/2}\left(x^2+\dfrac{4}{9}\eta^3\right)^{1/6}} + C_{11}$$

$$= C_{12} \int\limits_{0}^{\delta} \frac{d\tilde{\eta}}{(t^2+\tilde{\eta}^2)^{1/2}(x^2+\tilde{\eta}^2)^{1/6}} + C_{11}.$$

If $\tilde{x} = \max(x, t)$, it is easy to see that

$$\left|\frac{\partial H(t,\, x)}{\partial x}\right| \leqslant \frac{C_{13}}{\tilde{x}^{1/3}} \leqslant \frac{2C_{13}}{(x+t)^{1/3}} < \frac{C}{(x+t-2xt)^{1/3}} \quad \text{for} \quad 0 < x \leqslant \frac{1}{4}.$$

Evidently, the same estimate is valid for $3/4 \leqslant x < 1$.

LEMMA 2. *Let $\mu(x, y)$ be a bounded function in D^+. Then the function*

$$v(x,\, y) = \iint\limits_{D^+} \mu(\xi,\, \eta)\, q(x,\, y;\, \xi,\, \eta)\, d\xi\, d\eta$$

and its first-order partial derivatives are continuous in the half-plane $y \geqslant 0$. If $\mu(x, y)$ is continuous in $\overline{D}^+$ and has continuous first-order derivatives in the interior of D^+, then $v(x, y)$ is twice continuously differentiable in D^+ and satisfies the equation

$$yv_{xx} + v_{yy} = -\mu(x,\, y).$$

This lemma is proved in exactly the same way as the analogous theorem in potential theory.

LEMMA 3 ([4a]; see also [66g], Chapter II, §8). *Let $\mu(x, y)$ be a bounded function in $\overline{D}^+$. Then the function*

$$\Psi(x,\, y) = \iint\limits_{D^+} \mu(\xi,\, \eta)\, p(x,\, y;\, \xi,\, \eta)\, d\xi\, d\eta \tag{4.12}$$

has the following properties:

1) *It is continuous in $\overline{D}^+$.*
2) *Ψ_x, Ψ_y, Ψ_{xx} and Ψ_{yy} exist inside D^+ and are continuous there.*
3) *If $x_1(\delta)$ and $x_2(\delta)$ are the abscissas of the intersection points of the*

curve Γ *with a straight line* $y = \delta > 0$ (*where* δ *is small*) *and*

$$\hat{x} = \min\{x - x_1(\delta),\ x_2(\delta) - x_2\},$$

then

$$|\Psi_x(x, \delta)| \leqslant B \ln \frac{1}{\hat{x}}, \qquad |\Psi_y(x, \delta)| \leqslant B\delta^{1/2} \ln \frac{1}{\hat{x}}. \tag{4.13}$$

4) $\Psi_x(x, 0)$ *and* $\Psi_{xx}(x, 0)$ *exist and are continuous for* $0 < x < 1$; *moreover*, $|\Psi_x(x, 0)| \leqslant B_1$.

We now construct a solution $u(x, y)$ of problem N for equation (4.1). Let $u_0(x, y)$ be the solution of problem N for equation (*) with the same boundary data on the boundary of D^+. Then the solution of problem N for (4.1) is reduced to solution of the integral equation

$$u(x, y) - \iint\limits_{D^+} c(\xi, \eta)\, G(x, y; \xi, \eta)\, u(\xi, \eta)\, d\xi\, d\eta = u_0(x, y), \tag{4.14}$$

to which the Fredholm theory is applicable.

Assume that (4.14) has a bounded solution $u(x, y)$. Then, by Lemmas 2 and 3, $u(x, y)$ is continuous in $\overline{D}^+$, twice continuously differentiable in D^+, and satisfies (4.1). By the definition of the Green's function $G(x, y; \xi, \eta)$, we have $u(x, y) = u_0(x, y)$ on Γ and

$$\lim_{y \to 0} \frac{\partial u(x, y)}{\partial x} = \lim_{y \to 0} \frac{\partial u_0(x, y)}{\partial y},$$

i.e. $u(x, y)$ is a solution of problem N for equation (4.1).

The integral equation (4.14) is always solvable, since the corresponding homogeneous equation has only the trivial solution, as follows from the uniqueness theorem for problem N.

Let $R(x, y; \xi, \eta; \lambda)$ be the resolvent of the kernel

$$c(\xi, \eta)\, G(x, y; \xi, \eta);$$

then the solution of (4.14) is written as

$$u(x, y) = u_0(x, y) + \iint\limits_{D^+} R(x, y; \xi, \eta; 1)\, u_0(\xi, \eta)\, d\xi\, d\eta \tag{4.15}$$

or, by (4.3),

$$u(x, y) = -\int_0^1 \nu(t)\, \tilde{G}(t, 0; x, y)\, dt - \int_0^l \varphi(s)\, \tilde{\rho}(s; x, y)\, ds, \tag{4.16}$$

where

$$\tilde{G}(t, 0; x, y) = G(t, 0; x, y) + \iint_{D^+} R(x, y; \xi, \eta; 1)\, G(t, 0; \xi, \eta)\, d\xi\, d\eta = G(t, 0; x, y) + \tilde{H}(t, x, y), \tag{4.17}$$

$$\tilde{\rho}(s; x, y) = \rho(s; x, y) + \iint_{D^+} R(x, y; \xi, \eta; 1)\, \rho(s; \xi, \eta)\, d\xi\, d\eta. \tag{4.18}$$

It can be shown that

$$|\tilde{H}(t, x, y)| < C \tag{4.19}$$

for $0 \leqslant t \leqslant 1$ and $(x, y) \in \overline{D}^+$.

Set $y = 0$ in (4.16); then, in view of (4.17) and (4.10),

$$\tau(x) = -k \int_0^1 \left[\frac{1}{|x-t|^{1/3}} - \frac{1}{(x+t-2xt)^{1/3}}\right] \nu(t)\, dt - \int_0^1 \chi(t, x)\, \nu(t)\, dt - \int_0^l \varphi(s)\, \tilde{\rho}(s, x, 0)\, ds, \tag{4.20}$$

where

$$\chi(t, x) = H(t, x) + \tilde{H}(t, x, 0). \tag{4.21}$$

Formula (4.20) gives the *first functional equation* for $\tau(x)$ and $\nu(x)$; it follows from the condition that the solution $u(x, y)$ of (4.1) in D^+ must take prescribed values $\varphi(s)$ on Γ.

LEMMA 4 [4a]. *There exist constants H_1 and H_2 such that*

$$\left|\frac{\partial}{\partial x} \tilde{H}(t, x, 0)\right| \leqslant H_1 \ln \frac{1}{t-x} + H_2 (t + x - 2xt)^{-1/3}. \tag{4.22}$$

The proof is tedious and is therefore not given here.

2. CAUCHY-GOURSAT PROBLEM. *Find a solution of equation* (4.1) *in* D^-, *continuous in* $\overline{D}^-$ *and satisfying boundary conditions*

$$u|_{\xi=0} = \psi(\eta), \qquad \left.\frac{\partial u}{\partial y}\right|_{y=0} = \nu(x). \tag{4.23}$$

In characteristic coordinates equation (4.1) is

$$F(u) \equiv \frac{\partial^2 u}{\partial \xi\, \partial \eta} + \frac{1}{6(\eta - \xi)} \left(\frac{\partial u}{\partial \xi} - \frac{\partial u}{\partial \eta}\right) + C(\xi, \eta)\, u = 0, \tag{4.24}$$

where

$$C(\xi, \eta) = -\frac{1}{4}\left(\frac{4}{3}\right)^{2/3} \frac{c}{(\eta-\xi)^{2/3}}. \tag{4.25}$$

The domain D^- is transformed into a triangle ΔABC bounded by the characteristics $\xi = 0$, $\eta = 1$ and the straight line $\eta = \xi$. The boundary conditions (4.23) become

$$\begin{gathered} u\,|_{\xi=0} = \psi(\eta), \\ \lim_{\eta-\xi\to 0} \left(\frac{3}{4}\right)^{1/3} (\eta-\xi)^{1/3}\left(\frac{\partial u}{\partial \eta} - \frac{\partial u}{\partial \xi}\right) = \nu(\xi). \end{gathered} \tag{4.26}$$

The Riemann-Hadamard function(6) plays an essential part in solution of the Cauchy-Goursat problem.

The Riemann-Hadamard function $v(\xi, \eta; \xi_0, \eta_0)$ for the equation

$$E(u) \equiv \frac{\partial^2 u}{\partial \xi\, \partial \eta} + \frac{1}{6(\eta-\xi)}\left(\frac{\partial u}{\partial \xi} - \frac{\partial u}{\partial \eta}\right) = 0 \tag{4.27}$$

is defined by the following conditions:

1) It satisfies the adjoint equation $E^*(u) = 0$ as a function of (ξ, η).
2) It satisfies the equation $E(u) = 0$ as a function of (ξ_0, η_0).
3) $v(\xi_0, \eta_0; \xi_0, \eta_0) = 1$.
4) $v(\xi, \xi; \xi_0, \eta_0) = 0$.
5) $\partial[v]/\partial\xi + [v]/6(\eta - \xi) = 0$ for $\eta = \xi_0$, where

$$[v] = \lim_{\varepsilon\to 0}[v(\xi, \xi_0+\varepsilon; \xi_0, \eta_0) - v(\xi, \xi_0-\varepsilon; \xi_0, \eta_0)].$$

The Riemann-Hadamard function $v(\xi, \eta; \xi_0, \eta_0)$ was first constructed by Gellerstedt [24a]. It has the form

$$v(\xi, \eta; \xi_0, \eta_0) = \begin{cases} v_1(\xi, \eta; \xi_0, \eta_0) = \left(\dfrac{\eta-\xi}{\eta_0-\xi_0}\right)^{1/6} F\left(\dfrac{1}{6}, \dfrac{5}{6}, 1; s\right) & \text{for } \eta \geqslant \xi_0, \\ v_2(\xi, \eta; \xi_0, \eta_0) = k' \dfrac{(\eta-\xi)^{1/3}}{(\xi_0-\xi)^{1/6}(\eta_0-\eta)^{1/6}} F\left(\dfrac{1}{6}, \dfrac{1}{6}, \dfrac{1}{3}; \dfrac{1}{s}\right) & \text{for } \eta \leqslant \xi_0, \end{cases} \tag{4.28}$$

where

$$s = \frac{(\xi_0-\xi)(\eta_0-\eta)}{(\eta-\xi)(\eta_0-\xi_0)}, \qquad k' = \frac{\Gamma\left(\frac{1}{6}\right)}{\Gamma\left(\frac{5}{6}\right)\Gamma\left(\frac{1}{3}\right)}.$$

---

(6) On p. 27 we constructed the Riemann-Hadamard function for the Goursat problem.

Note that condition (5) can be written as

$$\frac{\partial (v_1 - v_2)}{\partial \xi} + \frac{1}{6}\,\frac{v_1 - v_2}{\eta - \xi} = 0 \quad \text{for} \quad \eta = \xi_0.$$

Let $u(\xi, \eta)$ be a twice continuously differentiable solution of (4.24) in $\triangle ABC$ which satisfies boundary conditions (4.26). We have $vE(u) - uE^*(v) = H_\xi + L_\eta$, where

$$H = \frac{1}{2}\left(v\,\frac{\partial u}{\partial \eta} - u\,\frac{\partial v}{\partial \eta}\right) + \frac{uv}{6(\eta - \xi)},$$

$$L = \frac{1}{2}\left(v\,\frac{\partial u}{\partial \xi} - u\,\frac{\partial v}{\partial \xi}\right) - \frac{uv}{6(\eta - \xi)}.$$

Thus, integrating the identity

$$H_\xi + L_\eta = -\,C(\xi, \eta)\,uv$$

over the triangle $\Delta A'B'C'$ (see Figure 6 on p. 28) bounded by a segment $A'B'$ of the line $\eta - \xi = \epsilon$, a segment $B'C'$ of the characteristic $\eta = \xi_0 - \epsilon$ and a segment $C'A'$ of the characteristic $\xi = 0$, and also over the rectangle $C''B''D''D'$ bounded by segments $C''B''$ and $D'D''$ of the characteristics $\eta = \xi_0 + \epsilon$ and $\eta = \eta_0$, and segments $C''D'$ and $B''D''$ of the characteristics $\xi = 0$ and $\xi = \xi_0 - 2\epsilon$, we see, letting $\epsilon \longrightarrow 0$, that

$$\begin{aligned} u(\xi_0, \eta_0) = k \int_0^{\xi_0} & \frac{\nu(t)\,dt}{(\xi_0 - t)^{1/6}(\eta_0 - t)^{1/6}} \\ & + \int_0^{\eta_0} \left[\psi'(\eta) + \frac{\psi(\eta)}{6\eta}\right] v(0, \eta; \xi_0, \eta_0)\,d\eta \\ & - \int_0^{\xi_0} d\xi \int_\xi^{\eta_0} C(\xi, \eta)\, v(\xi, \eta; \xi_0, \eta_0)\, u(\xi, \eta)\,d\eta, \end{aligned} \tag{4.29}$$

$$k = \frac{1}{2}\left(\frac{4}{3}\right)^{1/3} \frac{\Gamma\left(\frac{1}{6}\right)}{\Gamma\left(\frac{1}{3}\right)\Gamma\left(\frac{5}{6}\right)}.$$

Here we have used conditions 1)–5) and the boundary conditions (4.26).

According to Lemma 3 of §2, (4.29) remains valid for a generalized solution of (4.24) in $\mathfrak{R}_1$.

We consider the case in which the generalized solution $u(\xi, \eta) \in \mathfrak{R}_1$ vanishes on the characteristic $\xi = 0$. Then (4.29) yields an integral equation for $u(\xi, \eta)$:

$$u(\xi,\eta)+\int_0^\xi d\xi'\int_{\xi'}^\eta C(\xi',\eta')\,v(\xi',\eta';\xi,\eta)\,u(\xi',\eta')\,d\eta'$$
$$=k\int_0^\xi(\xi-t)^{-1/6}(\eta-t)^{-1/6}\nu(t)\,dt. \tag{4.30}$$

We look for a solution of (4.30) in the form

$$u(\xi,\eta)=\int_0^\xi K(\xi,\eta,t)\,\nu(t)\,dt. \tag{4.31}$$

The function $K(\xi, \eta, t)$ is defined by the equation

$$K(\xi,\eta,t)+\int_t^\xi d\xi'\int_{\xi'}^\eta C(\xi',\eta')\,v(\xi',\eta';\xi,\eta)\,K(\xi',\eta',t)\,d\eta' \tag{4.32}$$
$$=K_0(\xi,\eta,t),$$

where

$$K_0(\xi,\eta,t)=k(\xi-t)^{-1/6}(\eta-t)^{-1/6}.$$

Using (4.25) and setting $\lambda = \frac{1}{4}(4/3)^{2/3}$, we look for a solution of (4.32) in series form:

$$K(\xi,\eta,t)=\sum_{n=0}^{\infty}\lambda^n K_n(\xi,\eta,t). \tag{4.33}$$

Substituting (4.33) into (4.32), we get

$$K_{n+1}(\xi,\eta,t)$$
$$=\int_t^\xi d\xi'\int_{\xi'}^\eta \frac{\bar C(\xi',\eta')}{(\eta'-\xi')^{2/3}}\,v(\xi',\eta';\xi,\eta)\,K_n(\xi',\eta',t)\,d\eta', \tag{4.34}$$

where

$$\bar C(\xi,\eta)=c\left(\frac{\xi+\eta}{2},\ -\left(\frac{3}{4}\right)^{2/3}(\eta-\xi)^{2/3}\right).$$

We must prove that the series (4.33) is convergent. Consider (4.34) with $\xi = t + (x-t)z$ and $\eta = t + (x-t)\zeta$, and replace ξ' and η' by similar expressions with z' and ζ' in place of z and ζ; denoting

$$K_n(t+(x-t)z,\ t+(x-t)\zeta,\ t)=K_n(z,\zeta,x,t),$$
$$\bar C(\xi,\eta)=C(z,\zeta,x,t),$$
$$v(\xi',\eta';\xi,\eta)=v(z,\zeta;z',\zeta'),$$

we obtain

$$K_{n+1}(z, \zeta, x, t)$$
$$= (x-t)^{4/3} \int_0^z dz' \int_{z'}^{\zeta} \frac{C(z', \zeta', x, t)}{(z'-\zeta')^{2/3}} v(z, \zeta; z', \zeta') K_n(z', \zeta', x, t)\, d\zeta'.$$

Hence we readily obtain estimates for $K_n(z, \zeta, x, t)$. In fact, using the inequalities

$$F\left(\frac{1}{6}, \frac{1}{6}, \frac{1}{3}; \frac{1}{s}\right) < C_1\left(1 - \frac{1}{s}\right)^{-1/6},$$
$$F\left(\frac{1}{6}, \frac{5}{6}, 1; s\right) < C_1(1-s)^{-1/6} \tag{4.35}$$

(C_1 = const) in the estimate of the Riemann-Hadamard function, we get

$$|K_{n+1}(z, \zeta, x, t)|$$
$$\leqslant C_2(x-t)^{4/3} \int_0^z \frac{dz'}{(\zeta - z')^{1/6}} \int_{z'}^{\zeta} \frac{|K_n(z', \zeta', x, t)|}{(\zeta' - z')^{1/3}|\zeta' - z|^{1/6}} d\zeta'. \tag{4.36}$$

Then, since

$$|K_1(z, \zeta, x, t)| \leqslant C_3(x-t)\zeta^{1/3},$$

we obtain by induction from (4.26)

$$|K_{n+1}(z, \zeta, x, t)| < C_3^{n+1}(x-t)^{\frac{4n}{3}+1} \frac{z^n \zeta^{\frac{n+1}{3}}}{n!} \tag{4.37}$$
$$(n = 1, 2, \ldots).$$

Thus the series (4.33) is uniformly and absolutely convergent for $0 \leqslant z < \infty$ $0 \leqslant \zeta < \infty$ and $0 < x - t < \infty$.

Setting $\eta = \xi = x$ in (4.31), we get

$$u(x, x) = k \int_0^x (x-t)^{-1/3} \nu(t)\, dt + \int_0^x K(x, t)\nu(t)\, dt, \tag{4.38}$$

where

$$K(x, t) = \sum_{n=1}^{\infty} \lambda^n K_n(x, x, t)$$
$$= \frac{\Gamma\left(\frac{1}{6}\right)}{\Gamma\left(\frac{5}{6}\right)\Gamma\left(\frac{1}{3}\right)} \sum_{n=1}^{\infty} \lambda^n \int_t^x d\xi \int_{\xi}^x \frac{\bar{C}(\xi, \eta) K_{n-1}(\xi, \eta, t)\, d\eta}{(\eta - \xi)^{1/3}(x-\xi)^{1/6}(x-\eta)^{1/6}} \tag{4.39}$$
$$= \frac{1}{4}\left(\frac{4}{3}\right)^{2/3} \frac{\Gamma\left(\frac{1}{6}\right)}{\Gamma\left(\frac{5}{6}\right)\Gamma\left(\frac{1}{3}\right)} \int_t^x \frac{d\xi}{(x-\xi)^{1/6}} \int_{\xi}^x \frac{\bar{C}(\xi, \eta) K(\xi, \eta, t)}{(\eta - \xi)^{1/3}(x-\eta)^{1/6}} d\eta.$$

LEMMA 5. *There exists a constant C, depending on*

$$\max_{D^-}\{|c(x, y)|, |c'_x(x, y)|, |c'_y(x, y)|\},$$

such that

$$\left|\frac{\partial K(x, t)}{\partial x}\right| < C. \tag{4.40}$$

PROOF. We transform to new variables (z, ζ) in the integral (4.39):

$$\xi = t + (x - t)z, \qquad \eta = t + (x - t)\zeta.$$

Then (4.39) becomes

$$K(x, t) = \frac{1}{4}\left(\frac{4}{3}\right)^{2/3} \frac{\Gamma\left(\frac{1}{6}\right)(x - t)^{4/3}}{\Gamma\left(\frac{1}{3}\right)\Gamma\left(\frac{5}{6}\right)} \times \int_0^1 (1 - z)^{-1/6} dz \int_z^1 \frac{\bar{C}(z, \zeta, x, t) K(z, \zeta, x, t)}{(z - \zeta)^{1/3}(1 - \zeta)^{1/6}} d\zeta. \tag{4.41}$$

Differentiating the expression for $K_{n+1}(z, \zeta, x, t)$ with respect to x and proceeding in exactly the same way, we get estimates similar to (4.37) for $\partial K_{n+1}/\partial x$. Thus the series

$$\sum_{n=1}^{\infty} \lambda^n \frac{\partial K_n(z, \zeta, x, t)}{\partial x}$$

is absolutely and uniformly convergent for $0 \leqslant z < \infty$, $0 \leqslant \zeta < \infty$ and $0 \leqslant x - t < \infty$.

Now, since

$$K(z, \zeta, x, t) = k(x - t)^{-1/3} z^{-1/6} \zeta^{-1/6} + \sum_{n=1}^{\infty} \lambda^n K_n(z, \zeta, x, t),$$

it follows that

$$\left|\frac{\partial K(z, \zeta, x, t)}{\partial x}\right| \leqslant \frac{C_4}{(x - t)^{4/3}} z^{-1/6} \zeta^{-1/6}.$$

It follows from (4.41) that

$$\left|\frac{\partial K(x, t)}{\partial x}\right| < C.$$

Q.E.D.

We now consider the case that the generalized solution $u(\xi, \eta)$ has prescribed values $\psi(\eta)$ on the characteristic $\xi = 0$. Let $\psi(0) = 0$ and $\nu(x) \equiv 0$. Then, if $\psi'(\eta)$ exists and is bounded, we have from (4.29)

$$u(\xi, \eta) - \lambda \int_0^\xi d\xi' \int_{\xi'}^\eta \frac{\bar{C}(\xi', \eta') v(\xi', \eta'; \xi, \eta)}{(\eta' - \xi')^{2/3}} u(\xi', \eta') d\eta' = \int_0^\eta \psi_1(t) v(0, t; \xi, \eta) dt, \tag{4.42}$$

where $\psi_1(t) = \psi'(t) + \psi(t)/6t$.

We shall seek a solution of (4.42) in the form

$$u(\xi, \eta) = \int_0^\eta \psi_1(t) L(\xi, \eta, t) dt. \tag{4.43}$$

Then $L(\xi, \eta, t)$ is defined by

$$L(\xi, \eta, t) - \lambda \int_0^\xi d\xi' \int_{\max(t, \xi')}^\eta \frac{\bar{C}(\xi', \eta')}{(\eta' - \xi')^{2/3}} v(\xi', \eta'; \xi, \eta) L(\xi', \eta', t) d\eta' = L_0(\xi, \eta, t), \tag{4.44}$$

where $L_0(\xi, \eta, t) = v(0, t; \xi, \eta)$.

The solution of (4.44) is sought in series form

$$L(\xi, \eta, t) = \sum_{k=0}^\infty \lambda^k L_k(\xi, \eta, t). \tag{4.45}$$

Substituting (4.45) into (4.44), we get

$$L_{k+1}(\xi, \eta, t) = \int_0^\xi d\xi' \int_{\max(t, \xi')}^\eta \frac{\bar{C}(\xi', \eta')}{(\eta' - \xi')^{2/3}} v(\xi', \eta'; \xi, \eta) L_k(\xi', \eta', t) d\eta' \tag{4.46}$$

$$(k = 0, 1, 2, \ldots).$$

By (4.35),

$$| v(\xi', \eta'; \xi, \eta) | < C_1 \frac{(\eta' - \xi')^{1/6}}{| \eta' - \xi |^{1/6}}, \tag{4.47}$$

and so

$$| L_{k+1}(\xi, \eta, t) | \leqslant C_2 \int_0^\xi d\xi' \int_{\xi'}^\eta \frac{| L_k(\xi', \eta', t) | d\eta'}{(\eta' - \xi')^{1/2} | \eta' - \xi |^{1/6}}. \tag{4.48}$$

Hence

$$| L_{k+1}(\xi, \eta, t) | < C_3^{k+1} \frac{\xi^k \eta^{k/3}}{k!} t^{1/6} \qquad (k = 0, 1, 2, \ldots), \tag{4.49}$$

and therefore

$$|L(\xi, \eta, t)| < C_4 \frac{t^{1/6}}{|\xi - t|^{1/6}}. \tag{4.50}$$

Let

$$\tilde{L}(\xi, \eta, t) = \int_0^{\xi} d\xi' \int_{\max(t, \xi')}^{\eta} \frac{\bar{C}(\xi', \eta')}{(\eta' - \xi')^{2/3}} v(\xi', \eta'; \xi, \eta) L(\xi', \eta', t)\, d\eta'. \tag{4.51}$$

Differentiating with respect to η, we get

$$\frac{\partial \tilde{L}(\xi, \eta, t)}{\partial \eta} = \int_0^{\xi} \frac{\bar{C}(\xi', \eta) v(\xi', \eta; \xi, \eta)}{(\eta - \xi')^{2/3}} L(\xi', \eta, t)\, d\xi' + \int_0^{\xi} d\xi' \int_{\max(t, \xi')}^{\eta} \frac{\bar{C}(\xi', \eta') L(\xi', \eta', t)}{(\eta' - \xi')^{2/3}} \frac{\partial v}{\partial \eta}\, d\eta'.$$

Hence, in view of the inequalities

$$\left| \frac{\partial v(\xi', \eta'; \xi, \eta)}{\partial \eta} \right| < \begin{cases} C_5 \dfrac{v}{\eta - \xi} & \text{for} \quad \eta' > \xi, \\ C_6 \dfrac{v}{\eta - \eta'} & \text{for} \quad \eta' < \xi, \end{cases} \tag{4.52}$$

simple manipulations give

$$\left| \frac{\partial \tilde{L}(\xi, \eta, t)}{\partial \eta} \right| < C_7 \frac{t^{1/6}}{(\eta - \xi)^{1/6}}. \tag{4.53}$$

Using (4.47), (4.52) and (4.53), we infer from (4.44) that

$$\left| \frac{\partial L}{\partial \eta} \right| < C_8 \frac{t^{1/6}}{(\eta - \xi)\,|\xi - t|^{1/6}}. \tag{4.54}$$

Setting $\eta = \xi = x$ in (4.43) and noting (4.44) and (4.51), we obtain

$$u(x, x) = \frac{\Gamma\left(\frac{1}{6}\right)}{\Gamma\left(\frac{1}{3}\right)\Gamma\left(\frac{5}{6}\right)} \int_0^x \psi_1(t) \frac{t^{1/3}\, dt}{x^{1/6}(x - t)^{1/6}} + \lambda \int_0^x \psi_1(t) \tilde{L}(x, x, t)\, dt.$$

LEMMA 6. *The function $\tilde{L}(x, x, t)$ has a bounded partial derivative with respect to x; specifically,*

$$\left| \frac{\partial \tilde{L}(x, x, t)}{\partial x} \right| < C x^{1/6} t^{1/6}. \tag{4.55}$$

PROOF. Set $\eta = \xi = x$ in (4.51). Then, in view of (4.28),

$$\tilde{L}(x, x, t) = \frac{\Gamma\left(\frac{1}{6}\right)}{\Gamma\left(\frac{1}{3}\right)\Gamma\left(\frac{5}{6}\right)} \int_0^x (x-\xi)^{-1/6}\, d\xi \int_{\max(\xi, t)}^x \frac{\bar{C}(\xi, \eta) L(\xi, \eta, t)}{(x-\eta)^{1/6}(\eta-\xi)^{1/3}}\, d\eta$$
$$= \frac{\Gamma\left(\frac{1}{6}\right)}{\Gamma\left(\frac{1}{3}\right)\Gamma\left(\frac{5}{6}\right)} I(x, t). \tag{4.56}$$

By (4.54), we can integrate by parts in the inner integral. Then

$$I(x, t) = \frac{\Gamma\left(\frac{2}{3}\right)\Gamma\left(\frac{5}{6}\right)}{\Gamma\left(\frac{3}{2}\right)} \int_0^x (x-\xi)^{1/3} \bar{C}(\xi, x) L(\xi, x, t)\, d\xi$$
$$- \int_0^x (x-\xi)^{1/3}\, d\xi \int_{\max(\xi, t)}^x \frac{\partial(\bar{C}L)}{\partial \eta}\, d\eta \int_0^{\frac{\eta-\xi}{x-\xi}} \frac{dz}{z^{1/3}(1-z)^{1/6}}.$$

Differentiation with respect to x gives

$$\frac{\partial I(x, t)}{\partial x} = \frac{\Gamma\left(\frac{2}{3}\right)\Gamma\left(\frac{5}{6}\right)}{3\Gamma\left(\frac{3}{2}\right)} \int_0^x (x-\xi)^{-2/3} \bar{C}(\xi, x) L(\xi, x, t)\, d\xi$$
$$- \frac{1}{3} \int_0^x (x-\xi)^{-2/3}\, d\xi \int_{\max(\xi, t)}^x \frac{\partial(\bar{C}L)}{\partial \eta}\, d\eta \int_0^{\frac{\eta-\xi}{x-\xi}} \frac{dz}{z^{1/3}(1-z)^{1/6}}$$
$$+ \int_0^x (x-\xi)^{-7/6}\, d\xi \int_{\max(\xi, t)}^x (\eta-\xi)^{2/3}(x-\eta)^{-1/6} \frac{\partial(\bar{C}L)}{\partial \eta}\, d\eta.$$

Hence, in view of (4.54), we have

$$\left|\frac{\partial I(x, t)}{\partial x}\right| < C_9 x^{1/6} t^{1/6}.$$

Q.E.D.

In view of the preceding analysis, the solution of the Cauchy-Goursat problem can be expressed as the sum of (4.31) and (4.42):

$$u(\xi, \eta) = \int_0^\xi K(\xi, \eta, t)\, \nu(t)\, dt + \int_0^\eta \psi_1(t) L(\xi, \eta, t)\, dt.$$

Setting $\xi = \eta = x$, we get

$$\tau(x) = k \int_0^x (x-t)^{-1/3} \nu(t)\,dt + \int_0^x K(x, t)\,\nu(t)\,dt$$
$$+ \frac{\Gamma\left(\frac{1}{6}\right)}{\Gamma\left(\frac{5}{6}\right)\Gamma\left(\frac{1}{3}\right)} \int_0^x \frac{t^{1/3}\psi_1(t)\,dt}{x^{1/6}(x-t)^{1/6}} + \int_0^x \psi_1(t)\,l(x, t)\,dt, \tag{4.57}$$

where $K(x, t)$ is defined by (4.39), $l(x, t) = \lambda\widetilde{L}(x, x, t)$, and $\widetilde{L}(x, x, t)$ is defined by (4.56).

Formula (4.57) gives the *second functional equation* for $\tau(x)$ and $\nu(x)$, which results from the condition that the solution $u(x, y)$ of (4.1) in D^- must take the prescribed values $\psi(\eta)$ on the characteristic $\xi = 0$.

3. *Reduction of the Tricomi problem to a singular equation.* In view of the results of subsections 1 and 2, the problem of the existence of a solution to problem T for equation (4.1) is equivalent to the solvability of equations (3.20) and (4.57). Elimination of $\tau(x)$ from these equations gives

$$k \int_0^x (x-t)^{-1/3} \nu(t)\,dt + k \int_0^1 \left[\frac{1}{|x-t|^{1/3}} - \frac{1}{(x+t-2xt)^{1/3}}\right] \nu(t)\,dt$$
$$+ \int_0^x K(x, t)\,\nu(t)\,dt + \int_0^1 \chi(t, x)\,\nu(t)\,dt = f(x), \tag{4.58}$$

where

$$f(x) = -\int_0^l \varphi(s)\,\tilde{\rho}(s, x, 0)\,ds$$
$$- k' \int_0^x \frac{t^{1/3}\psi_1(t)\,dt}{x^{1/6}(x-t)^{1/6}} - \int_0^x \psi_1(t)\,l(x, t)\,dt. \tag{4.59}$$

Thus, if (4.58) has a solution, the system (4.20), (4.57) also has a solution, i.e. problem T is solvable.

Henceforth, we consider only solutions $\nu(x)$ which are continuous on $0 < x < 1$ and such that $\nu(x) \in L_p(0, 1)$, $p > 3/2$.

Using the inversion formula

$$\varphi(x) = \frac{\sin \pi\alpha}{\pi} \frac{d}{dx} \int_0^x \frac{\Phi(t)\,dt}{(x-t)^{1-\alpha}}$$

for the Abel integral equation

$$\int_0^x (x-t)^{-\alpha}\varphi(t)\,dt = \Phi(x), \qquad 0<\alpha<1,$$

we can rewrite (4.58) as

$$\begin{aligned} \nu(x) &+ \frac{\sqrt{3}}{2\pi}\frac{d}{dx}\int_0^1 \nu(t)\,dt\left\{\int_0^x \frac{d\xi}{|\xi-t|^{1/3}(x-\xi)^{2/3}}\right. \\ &\left.-\int_0^x \frac{d\xi}{(\xi+t-2\xi t)^{1/3}(x-\xi)^{2/3}} + \frac{1}{k}\int_0^x \frac{\chi(t,\xi)}{(x-\xi)^{2/3}}\,d\xi\right\} \\ &+\frac{\sqrt{3}}{2\pi k}\frac{d}{dx}\int_0^x \nu(t)\,dt\int_t^x (x-\xi)^{-2/3}K(\xi,t)\,d\xi = \frac{3}{2}F(x), \end{aligned} \tag{4.60}$$

where

$$F(x) = \frac{1}{\pi k\sqrt{3}}\frac{d}{dx}\int_0^x \frac{f(t)\,dt}{(x-t)^{2/3}}. \tag{4.61}$$

For x in the interior of $[0, 1]$, we can apply the substitution

$$\frac{x-\xi}{\xi-\dfrac{t}{2t-1}} = z^3 \quad \text{and} \quad \frac{x-\xi}{\dfrac{t}{2t-1}-\xi} = z_1^3$$

to the second integral term on the left of (4.60), rewriting it as

$$\begin{aligned} &\frac{d}{dx}\int_0^1 \nu(t)\,dt\int_0^x \frac{d\xi}{(\xi+t-2t\xi)^{1/3}(x-\xi)^{2/3}} \\ &\quad = 3\frac{d}{dx}\left[\int_0^{1/2}\frac{\nu(t)\,dt}{(1-2t)^{1/3}}\int_0^{\left[\frac{x(1-2t)}{t}\right]^{1/3}}\frac{dz}{1+z^3}\right. \\ &\quad\left.+\int_{1/2}^1 \frac{\nu(t)\,dt}{(2t-1)^{1/3}}\int_0^{\left[\frac{x(2t-1)}{t}\right]^{1/3}}\frac{dz_1}{1-z_1^3}\right] = \int_0^1\left(\frac{t}{x}\right)^{2/3}\frac{\nu(t)\,dt}{x+t-2xt}. \end{aligned} \tag{4.62}$$

Consider the integral

$$I(x)=\int_0^1 \nu(t)\,dt\int_0^x \frac{d\xi}{|\xi-t|^{1/3}(x-\xi)^{2/3}}=\lim_{\varepsilon\to 0} I_\varepsilon(x)$$

$$=\lim_{\varepsilon\to 0}\left\{\int_0^{x-\varepsilon}\nu(t)\,dt\left[\int_0^t \frac{d\xi}{(t-\xi)^{1/3}(x-\xi)^{2/3}}+\int_t^x \frac{d\xi}{(\xi-t)^{1/3}(x-\xi)^{2/3}}\right]\right. \tag{4.63}$$

$$\left.+\int_{x+\varepsilon}^1 \nu(t)\,dt\int_0^x \frac{d\xi}{(t-\xi)^{1/3}(x-\xi)^{2/3}}\right\}.$$

Using the substitution $\xi = t + (x-t)z$, we easily evaluate the integral

$$\int_t^x \frac{d\xi}{(\xi-t)^{1/3}(x-\xi)^{2/3}}=\int_0^1 z^{-1/3}(1-z)^{-2/3}\,dz=\frac{2\pi}{\sqrt{3}}\quad (x>t). \tag{4.64}$$

Next, the substitution $z^3 = (x-\xi)/(t-\xi)$ gives

$$\int_0^t \frac{d\xi}{(t-\xi)^{1/3}(x-\xi)^{2/3}}=-3\int_{\left(\frac{x}{t}\right)^{1/3}}^{\infty}\frac{dz}{1-z^3}$$

$$=-\ln(\zeta-1)+\ln\sqrt{1+\zeta+\zeta^2}+\sqrt{3}\arctan\frac{2\zeta+1}{\sqrt{3}}-\frac{3\tau}{2\sqrt{3}},\qquad x>t,$$

$$\int_0^x \frac{d\xi}{(t-\xi)^{1/3}(x-\xi)^{2/3}}=-3\int_{\left(\frac{x}{t}\right)^{1/3}}^{0}\frac{dz}{1-z^3} \tag{4.65}$$

$$=-\ln(1-\zeta)+\ln\sqrt{1+\zeta+\zeta^2}+\sqrt{3}\arctan\frac{2\zeta+1}{\sqrt{3}}-\frac{\pi}{2\sqrt{3}},\qquad x<t,$$

where $\zeta = (x/t)^{1/3}$.

Substituting (4.64) and (4.65) into (4.63), we get

$$I_\varepsilon(x)=\frac{\pi}{2\sqrt{3}}\left[\int_0^{x-\varepsilon}\nu(t)\,dt-\int_{x+\varepsilon}^1 \nu(t)\,dt\right]$$

$$+\left(\int_0^{x-\varepsilon}+\int_{x+\varepsilon}^1\right)\left[-\ln|1-\zeta|+\ln\sqrt{1+\zeta+\zeta^2}+\sqrt{3}\arctan\frac{2\zeta+1}{\sqrt{3}}\right]\nu(t)\,dt.$$

Differentiating with respect to x, we get

$$\frac{dI_\varepsilon(x)}{dx} = \frac{\pi}{2\sqrt{3}}\,[\nu(x-\varepsilon)+\nu(x+\varepsilon)] + \left(\int_0^{x-\varepsilon} + \int_{x+\varepsilon}^{1}\right)\left(\frac{t}{x}\right)^{2/3}\frac{\nu(t)}{t-x}\,dt$$
$$+\nu(x-\varepsilon)\left[-\ln(\zeta-1)+\ln\sqrt{1+\zeta+\zeta^2}+\sqrt{3}\arctan\frac{2\zeta+1}{\sqrt{3}}\right]\Bigg|_{\zeta=\left(\frac{x}{x-}\right)^{1/3}}$$
$$+\nu(x+\varepsilon)\left[\ln(1-\zeta)-\ln\sqrt{1+\zeta+\zeta^2}-\sqrt{3}\arctan\frac{2\zeta+1}{\sqrt{3}}\right.\Bigg|_{\zeta=\left(\frac{x}{x+\varepsilon}\right)^{1/3}}.$$

Letting $\epsilon \longrightarrow 0$, we obtain

$$\frac{dI(x)}{dx} = \lim_{\varepsilon\to 0}\frac{dI_\varepsilon(x)}{dx} = \frac{\pi}{\sqrt{3}}\,\nu(x) + \int_0^1 \left(\frac{t}{x}\right)^{2/3}\frac{\nu(t)\,dt}{t-x}. \tag{4.66}$$

Now consider the integral

$$\begin{aligned}\chi_1(x,\,t) &= \int_0^x \frac{\chi(t,\,\xi)\,d\xi}{(x-\xi)^{2/3}} \\ &= 3x^{1/3}\chi(t,\,0) + 3\int_0^x (x-\xi)^{1/3}\,\frac{\partial\chi(t,\,\xi)}{\partial\xi}\,d\xi.\end{aligned} \tag{4.67}$$

The existence of $\partial\chi(t,\,\xi)/\partial\xi$ is guaranteed by Lemmas 1 and 4. Since $\chi(t,\,0)=0$, we can write

$$\chi_1(x,\,t) = 3\int_0^x (x-\xi)^{1/3}\,\frac{\partial\chi(t,\,\xi)}{\partial\xi}\,d\xi.$$

Differentiation with respect to x gives

$$\frac{\partial\chi_1(x,\,t)}{\partial x} = \int_0^x (x-\xi)^{-2/3}\,\frac{\partial\chi(t,\,\xi)}{\partial\xi}\,d\xi$$

and so

$$\begin{aligned}\frac{d}{dx}\int_0^1 \chi_1(x,\,t)\,\nu(t)\,dt &= \int_0^1 \frac{\partial\chi_1(x,\,t)}{\partial x}\,\nu(t)\,dt \\ &= \int_0^1 \nu(t)\,dt\int_0^x (x-\xi)^{-2/3}\,\frac{\partial\chi(t,\,\xi)}{\partial\xi}\,d\xi.\end{aligned} \tag{4.68}$$

Finally, by Lemma 5,

$$\begin{aligned}&\int_0^x \nu(t)\,dt\int_t^x (x-\xi)^{-2/3}K(\xi,\,t)\,d\xi \\ &\qquad = \int_0^x \nu(t)\,dt\left[3(x-t)^{1/3}K(t,\,t) + 3\int_t^x (x-\xi)^{1/3}\,\frac{\partial K}{\partial\xi}\,d\xi\right].\end{aligned}$$

But it follows from (4.39) that $K(t, t) = 0$. Consequently,

$$\int_0^x \nu(t)\,dt \int_t^x (x-\xi)^{-2/3} K(\xi, t)\,d\xi = 3\int_0^x \nu(t)\,dt \int_t^x (x-\xi)^{1/3} \frac{\partial K(\xi, t)}{\partial \xi}\,d\xi.$$

Differentiating with respect to x, we get

$$\frac{d}{dx}\int_0^x \nu(t)\,dt \int_t^x (x-\xi)^{-2/3} K(\xi, t)\,d\xi = \int_0^x \nu(t)\,dt \int_t^x (x-\xi)^{-2/3} \frac{\partial K(\xi, t)}{\partial \xi}\,d\xi. \tag{4.69}$$

Thanks to (4.62) and (4.66)–(4.69), equation (4.60) can be written as

$$\nu(x) + \frac{1}{\pi\sqrt{3}}\int_0^1 \left(\frac{t}{x}\right)^{2/3}\left(\frac{1}{t-x} - \frac{1}{t+x-2xt}\right)\nu(t)\,dt + \int_0^1 L(x, t)\,\nu(t)\,dt = F(x), \tag{4.70}$$

where

$$L(x, t) = \frac{1}{k\pi\sqrt{3}}\left[\int_t^x (x-\xi)^{-2/3}\frac{\partial K(\xi, t)}{\partial \xi}\,d\xi + \int_0^x (x-\xi)^{-2/3}\frac{\partial \chi(t, \xi)}{\partial \xi}\,d\xi\right], \quad 0 < t \leqslant x, \tag{4.71}$$

$$L(x, t) = \frac{1}{k\pi\sqrt{3}}\int_0^x (x-\xi)^{-2/3}\frac{\partial \chi(t, \xi)}{\partial \xi}\,d\xi, \quad x \leqslant t < 1.$$

4. *Investigation of the function $F(x)$.* We now study the behavior of $F(x)$ as $x \longrightarrow 0$ and $x \longrightarrow 1$. Set

$$f_1(x) = \int_0^l \varphi(s)\,\tilde{\rho}(s, x, 0)\,ds, \tag{4.72}$$

$$f_2(x) = \frac{\Gamma\left(\frac{1}{6}\right)}{\Gamma\left(\frac{1}{3}\right)\Gamma\left(\frac{5}{6}\right)}\int_0^x \frac{t^{1/3}\psi_1(t)\,dt}{x^{1/6}(x-t)^{1/6}} + \int_0^x l(x, t)\,\psi_1(t)\,dt. \tag{4.73}$$

Then $f(x) = -f_1(x) - f_2(x)$.

LEMMA 7. *If $\varphi(s)$ satisfies a Hölder condition with exponent α, and*

$$\varphi(s) = \varphi(l) + \varphi_1(s), \qquad \varphi(s) = \varphi(0) + \varphi_2(s), \tag{4.74}$$

where

$$|\varphi_1(s)| < C(l-s)^{4/3}, \qquad |\varphi_2(s)| < C's, \tag{4.75}$$

then

$$|f_1'(x)| \leqslant C_1 (x^{-1/9} + (1-x)^{-1/3}). \tag{4.76}$$

PROOF. In view of (4.18), it follows from (4.72) that

$$\begin{aligned} f_1(x) = \int_0^l \varphi(s)\rho(s, x, 0)\,ds \\ + \iint_{D^+} R(x, 0; \xi, \eta; 1)\,d\xi\,d\eta \int_0^l \varphi(s)\rho(s, \xi, \eta)\,ds \\ = f_{11}(x) + f_{12}(x). \end{aligned} \tag{4.77}$$

Let $\bar{u}(x, y)$ be a bounded solution of (4.1) such that

$$\bar{u}|_\Gamma = \varphi(s), \qquad \lim_{y\to 0} \frac{\partial \bar{u}}{\partial y} = 0 \quad \text{for} \quad 0 < x < 1.$$

This solution may be expressed in the form

$$\bar{u}(x, y) = -\int_0^l \varphi(s)\rho(s, x, y)\,ds + \iint_{D^+} c(\xi, \eta) G(x, y; \xi, \eta)\bar{u}(\xi, \eta)\,d\xi\,d\eta.$$

Hence

$$\bar{u}(x, 0) = -f_1(x),$$

$$f_{12}(x) = -\iint_{D^+} c(\xi, \eta) G(x, 0; \xi, \eta)\bar{u}(\xi, \eta)\,d\xi\,d\eta.$$

Consequently, by Lemmas 2 and 3 of subsection 1,

$$|f_{12}'(x)| < C_2 \max|\varphi|, \qquad 0 < x < 1. \tag{4.78}$$

Consider $f_{11}(x)$. By (4.74) and (4.5), we have

$$\begin{aligned} f_{11}(x) = \int_0^l \varphi(s)\rho(s, x, 0)\,ds = -\varphi(l) + \int_0^l \varphi_1(s)\rho(s, x, 0)\,ds \\ = -\varphi(l) + \int_0^l \bar{\varphi}_1(s) A(s, x, 0)\,ds, \end{aligned}$$

where

$$\bar{\varphi}_1(s) = 2\varphi_1(s) + 4\int_0^l R(s, t; 2)\varphi_1(t)\,dt,$$

i.e. the function $\bar{\varphi}_1(s)$ satisfies

$$\bar{\varphi}_1(s) - 2\int_0^l K(s, t)\,\bar{\varphi}_1(t)\,dt = 2\varphi_1(s).$$

Thus,

$$f_{11}(x) = -\varphi(l) - \bar{\varphi}_1(l) + \int_0^l [\bar{\varphi}_1(s) - \bar{\varphi}_1(l)]\,A(s;\ x,\ 0)\,ds.$$

One can show (see, for example, [66g]) that

$$\int_0^{\tilde{l}} K(\tilde{s}, \tilde{t})\,\bar{\varphi}_1(\tilde{t})\,dt$$

satisfies a Lipschitz condition, where $\tilde{s}$ is arc-length on the curve $\tilde{\Gamma}$ and $\tilde{l}$ is the length of $\tilde{\Gamma}$ in the plane $(x, \tilde{y} = 2y^{3/2}/3)$. Hence, by (4.75),

$$|\bar{\varphi}_1(l) - \bar{\varphi}_1(s)| < C_3(l - s)^{1/3}.$$

Thus, for sufficiently small x,

$$|f'_{11}(x)| < C_3 \int_{l-\varepsilon}^{l} (l - s)^{4/3} |A'_x(s, x, 0)|\,ds + O(1)$$

or, replacing $A'_x(s, x, 0)$ by the corresponding expression,

$$|f'_{11}(x)| < C_4 \int_{l-\varepsilon}^{l} (l - s)^{4/3} \frac{\eta \left|\frac{d\eta}{ds}\right| ds}{\left[(\xi - x)^2 + \frac{4}{9}\eta^3\right]^{7/6}} + O(1).$$

But, for sufficiently small ϵ,

$$\frac{1}{(\xi - x)^2 + \frac{4}{9}\eta^3} < C_5 \frac{1}{x^2 + \frac{4}{9}\eta^3}.$$

Therefore,

$$|f'_{11}(x)| < C_6 \int_{l-\varepsilon}^{l} \frac{(l - s)^{4/3}\,\eta \left|\frac{d\eta}{ds}\right| ds}{\left(x^2 + \frac{4}{9}\eta^3\right)^{7/6}} + O(1) < C_7 x^{-1/9}. \tag{4.79}$$

Thus by (4.77)–(4.79) we have $|f'_1(x)| < C_8 x^{-1/9}$.

If $(1 - x)$ is small, then, using the second inequality of (4.75), we find in the same way that

$$|f'_1(x)| < C_9(1 - x)^{-1/3}.$$

Q.E.D.

LEMMA 8. *If $\varphi(s)$ satisfies the conditions of Lemma 7 and $\psi'(\eta)$ satisfies* *a Hölder condition with exponent β, then F(x) satisfies a Hölder condition with exponent* $1/6+\epsilon$, $\epsilon>0$, *for* $0 \leqslant x < 1$.

PROOF. Indeed,

$$F(x) = -\frac{1}{\pi k\sqrt{3}}\frac{d}{dx}\int_0^x (x-t)^{-2/3}[f_1(t)+f_2(t)]\,dt.$$

Next,

$$f_2(x) = \frac{\Gamma\left(\frac{1}{6}\right)}{\Gamma\left(\frac{1}{3}\right)\Gamma\left(\frac{5}{6}\right)}\int_0^x \left[\psi'(t) + \frac{\psi(t)-\psi(0)}{6t}\right]\frac{t^{1/3}\,dt}{x^{1/6}(x-t)^{1/6}}$$
$$+\int_0^x \psi_1(t)\,l(x,t)\,dt + \psi(0) = \psi_{11}(x)+\psi_{12}(x)+\psi(0).$$

It can be shown that

$$\frac{d}{dx}\int_0^x (x-t)^{-2/3}\psi_{11}(t)\,dt = \int_0^x \left[\psi'(t)+\frac{\psi(t)-\psi(0)}{6t}\right]\frac{t\,dt}{x^{5/6}(x-t)^{5/6}}$$
$$= \int_0^x x^{1/6}(x-t)^{-5/6}\psi'(t)\,dt. \tag{4.80}$$

Since $\psi'(x)$ satisfies a Hölder condition with exponent β on $[0, 1)$, the integral (4.80) satisfies a Hölder condition with exponent $1/6+\delta<1$, $\delta<\beta$.

By Lemma 6, $|\psi'_{12}(x)| < C_{10}$. Hence

$$F(x) = -\frac{1}{\pi k\sqrt{3}}\left\{\int_0^x (x-t)^{-2/3} f_1'(t)\,dt\right.$$
$$\left.+\frac{d}{dx}\int_0^x (x-t)^{-2/3}\psi_{11}(t)\,dt + \int_0^x (x-t)^{-2/3}\psi'_{12}(t)\,dt\right\}. \tag{4.81}$$

By Lemma 7, the first integral in (4.81) satisfies a Hölder condition with exponent $\delta < 2/9$ for $0 \leqslant x < 1$, while the third integral satisfies a Hölder condition with exponent $1/3$. Thus, $F(x)$ satisfies a Hölder condition with exponent $1/6+\epsilon<2/9$ if $0 \leqslant x < 1$. Q.E.D.

5. *Regularization of the singular equation.* The singular equation (4.70) can be regularized with help of the following lemma.

LEMMA 9 (TRICOMI). *If* $g(x)$ *satisfies a Hölder condition for* $0<x<1$ *and* $g(x)\in L_p(0,1)$, $p>1$, *then the solution of the equation*

$$\nu(x) + \frac{1}{\pi\sqrt{3}}\int_0^1 \left(\frac{t}{x}\right)^{2/3}\left(\frac{1}{t-x} - \frac{1}{t+x-2xt}\right)\nu(t)\,dt = g(x) \tag{4.82}$$

in the class of functions $\nu(x)$ such that $x^{2/3}\nu(x) \in L_p(0, 1)$ is given by

$$\nu(x) = \frac{3}{4}\left\{ g(x) - \frac{1}{\pi\sqrt{3}} \int_0^1 \left[\frac{t(1-t)}{x(1-x)}\right]^{1/3} \right. \tag{4.83}$$
$$\left. \times \left(\frac{1}{t-x} - \frac{1}{t+x-2xt}\right) g(t)\,dt \right\} + Ax^{-4/3}(1-x)^{-1/3},$$

where A is an arbitrary constant.

PROOF (see [55a]). Set $x^{2/3}\nu(x) = \varphi(x)$ and $x^{2/3}g(x) = f(x)$; then (4.82) is reduced to

$$\varphi(x) + \frac{1}{\pi\sqrt{3}} \int_0^1 \left(\frac{1}{t-x} - \frac{1}{x+t-2xt}\right) \varphi(t)\,dt = f(x). \tag{4.84}$$

Let z be an arbitrary point of the complex plane. Following Carleman, we set

$$\Phi(z) = \frac{1}{2\pi i} \int_0^1 \left(\frac{1}{t-z} - \frac{1}{t+z-2tz}\right) \varphi(t)\,dt. \tag{4.85}$$

Obviously, $\Phi(z)$ is holomorphic in both the upper and lower half-planes, and for $0 < x < 1$

$$\Phi^+(x) - \Phi^-(x) = \varphi(x),$$

$$\Phi^+(x) + \Phi^-(x) = \frac{1}{\pi i} \int_0^1 \left(\frac{1}{t-x} - \frac{1}{x+t-2xt}\right) \varphi(t)\,dt.$$

We can thus reduce (4.84) to the equation

$$\left(1 + \frac{i}{\sqrt{3}}\right)\Phi^+(x) - \left(1 - \frac{i}{\sqrt{3}}\right)\Phi^-(x) = f(x) \qquad (0 < x < 1). \tag{4.86}$$

It is readily seen from (4.85) that

$$\Phi\left(\frac{z}{2z-1}\right) = (2z-1)\,\Phi(z). \tag{4.87}$$

The Möbius transformation $\zeta = z/(2z-1)$ maps the upper half-plane onto the lower, and conversely; the interval $(0, 1)$ is mapped onto the two rays $(0, \infty)$ and $(\infty, 1)$. Replacing x by $x/(2x-1)$ in (4.86) and using (4.87), we get

$$\left(1 + \frac{i}{\sqrt{3}}\right)\Phi^-(x) - \left(1 - \frac{i}{\sqrt{3}}\right)\Phi^+(x) = \frac{1}{2x-1} f\left(\frac{x}{2x-1}\right) \tag{4.88}$$
$$(-\infty < x < 0,\ 1 < x < \infty).$$

Set

$$G(x)=\begin{cases}\dfrac{1-i\sqrt{3}}{2}, & 0<x<1,\\[2mm] \dfrac{1+i\sqrt{3}}{2}, & -\infty<x<0,\quad 1<x<\infty,\end{cases}$$

$$h(x)=\begin{cases}\dfrac{\sqrt{3}}{\sqrt{3}+i}f(x), \quad 0<x<1,\\[2mm] -\dfrac{\sqrt{3}}{\sqrt{3}-i}\,\dfrac{1}{2x-1}\,f\left(\dfrac{x}{2x-1}\right),\\[2mm] -\infty<x<0,\ 1<x<\infty.\end{cases}$$

Then (4.86) and (4.88) can be combined into a single equation:

$$\Phi^+(x)-G(x)\,\Phi^-(x)=h(x),\qquad -\infty<x<\infty,\quad x\neq 0,\quad x\neq 1. \tag{4.89}$$

Thus, solution of the integral equation (4.84) is reduced to the following problem of complex analysis: *Find a function* $\Phi(z)$, *holomorphic in both upper and lower half-planes and satisfying the boundary condition* (4.89).

The solution of this problem can be obtained in explicit form. In fact, it is readily seen that the function

$$X(z)=\exp\left[-\frac{1}{6}\int_0^1\left(\frac{1}{t-z}-\frac{2z-1}{t+z-2tz}\right)dt\right] \tag{4.90}$$

is a particular solution of the corresponding homogeneous problem $X^+(x)=G(x)X^-(x)$ and satisfies the condition

$$X\left(\frac{z}{2z-1}\right)=X(z). \tag{4.91}$$

From (4.90) we have

$$X^+(x)=e^{-\pi i/6}\left(\frac{x}{1-x}\right)^{1/3},\quad X^-(x)=e^{\pi i/6}\left(\frac{x}{1-x}\right)^{1/3},\qquad 0<x<1, \tag{4.92}$$

and, by (4.91) and (4.92),

$$X^+(x)=e^{\pi i/6}\left(\frac{x}{1-x}\right)^{1/3},\quad X^-(x)=e^{-\pi i/6}\left(\frac{x}{1-x}\right)^{1/3}\qquad (-\infty<x<0,\ 1<x<\infty). \tag{4.93}$$

Now replace $G(x)$ by $X^+(x)/X^-(x)$ in (4.89) and divide by $X^+(x)$. Then

$$\frac{\Phi^+(x)}{X^+(x)}-\frac{\Phi^-(x)}{X^-(x)}=\frac{h(x)}{X^+(x)}. \tag{4.94}$$

One solution of this equation is

$$\Phi(z) = \frac{X(z)}{2\pi i} \int_{-\infty}^{\infty} \frac{h(t)}{X^+(x)} \frac{dt}{t-z}. \tag{4.95}$$

In order to find the general solution, we consider the homogeneous equation

$$\frac{\Phi^+(x)}{X^+(x)} - \frac{\Phi^-(x)}{X^-(x)} = 0.$$

This equation shows that the function $F(z) = \Phi(z)/X(z)$ is holomorphic in the whole plane, with the possible exception of $z = 0$ and $z = 1$, which can only be poles. $F(z)$ vanishes at infinity. Assuming that $\Phi(z)$ may have poles of order less than one as $z \longrightarrow 0$ or $z \longrightarrow 1$, we see that $F(z) = a/z$, where a is an arbitrary constant. Thus the general solution of (4.89) is

$$\Phi(z) = \frac{X(z)}{2\pi i} \int_{-\infty}^{\infty} \frac{h(t)}{X^+(t)} \frac{dt}{t-z} + a\frac{X(z)}{z}. \tag{4.96}$$

We split the integral (4.96) into three parts: over the intervals $(-\infty, 0)$, $(1, \infty)$ and $(0, 1)$; in the first two we replace t by $t/(2t-1)$. Then, in view of (4.91), we get

$$\Phi(z) = \frac{X(z)}{2\pi i} \int_0^1 \frac{\sqrt{3}}{\sqrt{3}+i} \frac{f(t)}{X^+(t)} \left(\frac{1}{t-z} - \frac{1}{t+z-2zt}\right) dt + a\frac{X(z)}{z}.$$

Next, define $\varphi(x) = \Phi^+(x) - \Phi^-(x)$, $0 < x < 1$. Performing the calculations, we get

$$\varphi(x) = \frac{3}{4}\left[f(x) - \frac{1}{\pi\sqrt{3}}\left(\frac{x}{1-x}\right)^{1/3} \int_0^1 \left(\frac{1-t}{t}\right)^{1/3} \left(\frac{1}{t-x} - \frac{1}{t+x-2xt}\right) f(t)\, dt\right]$$
$$+ Ax^{-2/3}(1-x)^{-1/3},$$

where A is an arbitrary constant. Hence, returning to the previous functions, we get (4.83), proving the lemma.

We now return to the singular integral equation (4.70), writing it as

$$\nu(x) + \frac{1}{\pi\sqrt{3}} \int_0^1 \left(\frac{t}{x}\right)^{2/3} \left(\frac{1}{t-x} - \frac{1}{t+x-2xt}\right) \nu(t)\, dt$$
$$= F(x) - \int_0^1 L(x, t)\, \nu(t)\, dt.$$

Then, by Lemma 9,

$$\nu(x) + \int_0^1 K_1(x, t)\, \nu(t)\, dt = g_1(x) + Ax^{-4/3}(1-x)^{-1/3}, \tag{4.97}$$

where

$$K_1(x, t) = \frac{3}{4}\Bigg[L(x, t) - \frac{1}{\pi\sqrt{3}}\int_0^1 \left[\frac{y(1-y)}{x(1-x)}\right]^{1/3} \times \left(\frac{1}{y-x} - \frac{1}{y+x-2xy}\right) L(y, t)\, dy\Bigg], \tag{4.98}$$

$$g_1(x) = \frac{3}{4}\left[F(x) - \frac{1}{\pi\sqrt{3}}\int_0^1 \left[\frac{t(1-t)}{x(1-x)}\right]^{1/3}\left(\frac{1}{t-x} - \frac{1}{t+x-2xt}\right) F(t)\, dt\right]. \tag{4.99}$$

Consider the kernel $K_1(x, t)$. Since, by Lemmas 4 and 5,

$$\left|\frac{\partial\chi(t, y)}{\partial y}\right| < C_1\left[\ln\frac{1}{|t-y|} + (t+y-2yt)^{-1/3}\right], \quad \left|\frac{\partial K(y, t)}{\partial y}\right| < C_2, \tag{4.100}$$

it follows from (4.71) that

$$|t^{\varepsilon}(1-t)^{\varepsilon} L(x, t)| < C_3,$$

where $\epsilon > 0$ is arbitrarily small. Then it follows easily from (4.98) that

$$|t^{\varepsilon}(1-t)^{\varepsilon} K_1(x, t)| < C_4(1-x)^{-\frac{1}{3}-\delta}, \tag{4.101}$$

where $\delta > 0$ is arbitrarily small.

We now investigate the function $g_1(x)$. According to Lemma 8, $g_1(x)$ is continuous for $0 \leqslant x < 1$, and since, by Lemma 7,

$$|F(x)| < C_5 + C_6 \ln\frac{1}{1-x},$$

it follows from (4.99) that $x^{1/3}(1-x)^{1/3} g_1(x) \in L_p(0, 1)$ for $p > 1$, i.e. $g_1(x) \in L_r(0, 1)$, $r < 3$.

However, we are looking for solutions $\nu(x)$ of (4.70) which are integrable to some power $p > 3/2$, so we must set $A = 0$ in (4.97). Thus we have obtained the equation

$$\nu(x) + \int_0^1 K_1(x, t)\,\nu(t)\, dt = g_1(x), \tag{4.102}$$

to which Fredholm theory is applicable.

We claim that the homogeneous equation

$$\nu(x) + \int_0^1 K_1(x, t)\,\nu(t)\, dt = 0 \tag{4.103}$$

has no nontrivial solutions $\nu(x) \in L_p(0, 1)$, $p > 3/2$. Suppose the contrary. Thanks to (4.101), it follows from (4.103) that $\nu(x) \in L_2(0, 1)$. In view of

(4.98), equation (4.103) can be written as

$$\nu(x)=\frac{3}{4}\left\{\varkappa(x)-\frac{1}{\pi\sqrt{3}}\int_0^1\left[\frac{t(1-t)}{x(1-x)}\right]^{1/3}\left(\frac{1}{t-x}-\frac{1}{t+x-2xt}\right)\varkappa(t)\,dt\right\}, \tag{4.104}$$

where

$$\varkappa(x)=-\int_0^1 L(x,t)\,\nu(t)\,dt$$

$$=-\frac{1}{\pi k\sqrt{3}}\int_0^x(x-y)^{-2/3}\,dy\left[\int_0^y\frac{\partial K(y,t)}{\partial y}\nu(t)\,dt+\int_0^1\frac{\partial\chi(y,t)}{\partial y}\nu(t)\,dt\right]$$

$$=-\frac{\sqrt{3}}{2\pi k}\int_0^x(x-y)^{-2/3}\varkappa_1(y)\,dy.$$

Using the Schwarz-Bunjakovskiĭ inequality and (4.100), we see that $|\kappa_1(y)|<C_7$. Then, by the Hardy-Littlewood theorem (see Appendix 2 at the end of this section), $\kappa(x)$ satisfies a Hölder condition with exponent $1/3$ for $0\leqslant x<1$.

In order to complete the proof, we need

LEMMA 10. *If $\kappa(x)$ satisfies a Hölder condition with exponent α for $0\leqslant x<1$, then*

$$\theta(x)=\int_0^1\left[\frac{t(1-t)}{x(1-x)}\right]^{1/3}\left(\frac{1}{t-x}-\frac{1}{t+x-2xt}\right)\varkappa(t)\,dt$$

satisfies a Hölder condition with exponent $\alpha-\epsilon$, $\epsilon>0$, for $0\leqslant x<1$.

PROOF. In fact,

$$\theta(x)=\int_0^1\left[\frac{t(1-t)}{x(1-x)}\right]^{1/3}\frac{2x(1-t)}{t+x-2xt}\,\frac{\varkappa(t)-\varkappa(x)}{t-x}\,dt$$

$$+\varkappa(x)\int_0^1\left[\frac{t(1-t)}{x(1-x)}\right]^{1/3}\left(\frac{1}{t-x}-\frac{1}{t+x-2xt}\right)dt=\theta_1(x)+\theta_2(x).$$

Using Tricomi's calculations [69a] (see Appendix 1 at the end of this section), we get

$$\theta_2(x) = \varkappa(x)\left[-\frac{\pi}{\sqrt{3}} + \frac{2}{3}\,\frac{\Gamma\left(\frac{1}{3}\right)\Gamma\left(-\frac{2}{3}\right)}{\Gamma\left(\frac{2}{3}\right)}\,x^{2/3}F\left(\frac{7}{3},\frac{2}{3},\frac{5}{3},x\right)\right] + \varkappa(x)\,\frac{2\Gamma^2\left(\frac{1}{3}\right)}{5\Gamma\left(\frac{2}{3}\right)}(1-x)^{-5/3}F\left(\frac{1}{3},\frac{5}{3},\frac{8}{3};\frac{1-2x}{1-x}\right).$$

Hence it is evident that $\theta_2(x)$ satisfies a Hölder condition with exponent α for $0 \leqslant x < 1$. Standard techniques show that $\theta_1(x)$ satisfies a Hölder condition with exponent $\alpha - \epsilon$ $(\epsilon > 0)$ for $0 \leqslant x < 1$. Q.E.D.

According to Lemma 10, the solution $\nu(x)$ of (4.103) satisfies a Hölder condition with exponent $1/3 - \epsilon$ for $0 \leqslant x < 1$. Substituting $\nu(x)$, say, into equation (4.57) with $\psi_1(t) \equiv 0$, we find that $\tau(x)$ satisfies a Hölder condition with exponents $1 - \epsilon$ for $0 \leqslant x < 1$ (see Appendix 2 at the end of this section). Hence the functions $\tau(x)$ and $\nu(x)$ determine a solution $u(x, y)$ of equation (4.1) in class $\Re_1$ which vanishes on Γ and on the characteristic $\xi = 0$. However, if the conditions of either Theorem 2.2 or Theorem 2.5 are satisfied, then $u(x, y) \equiv 0$, and so $\nu(x) \equiv 0$. Therefore, equation (4.102) is solvable whatever its right-hand member, and $\nu(x) \in L_2(0, 1)$.

By Lemmas 8 and 10, it follows from (4.99) that $g_1(x)$ in (4.102) satisfies a Hölder condition with exponent $1/6 + \epsilon_1$ $(\epsilon_1 > 0)$ for $0 \leqslant x < 1$. Applying Lemma 10 to the integral

$$\int_0^1 K_1(x, t)\,\nu(t)\,dt,$$

we find that $\nu(x)$ satisfies a Hölder condition with exponent $1/6 + \epsilon_1$ for $0 \leqslant x < 1$. From (4.57) it follows that $\tau(x)$ satisfies a Hölder condition with exponent $\alpha > 5/6$ for $0 \leqslant x < 1$. Thus we have

THEOREM 2.9. *If $\varphi(s)$ satisfies the conditions of Lemma 7 and $\psi(\eta)$ those of Lemma 8, if the conditions of Theorem 2.2 or Theorem 2.5 are satisfied, and if Γ satisfies the conditions stated at the beginning of §4, then there exists a solution of equation (2.1) in D of class $\Re_1$ such that*

$$u|_\Gamma = \varphi(s), \qquad u|_{\xi=0} = \psi(\eta).$$

6. *Special case.* The solution of the Tricomi problem for the equation

$$y\frac{\partial^2 u}{\partial x^2} + \frac{\partial^2 u}{\partial y^2} = 0, \tag{*}$$

when Γ is the "normal" curve Γ_0:

$$\left(x-\frac{1}{2}\right)^2+\frac{4}{9}y^3=\frac{1}{4}, \qquad y\geqslant 0,$$

can be expressed in closed form. In fact, the Green's function $G_0(\xi, \eta; x, y)$ of problem N for equation (*) in a normal domain D_0^+ is given by an explicit formula (see (4.7)), and the solution of problem N is (see [66g])

$$u(x, y) = -\int_0^1 \nu(t)\, G_0(t, 0; x, y)\, dt - \int_0^l \varphi(s)\, A_s[G_0(\xi, \eta; x, y)]\, ds. \tag{4.105}$$

Set $y = 0$; then, in view of (4.4) and (4.7), we get

$$\tau(x) + k\int_0^1 \left[\frac{1}{|x-t|^{1/3}} - \frac{1}{(x+t-2xt)^{1/3}}\right]\nu(t)\, dt = \Phi(x), \tag{4.106}$$

where

$$\Phi(x) = \frac{k}{2}\left(\frac{2}{3}\right)^{2/3} x(1-x)\int_0^1 \varphi(\xi)\,\xi^{-1/3}(1-\xi)^{-1/3}[x^2-(2x-1)\xi]^{-7/6}\, d\xi. \tag{4.107}$$

The solution of equation (*) in the hyperbolic half-plane, satisfying the Cauchy data

$$u(x, 0) = \tau(x), \qquad \lim_{y\to 0}\frac{\partial u}{\partial y} = \nu(x),$$

is given by

$$\begin{aligned} u(x, y) = \gamma_1 \int_0^1 \tau\left[x + \frac{2}{3}(-y)^{3/2}(2t-1)\right] t^{-5/6}(1-t)^{-5/6}\, dt \\ + \left(\frac{4}{3}\right)^{2/3} \gamma_2 y \int_0^1 \nu\left[x + \frac{2}{3}(-y)^{3/2}(2t-1)\right] t^{-1/6}(1-t)^{-1/6}\, dt, \end{aligned} \tag{4.108}$$

where

$$\gamma_1 = \frac{\Gamma\left(\frac{1}{3}\right)}{\Gamma^2\left(\frac{1}{6}\right)}, \qquad \gamma_2 = \left(\frac{3}{4}\right)^{2/3} \frac{\Gamma\left(\frac{5}{3}\right)}{\Gamma^2\left(\frac{5}{6}\right)}.$$

Equating (4.108) to $\psi(x)$ on the characteristic arc AC, we have

$$\gamma_1 \int_0^1 \tau(2tx)\, t^{-5/6}(1-t)^{-5/6}\, dt - \gamma_2 (2x)^{2/3} \int_0^1 \nu(2tx)\, t^{-1/6}(1-t)^{-1/6}\, dt = \psi(x)$$

$(0 \leqslant x \leqslant 1/2)$ or

$$\int_0^x (x-\xi)^{-5/6} \xi^{-5/6} \tau(\xi)\, d\xi - \frac{\gamma_2}{\gamma_1} x^{-2/3} \int_0^x \xi^{-1/6} (x-\xi)^{-1/6} \nu(\xi)\, d\xi$$
$$= \frac{1}{\gamma_1} x^{-2/3} \psi\left(\frac{x}{2}\right).$$

Hence, using the inversion formula for the Abel integral equation (see p. 63), we get

$$\tau(x) = \psi_1(x) + k \int_0^x (x-t)^{-1/3} \nu(t)\, dt, \tag{4.109}$$

where

$$\psi_1(x) = \frac{1}{2\pi\gamma_1} x^{5/6} \frac{d}{dx} \int_0^x \frac{\bar{\psi}(t)\, dt}{t^{2/3}(x-t)^{1/6}}, \qquad \bar{\psi}(t) = \psi\left(\frac{t}{2}\right),$$

$$k = \frac{\gamma_2}{\gamma_1} \frac{\Gamma\left(\frac{5}{6}\right)}{\Gamma\left(\frac{1}{6}\right)\Gamma\left(\frac{2}{3}\right)} = \left(\frac{4}{3}\right)^{1/3} \frac{\Gamma^2\left(\frac{1}{6}\right)}{4\pi\Gamma\left(\frac{1}{3}\right)}. \tag{4.110}$$

Elimination of $\tau(x)$ from (4.106) and (4.109) gives

$$\int_0^x \frac{\nu(t)\, dt}{(x-t)^{1/3}} + \int_0^1 \left[\frac{1}{|x-t|^{1/3}} - \frac{1}{(x+t-2xt)^{1/3}}\right] \nu(t)\, dt = \frac{1}{k} f(x), \tag{4.111}$$

where

$$f(x) = \Phi(x) - \psi_1(x). \tag{4.112}$$

By the same manipulations as at the beginning of §4.3, we reduce (4.111) to the form

$$\nu(x) + \frac{1}{\pi\sqrt{3}} \int_0^1 \left(\frac{t}{x}\right)^{2/3} \left(\frac{1}{t-x} - \frac{1}{x+t-2xt}\right) \nu(t)\, dt = F(x), \tag{4.113}$$

where

$$F(x) = \frac{1}{k\pi\sqrt{3}} \frac{d}{dx} \int_0^x \frac{f(t)\, dt}{(x-t)^{2/3}}. \tag{4.114}$$

Let us investigate the behavior of the free term $F(x)$ in (4.113) as $x \to 0$ and $x \to 1$, assuming that $\varphi(s)$ is a continuous function and $\psi(x)$ has a bounded derivative satisfying a Hölder condition with exponent β. It is readily seen that the function $\Phi(x)$ displays the same behavior as $x \to 0$ and $x \to 1$. We therefore confine ourselves to its behavior as $x \to 0$. Applying the first mean-value theorem to the integral (4.107), we get

$$\Phi(x)=\frac{k}{2}\left(\frac{2}{3}\right)^{2/3}\varphi(\xi)\,x(1-x)\int_0^1 \xi^{-1/3}(1-\xi)^{-1/3}\left[x^2-(2x-1)\xi\right]^{-7/6}d\xi$$

or

$$\Phi(x)=\frac{k}{2}\left(\frac{2}{3}\right)^{2/3}\varphi(\xi)\,x^{-4/3}(1-x)\int_0^1 \xi^{-1/3}(1-\xi)^{-1/3}\left[1-\frac{2x-1}{x^2}\xi\right]^{-7/6}d\xi.$$

But the integral in the last formula can be expressed in terms of a hypergeometric function. Using a standard formula,(7) we get

$$\Phi(x)=\frac{k}{2}\left(\frac{2}{3}\right)^{2/3}\varphi(\xi)\,x^{-4/3}(1-x)\frac{\Gamma^2\left(\frac{2}{3}\right)}{\Gamma\left(\frac{4}{3}\right)}F\left(\frac{2}{3},\frac{7}{6},\frac{4}{3};\frac{2x-1}{x^2}\right).$$

Hence, using the relation between hypergeometric functions with arguments z and $1/(1-z)$ (see [49], formula (9.5.8)), we have

$$\Phi(x)=-3k\sqrt{\pi}\left(\frac{2}{3}\right)^{2/3}\frac{\Gamma\left(\frac{2}{3}\right)}{\Gamma\left(\frac{1}{6}\right)}\varphi(\xi)(1-x)^{-1/3}F\left(\frac{2}{3},\frac{1}{6},\frac{1}{2},\left(\frac{x}{1-x}\right)^2\right)$$

$$+k\sqrt{\pi}\left(\frac{2}{3}\right)^{2/3}\frac{\Gamma\left(\frac{2}{3}\right)}{\Gamma\left(\frac{1}{6}\right)}\varphi(\xi)\,x(1-x)^{-4/3}F\left(\frac{7}{6},\frac{2}{3},\frac{3}{2};\left(\frac{x}{1-x}\right)^2\right).$$

This shows that $\Phi(x)$ has a finite limit as $x\longrightarrow 0$. By analogous reasoning one easily shows that $\Phi'(x)$ has a finite limit as $x\longrightarrow 0$. The same is true of the behavior of $\Phi(x)$ and $\Phi'(x)$ as $x\longrightarrow 1$.

Now, from (4.110) we deduce that

$$\psi_1(x)=\frac{\Gamma\left(\frac{1}{6}\right)}{\Gamma\left(\frac{1}{3}\right)\Gamma\left(\frac{5}{6}\right)}\int_0^x\left[\bar\psi'(t)+\frac{\bar\psi(t)-\bar\psi(0)}{6t}\right]\frac{t^{1/3}\,dt}{x^{1/6}(x-t)^{1/6}}+\bar\psi(0)$$

$$=\psi_{11}(x)+\bar\psi(0).$$

It is easily seen that

$$\frac{d}{dx}\int_0^x (x-t)^{-2/3}\psi_{11}(t)\,dt=\int_0^x\left[\bar\psi'(t)+\frac{\bar\psi(t)-\bar\psi(0)}{6t}\right]\frac{t\,dt}{x^{5/6}(x-t)^{5/6}}$$

$$=\int_0^x x^{1/6}(x-t)^{-5/6}\bar\psi'(t)\,dt.$$

(7) See footnote (1) on p. 23.

Thus

$$F(x) = \frac{1}{k\pi\sqrt{3}} \int_0^x (x-t)^{-2/3}\Phi'(t)\,dt - \frac{x^{1/6}}{k\pi\sqrt{3}} \int_0^x (x-t)^{-5/6}\overline{\psi}'(t)\,dt.$$

Since $|\Phi'(x)| < C$, the first integral satisfies a Hölder condition with exponent $1/3$, while the second satisfies a Hölder condition with exponent $1/6 + \beta \leqslant 1$ (since $\overline{\psi}'(x)$ satisfies a Hölder condition with exponent β on $(0, 1)$). Thus, $F(x)$ satisfies a Hölder condition with exponent $\min(1/3, 1/6 + \beta)$ for $0 \leqslant x < 1$.

According to Lemma 9 of §4, the solution of (4.113) is given by (4.83) with $A = 0$, since $\nu(x)$ may have a pole of order less than $2/3$ as $x \longrightarrow 0$. Thus

$$\nu(x) = \frac{3}{4}\left\{F(x) - \frac{1}{\pi\sqrt{3}} \int_0^1 \left[\frac{t(1-t)}{x(1-x)}\right]^{1/3} \left(\frac{1}{t-x} - \frac{1}{t+x-2xt}\right) F(t)\,dt\right\}. \tag{4.115}$$

By Lemma 10 of §4, $\nu(x)$ satisfies a Hölder condition with exponent $1/6 + \epsilon_1$ ($\epsilon_1 > 0$) for $0 \leqslant x < 1$. It follows from (4.109) that $\tau(x)$ satisfies a Hölder condition with exponent $\alpha > 5/6$ for $0 \leqslant x < 1$.

Substituting (4.115) into (4.109), we find $\tau(x)$. Knowing $\tau(x)$ and $\nu(x)$, we can utilize (4.105) and (4.108) to obtain the solution $u(x, y)$ in the domains D^+ and D^-, respectively. This solution is in class $\mathfrak{R}_1$.

7. *Appendix* 1 [69a]. Let the function $A(t)$ be continuous on $[a, b]$ and have a continuous first derivative on $[a, b]$, which may have poles of order less than 1 at the endpoints of $[a, b]$; let c be an interior point of $[a, b]$. Then

$$\int_a^b \frac{A(t)\,dt}{t-c} = A(b)\ln(b-c) - A(a)\ln(c-a) - \int_a^b A'(t)\ln|t-c|\,dt. \tag{4.116}$$

Under the same conditions, we have

$$\int_a^b \frac{A(t)\,dt}{t-c} = \lim_{\varepsilon\to 0}\left[-\int_a^c \frac{A(t)\,dt}{(c-t)^{1-\varepsilon}} + \int_c^b \frac{A(t)\,dt}{(t-c)^{1-\varepsilon}}\right], \tag{4.117}$$

as is easily verified using (4.116) and recalling that

$$\lim_{\varepsilon\to 0} \frac{x^{\varepsilon}-1}{\varepsilon} = \ln x.$$

We now evaluate the integrals

$$\begin{aligned} T_1(\alpha, \beta; x) &= \int_0^\infty \frac{t^{\alpha-1}\,dt}{[x+(1-x)t]^\beta(1-t)}, \\ T_2(\alpha, \beta; x) &= \int_0^\infty \frac{t^{\alpha-1}\,dt}{[x+(1-x)t]^\beta(1+t)}, \end{aligned} \tag{4.118}$$

where α and β are constants such that $0 < \alpha < \beta + 1$ and $\alpha > \beta$. In order to evaluate T_1, we use (4.117). Then

$$T_1(\alpha, \beta; x) = \lim_{\varepsilon\to 0}\Bigg[x^{-\beta}\int_0^1 t^{\alpha-1}(1-t)^{\varepsilon-1}\left(1-\frac{x-1}{x}t\right)^{-\beta}dt$$
$$-(1-x)^{-\beta}\int_1^\infty t^{\alpha-1}(t-1)^{\varepsilon-1}\left(t-\frac{x}{x-1}\right)^{-\beta}dt\Bigg].$$

Set $t = 1/z$ in the second integral (thus converting it into a hypergeometric function); then

$$T_1(\alpha, \beta; x) = \lim_{\varepsilon\to 0}\Bigg[\frac{\Gamma(\alpha)\Gamma(\varepsilon)}{\Gamma(\alpha+\varepsilon)}F\left(\alpha, \beta, \alpha+\varepsilon; \frac{x-1}{x}\right)x^{-\beta}$$
$$-(1-x)^{-\beta}\frac{\Gamma(\beta-\alpha+1-\varepsilon)\Gamma(\varepsilon)}{\Gamma(\beta-\alpha+1)}$$
$$\times F\left(\beta, \beta-\alpha+1-\varepsilon, \quad \beta-\alpha+1; \frac{x}{x-1}\right)\Bigg]$$

or, using standard formulas (see [49], §9.5) for the hypergeometric functions,

$$T_1(\alpha, \beta; x)$$
$$=\lim_{\varepsilon\to 0}\Bigg\{\Gamma(\varepsilon)\left[\frac{\Gamma(\alpha-\beta)}{\Gamma(\alpha-\beta-\varepsilon)}-\frac{\Gamma(\beta-\alpha+1-\varepsilon)}{\Gamma(\beta-\alpha+1)}\right]F(\beta, \varepsilon, \beta-\alpha-1; x)$$
$$+x^{\alpha-\beta}\frac{\Gamma(\alpha)\Gamma(\beta-\alpha)}{\Gamma(\beta)}F(\alpha, \alpha-\beta-\varepsilon, \alpha-\beta+1; x)\Bigg\}.$$

Hence, passing to the limit, we get

$$\begin{aligned} T_1(\alpha, \beta{:}\ x) &= \pi\cot(\alpha-\beta)\pi \\ &+ x^{\alpha-\beta}\frac{\Gamma(\alpha)\Gamma(\beta-\alpha)}{\Gamma(\beta)}F(\alpha, \alpha-\beta, \alpha-\beta+1; x). \end{aligned} \tag{4.119}$$

The integral T_2 is easier to evaluate; indeed, setting $z = 1/(1+t)$, we get

$$T_2(\alpha, \beta; x) = \int_0^1 z^{\beta-\alpha}(1-z)^{\alpha-1}[(1-x)-(1-2x)z]^{-\beta}\,dz,$$

whence

$$\begin{aligned} &T_2(\alpha, \beta; x) \\ &= (1-x)\frac{-\beta\,\Gamma(\alpha)\Gamma(\beta-\alpha+1)}{\Gamma(1+\beta)}F\left(1-\alpha+\beta, \beta, 1+\beta; \frac{1-2x}{1-x}\right). \end{aligned} \tag{4.120}$$

We now evaluate the integral

$$I(x)=\int_0^1\left[\frac{t(1-t)}{x(1-x)}\right]^{1/3}\left(\frac{1}{t-x}-\frac{1}{x+t-2xt}\right)dt.$$

Replacing t by the new variable $\xi = x(1-t)/(1-x)t$, we obtain

$$I(x)=\int_0^\infty\frac{2\xi^{4/3}\,d\xi}{[x+(1-x)\xi]^{5/3}(1-\xi^2)}=\int_0^\infty\frac{\xi^{4/3}\,d\xi}{[x+(1-x)\xi]^{5/3}(1-\xi)}$$

$$+\int_0^\infty\frac{\xi^{4/3}\,d\xi}{[x+(1-x)\xi]^{5/3}(1+\xi)}=T_1\left(\frac{7}{3},\frac{5}{3};x\right)+T_2\left(\frac{7}{3},\frac{5}{3};x\right).$$

Hence, by (4.119) and (4.120),

$$I(x)=-\frac{\pi}{\sqrt{3}}+\frac{2}{3}\frac{\Gamma\left(\frac{1}{3}\right)\Gamma\left(-\frac{2}{3}\right)}{\Gamma\left(\frac{2}{3}\right)}F\left(\frac{7}{3},\frac{2}{3},\frac{5}{3};x\right)$$

$$+\frac{2}{5}\frac{\Gamma^2\left(\frac{1}{3}\right)}{\Gamma\left(\frac{2}{3}\right)}\Gamma\left(\frac{1}{3},\frac{5}{3},\frac{8}{3};\frac{1-2x}{1-x}\right)(1-x)^{-5/3}.$$

8. *Appendix* 2. Consider the following integral of fractional order α:

$$f_\alpha(x)=\frac{1}{\Gamma(\alpha)}\int_a^x(x-t)^{\alpha-1}f(t)\,dt\qquad(a\leqslant x\leqslant b).$$

The following theorems are proved in [30] and [31]:

THEOREM 1. *If $f(x)\in L_p(a,b)$, $p>1$, $-\infty<a<b\leqslant\infty$, and if $0<\alpha<1/p$ and $q=p/(1-p\alpha)$, then $f_\alpha(x)\in L_q(a,b)$, and*

$$\left(\int_a^b|f_\alpha(x)|^q\,dx\right)^{1/q}\leqslant K\left(\int_a^b|f(x)|^p\,dx\right),$$

where $K=K(p,\alpha)$ depends only on p and α.

COROLLARY. *If $p>1$ and $\alpha=1/p$, then $f_\alpha(x)$ is bounded.*

THEOREM 2. *If $p>1$ and $1/p<\alpha<1/p+1$ or $p=1$ and $1\leqslant\alpha<2$, and moreover $f(x)\in L_p(a,b)$, then $f_\alpha(x)$ satisfies a Hölder condition with exponent $\alpha-1/p$ on (a,b).*

THEOREM 3. *Let $k\geqslant 0$, $\alpha>0$ and $k+\alpha<1$. If $f(x)$ satisfies a Hölder condition with exponent k on (a,b) and $f(x)=O((x-a)^k)$ for small $(x-a)$, then $f_\alpha(x)$ satisfies a Hölder condition with exponent $k+\alpha$ on (a,b), and*

$$f_a(x) = O\left((x-a)^{k+\alpha}\right)$$

for small $(x-a)$.

§5. Further investigation of the Čaplygin equation

1. *Definition of a generalized solution of class* $\mathfrak{R}$. Consider the equation

$$K(u) \equiv K(\sigma)\frac{\partial^2 u}{\partial\theta^2} + \frac{\partial^2 u}{\partial\sigma^2} = 0, \quad K(0) = 0, \quad K'(\sigma) > 0, \tag{5.1}$$

in the domain D_σ bounded by a curve Γ_σ with endpoints A (0, 0) and B (1, 0), lying in the upper half-plane, and by two arcs of characteristics

$$\theta = -\int_0^\sigma \sqrt{-K(t)}\,dt, \qquad \theta - 1 = \int_0^\sigma \sqrt{-K(t)}\,dt.$$

We introduce new variables:

$$x = \theta, \qquad y^{3/2} = \frac{3}{2}\int_0^\sigma \sqrt{K(t)}\,dt. \tag{5.2}$$

Equation (5.1) becomes

$$yu_{xx} + u_{yy} + b(y)\,u_y = 0, \tag{5.3}$$

where

$$b(y) = \frac{d^2y}{d\sigma^2} : \left(\frac{dy}{d\sigma}\right)^2.$$

Setting

$$u = ze^{-\frac{1}{2}\int_0^y b(y)\,dy}, \tag{5.4}$$

we bring (5.3) to the form

$$F(z) \equiv yz_{xx} + z_{yy} + c(y)\,z = 0, \tag{5.5}$$

where $c(y) = -b^2/4 - b'/2$.

The domain D_σ is mapped onto a domain D in the (x, y)-plane, the curve Γ_σ onto a curve Γ. We shall assume that Γ satisfies the conditions of §4.

Consider the following class $\mathfrak{R}$ of generalized solutions of equation (5.5) (hence also of (5.1)), introduced by Babenko [4a].

A solution $z(x, y)$ of equation (5.5) is in *class* $\mathfrak{R}$ if it can be represented in D^+ as (see (4.16))

$$z(x, y) = -\int_0^1 \nu(t)\, \tilde{G}(t, 0; x, y)\, dt - \int_0^l \varphi(s)\, \tilde{\rho}(s, x, y)\, ds, \tag{5.6}$$

and in D^- as (see (2.5))

$$z(\xi, \eta) = \int_\xi^\eta \tau(t)\, g(\xi, \eta, t)\, dt - \int_\xi^\eta \nu(t)\, h(\xi, \eta, t)\, dt \tag{5.7}$$

and the function $\nu(x)$ is such that

$$\nu(x) = x^{1/6} \int_0^x (x-t)^{-5/6}\, t^{-1/6} \mu(t)\, dt,$$
$$\int_0^1 (1-t)\, \mu^2(t)\, dt < \infty. \tag{5.8}$$

The function $\varphi(s) = z|_\Gamma$ is required to be continuous for $0 \leqslant s \leqslant l$ and to satisfy a Hölder condition for $0 \leqslant s \leqslant \epsilon$ and $l - \epsilon \leqslant s \leqslant l$.

Thus a solution $z \in \mathfrak{R}$ is twice continuously differentiable in D^+ and satisfies equation (5.5) there. It may increase without bound as the point (x, y) approaches $(1, 0)$.

It can be shown that $\nu(x)$ is an integral of fractional order 1/6 of a function which is square-integrable over any interval $(0, \theta)$, $\theta < 1$, i.e.

$$\nu(x) = \int_0^x (x-t)^{-5/6}\, \psi(t)\, dt.$$

We shall need the following

THEOREM (Hardy; see [81]). *Let $p > 1$, $n < p - 1$, $f(x) \geqslant 0$ for $0 < x < \infty$, and $F(x) = \int_0^x f(t)\, dt$. If $f^p(x) x^n$ is integrable over $(0, \infty)$, then the same is true for $\{F(x)/x\}^p x^n$, and*

$$\int_0^\infty \left\{\frac{F(x)}{x}\right\}^p x^n\, dx \leqslant \left(\frac{p}{p-n-1}\right)^p \int_0^\infty f^p(x)\, x^n\, dx.$$

LEMMA 1. *If $\nu(x)$ is expressible in the form* (5.8), *then*

$$\left(\int_0^1 (1-x)^{3/2} |\nu(x)|^3\, dx\right)^{1/3} \leqslant A \left(\int_0^1 (1-x)\, \mu^2(x)\, dx\right)^{1/2}. \tag{5.9}$$

PROOF. Setting $g(x) = (1-x)^{1/2}\,|\mu(x)|$, we obtain

$$
\begin{aligned}
(1-x)^{1/2}\,|\nu(x)| &\leqslant x^{1/6}\int_0^x (x-t)^{-5/6}\,t^{-1/6}g(t)\,dt \\
&\leqslant 2^{5/6}x^{-2/3}\int_0^{x/2} t^{-1/6}\,g(t)\,dt + 2^{1/6}\int_{x/2}^{x}(x-t)^{-5/6}\,g(t)\,dt \qquad (5.10)\\
&\leqslant 2^{5/6}x^{-2/3}\int_0^{x} t^{-1/6}g(t)\,dt + 2^{1/6}\int_{0}^{x}(x-t)^{-5/6}\,g(t)\,dt \\
&= 2^{5/6}f_1(x) + 2^{1/6}f_2(x).
\end{aligned}
$$

By the Hardy-Littlewood theorem (see §4.8),

$$
\left(\int_0^1 f_2^3(x)\,dx\right)^{1/3} \leqslant B\left(\int_0^1 g^2(x)\,dx\right)^{1/2}.
$$

Now let $G(x) = \int_0^x t^{-1/6}g(t)\,dt$. By the Schwarz-Bunjakovskiĭ inequality,

$$
G(x) \leqslant \left(\frac{3}{2}\right)^{1/2} x^{1/3}\left(\int_0^1 g^2(x)\,dx\right)^{1/2}
$$

and therefore

$$
\begin{aligned}
\int_0^1 f_1^3(x)\,dx &= \int_0^1 x^{-2}G^3(x)\,dx \\
&\leqslant \left(\frac{3}{2}\right)^{1/2}\int_0^1 x^{-2}\left(x^{1/6}G(x)\right)^2dx\left(\int_0^1 g^2(x)\,dx\right)^{1/2},
\end{aligned}
$$

or

$$
\int_0^1 f_1^3(x)\,dx \leqslant B\left(\int_0^1 g^2(x)\,dx\right)^{3/2}.
$$

Applying the Minkowski inequality to (5.10), we get (5.9). Q.E.D.

COROLLARY. *If $p < 6/5$, then $\int_0^1 |\nu(x)|^p\,dx < \infty$.*

2. *Functional relation between $\lambda(x)$ and $\mu(x)$.* Let $z(x, y)$ be a solution of equation (5.5) in class $\mathfrak{R}$, vanishing on Γ. Then it follows from (5.6) that, in D^+,

$$z(x, y) = -\int_0^1 \nu(t)\tilde{G}(t, 0;\ x, y)\,dt$$
$$= -\int_0^1 \nu(t)G(t, 0;\ x, y)\,dt - \int_0^1 \nu(t)\tilde{H}(t;\ x, y)\,dt.$$

Hence, for $0 < x < 1$,

$$\lim_{y\to 0} z(x, y) = \tau(x)$$
$$= -\int_0^1 \nu(t)G(t, 0;\ x, 0)\,dt - \int_0^1 \nu(t)\tilde{H}(t;\ x, 0)\,dt, \tag{5.11}$$

and, in view of (4.21),

$$\tau(x) = -\int_0^1 \nu(t)G_0(t, 0;\ x, 0)\,dt - \int_0^1 \nu(t)\chi(t, x)\,dt, \tag{5.12}$$

where

$$\chi(t, x) = H(t, x) + \tilde{H}(t;\ x, 0),$$
$$G_0(t, 0;\ x, 0) = k\left[\frac{1}{|t-x|^{1/3}} - \frac{1}{(x+t-2xt)^{1/3}}\right]. \tag{5.13}$$

Let us replace x by ξ in (5.12), multiply by $(x-\xi)^{-2/3}$, and integrate with respect to ξ from 0 to x. Then, differentiating with respect to x, we obtain

$$\frac{d}{dx}\int_0^x (x-\xi)^{-2/3}\tau(\xi)\,d\xi = -k\frac{d}{dx}\int_0^1 \nu(t)\,dt$$
$$\times\left[\int_0^x \frac{d\xi}{|t-\xi|^{1/3}(x-\xi)^{2/3}} - \int_0^x \frac{d\xi}{(t+\xi-2\xi t)^{1/3}(x-\xi)^{2/3}}\right] \tag{5.14}$$
$$-\frac{d}{dx}\int_0^1 \nu(t)\,dt\int_0^x \chi(t, \xi)(x-\xi)^{-2/3}\,d\xi.$$

Submitting the right-hand side of (5.14) to the same treatment as in §4.3, we get

$$\frac{\sqrt{3}}{2\pi k}\frac{d}{dx}\int_0^x (x-\xi)^{-2/3}\tau(\xi)\,d\xi$$
$$= -\frac{1}{2}\nu(x) - \frac{\sqrt{3}}{2\pi}\int_0^1 \left(\frac{t}{x}\right)^{2/3}\left(\frac{1}{t-x} - \frac{1}{t+x-2xt}\right)\nu(t)\,dt$$
$$-\frac{\sqrt{3}}{2\pi k}\int_0^1 \nu(t)\,dt\,\frac{d}{dx}\int_0^x (x-\xi)^{-2/3}\chi(t, \xi)\,d\xi$$

or, setting

$$\frac{\sqrt{3}}{2\pi}\frac{d}{dx}\int_0^x (x-\xi)^{-2/3}\tau(\xi)\,d\xi = \varphi_1(x), \tag{5.15}$$

we have

$$\frac{1}{k}\varphi_1(x)$$
$$= -\frac{1}{2}\nu(x) - \frac{\sqrt{3}}{2\pi}\int_0^1 \left(\frac{t}{x}\right)^{2/3}\left(\frac{1}{t-x} - \frac{1}{t+x-2xt}\right)\nu(t)\,dt$$
$$- \frac{\sqrt{3}}{2\pi k}\int_0^1 \nu(t)\,dt\,\frac{d}{dx}\int_0^x (x-\xi)^{-2/3}\chi(t,\xi)\,d\xi.$$

For convenience, we put

$$\begin{aligned} &\frac{1}{k}\varphi_1(x) - \nu(x) = \tilde{\varphi}(x),\\ &\frac{1}{\pi\sqrt{3}\,k}\frac{d}{dx}\int_0^x (x-\xi)^{-2/3}\chi(t,\xi)\,d\xi = \chi_1(x,t). \end{aligned} \tag{5.16}$$

Then

$$\begin{aligned} \frac{2}{3}\tilde{\varphi}(x) + \nu(x) + \frac{1}{\pi\sqrt{3}}\int_0^1 \left(\frac{t}{x}\right)^{2/3}\left(\frac{1}{t-x} - \frac{1}{t+x-2xt}\right)\nu(t)\,dt&\\ + \int_0^1 \chi_1(x,t)\,\nu(t)\,dt = 0.& \end{aligned} \tag{5.17}$$

Multiply (5.17) by $(\zeta - x)^{-1/6}x^{-1/6}$, and integrate with respect to x from 0 to ζ. Changing the order of integration (that this is legitimate will be proved below) gives

$$\frac{2}{3}\int_0^\zeta (\zeta-x)^{-1/6}x^{-1/6}\tilde{\varphi}(x)\,dx + \int_0^\zeta (\zeta-x)^{-1/6}x^{-1/6}\nu(x)\,dx$$
$$+ \frac{1}{\pi\sqrt{3}}\int_0^1 t^{2/3}\nu(t)\,dt\int_0^\zeta (\zeta-x)^{-1/6}x^{-5/6}\left(\frac{1}{t-x} - \frac{1}{t+x-2xt}\right)dx$$
$$+ \int_0^1 \nu(t)\,dt\int_0^\zeta (\zeta-x)^{-1/6}x^{-1/6}\chi_1(x,t)\,dx = 0.$$

It is readily seen that

$$\int_0^\zeta (\zeta-x)^{-1/6}x^{-5/6}\frac{dx}{t-x} = \begin{cases} 2\pi t^{-5/6}(t-\zeta)^{-1/6} & \text{for} \quad t>\zeta,\\ \pi\sqrt{3}\,t^{-5/6}(\zeta-t)^{-1/6} & \text{for} \quad t<\zeta \end{cases}$$

and

$$\int_0^\zeta (\zeta - x)^{-1/6} x^{-5/6} (x + t - 2xt)^{-1} \, dx = 2\pi t^{-5/6} (t + \zeta - 2t\zeta)^{-1/6}.$$

Therefore

$$\begin{aligned} \frac{2}{3} \int_0^\zeta (\zeta - x)^{-1/6} x^{-1/6} \tilde{\varphi}(x) \, dx + 2 \int_0^\zeta (\zeta - x)^{-1/6} x^{-1/6} \nu(x) \, dx \\ + \frac{2}{\sqrt{3}} \int_\zeta^1 (t - \zeta)^{-1/6} t^{-1/6} \nu(t) \, dt \\ - \frac{2}{\sqrt{3}} \int_0^1 t^{-1/6} (t + \zeta - 2t\zeta)^{-1/6} \nu(t) \, dt \\ + \int_0^1 \nu(t) \, dt \int_0^\zeta (\zeta - x)^{-1/6} x^{-1/6} \chi_1(x, t) \, dx = 0. \end{aligned} \tag{5.18}$$

We now consider the individual terms in (5.18) in greater detail. By (5.8),

$$\begin{aligned} \int_0^\zeta (\zeta - x)^{-1/6} x^{-1/6} \nu(x) \, dx \\ = \int_0^\zeta (\zeta - x)^{-1/6} dx \int_0^x (x - t)^{-5/6} t^{-1/6} \mu(t) \, dt \\ = \int_0^\zeta t^{-1/6} \mu(t) \, dt \int_t^\zeta (\zeta - x)^{-1/6} (x - t)^{-5/6} \, dx \\ = \int_0^\zeta t^{-1/6} \mu(t) \, dt \int_0^1 v^{-5/6} (1 - v)^{-1/6} dv = 2\pi \int_0^\zeta t^{-1/6} \mu(t) \, dt. \end{aligned}$$

Differentiating with respect to ζ, we get

$$\frac{d}{d\zeta} \int_0^\zeta (\zeta - x)^{-1/6} x^{-1/6} \nu(x) \, dx = 2\pi \zeta^{-1/6} \mu(\zeta). \tag{5.19}$$

In view of (5.19), we have

$$\begin{aligned}\int_\zeta^1 (t-\zeta)^{-1/6}\, t^{-1/6}\nu(t)\,dt &= \int_\zeta^1 (t-\zeta)^{-1/6}\,dt \int_0^t (t-x)^{-5/6}\, x^{-1/6}\mu(x)\,dx\\ &= \int_0^\zeta x^{-1/6}\mu(x)\,dx \int_\zeta^1 (1-\zeta)^{-1/6}(t-x)^{-5/6}\,dt\\ &\quad+ \int_\zeta^1 x^{-1/6}\mu(x)\,dx \int_x^1 (t-\zeta)^{-1/6}(t-x)^{-5/6}\,dt.\end{aligned}$$

Set

$$K_1(x,\zeta)=\begin{cases}\displaystyle\int_\zeta^1 (t-\zeta)^{-1/6}(t-x)^{-5/6}\,dt & \text{for } x<\zeta,\\[2ex] \displaystyle\int_x^1 (t-\zeta)^{-1/6}(t-x)^{-5/6}\,dt & \text{for } x>\zeta\end{cases}$$

or, making the obvious substitution,

$$K_1(x,\zeta)=\begin{cases}\displaystyle 6\int_0^{\left(\frac{1-x}{1-\zeta}\right)^{1/6}} \frac{dv}{1-v^6} & \text{for } x>\zeta,\\[2ex] \displaystyle 6\int_{\left(\frac{1-x}{1-\zeta}\right)^{1/6}}^{\infty} \frac{dv}{v^6-1} & \text{for } x<\zeta.\end{cases} \tag{5.20}$$

Then

$$\int_\zeta^1 (t-\zeta)^{-1/6}\, t^{-1/6}\nu(t)\,dt = \int_0^\zeta x^{-1/6}\mu(x)\,K_1(x,\zeta)\,dx + \int_\zeta^1 x^{-1/6}K_1(x,\zeta)\,\mu(x)\,dx. \tag{5.21}$$

After evaluating the integrals (5.20) and substituting them into (5.21), we get

$$\int_\zeta^1 (t-\zeta)^{-1/6} t^{-1/6} \nu(t)\, dt$$

$$= -\sqrt{3}\,\pi \int_0^\zeta t^{-1/6}\mu(t)\, dt + \int_0^1 \ln\left|\frac{1+w}{1-w}\right| t^{-1/6}\mu(t)\, dt$$

$$-\frac{1}{2}\int_0^1 \ln\frac{w^2-w+1}{w^2+w+1}\, t^{-1/6}\mu(t)\, dt$$

$$+\sqrt{3}\int_0^1 t^{-1/6}\left[\arctan\frac{2w-1}{\sqrt{3}} + \arctan\frac{2w+1}{\sqrt{3}}\right]\mu(t)\, dt$$

$$\left(w = \left(\frac{1-t}{1-\zeta}\right)^{1/6}\right).$$

Hence, differentiating with respect to ζ, we see that for almost all ζ $(0 < \zeta < 1)$

$$\frac{d}{d\zeta}\int_\zeta^1 (t-\zeta)^{-1/6} t^{-1/6}\nu(t)\, dt$$
$$= -\pi\sqrt{3}\,\zeta^{-1/6}\mu(\zeta) + \int_0^1 \left(\frac{1-t}{1-\zeta}\right)^{1/6} t^{-1/6}\,\frac{\mu(t)\, dt}{t-\zeta}. \tag{5.22}$$

Consider the integral

$$\int_0^1 t^{-1/6}(t+\zeta-2t\zeta)^{-1/6}\,\nu(t)\, dt.$$

Using (5.8), we have

$$\int_0^1 t^{-1/6}(t+\zeta-2t\zeta)^{-1/6}\,\nu(t)\, dt$$
$$= \int_0^1 x^{-1/6}\mu(x)\, dx \int_x^1 (t-x)^{-5/6}(t+\zeta-2t\zeta)^{-1/6}\, dt \tag{5.23}$$
$$= \int_0^1 x^{-1/6} K_2(x,\zeta)\,\mu(x)\, dx,$$

where

$$K_2(x,\zeta) = \int_x^1 (t-x)^{-5/6}(t+\zeta-2t\zeta)^{-1/6}\, dt$$
$$= \int_0^{\frac{1-x}{x+\zeta-2x\zeta}} v^{-5/6}\,[1+(1-2\zeta)\,v]^{-1/6}\, dv.$$

Differentiating with respect to ζ, we obtain

$$\begin{aligned}\frac{\partial K_2(x,\zeta)}{\partial \zeta} &= -\left(\frac{1-x}{1-\zeta}\right)^{1/6}\frac{1-2x}{x+\zeta-2x\zeta}\\ &-\left(\frac{1-x}{1-\zeta}\right)^{1/6}\frac{2}{1-2\zeta}\left\{1-F\left(1,\frac{1}{6},\frac{7}{6},\frac{(1-x)(1-2\zeta)}{1-\zeta}\right)\right\}\\ &= -\left(\frac{1-x}{1-\zeta}\right)^{1/6}\frac{1-2x}{x+\zeta-2x\zeta}-\left(\frac{1-x}{1-\zeta}\right)^{1/6}\frac{\sqrt{3}}{2}K_3(x,\zeta).\end{aligned} \tag{5.24}$$

It is easy to see that

$$|K_3(x,\zeta)|\leqslant C\left(\ln\left|\frac{1}{x+\zeta-2x\zeta}\right|+1\right). \tag{5.25}$$

By Lemmas 1 and 2 of §4, from (5.16) we have

$$\chi_1(x,t)=\frac{1}{\pi\sqrt{3}k}\int_0^x (x-y)^{-2/3}\frac{\partial\chi(t,y)}{\partial y}\,dy.$$

Then we get the equality

$$\begin{aligned}&\int_0^\zeta (\zeta-x)^{-1/6}x^{-1/6}\chi_1(x,t)\,dx\\ &=\frac{1}{\pi\sqrt{3}k}\int_0^\zeta \frac{\partial\chi(t,y)}{\partial y}\,dy\int_y^\zeta (\zeta-x)^{-1/6}(x-y)^{-2/3}x^{-1/6}\,dx.\end{aligned} \tag{5.26}$$

The substitution $x=\zeta-(\zeta-y)\xi$ converts the inner integral in (5.26) into a hypergeometric integral, and thus

$$\begin{aligned}\int_0^\zeta (\zeta-x)^{-1/6}x^{-1/6}\chi_1(x,t)\,dx &= \frac{1}{\pi\sqrt{3}k}\,\frac{\Gamma\left(\frac{5}{6}\right)\Gamma\left(\frac{1}{3}\right)}{\Gamma\left(\frac{7}{6}\right)}\\ &\times\int_0^\zeta\left(\frac{\zeta-y}{\zeta}\right)^{1/6}F\left(\frac{5}{6},\frac{1}{6},\frac{7}{6};\frac{\zeta-y}{\zeta}\right)\frac{\partial\chi(t,y)}{\partial y}\,dy.\end{aligned}$$

Differentiating with respect to ζ, we get

$$\frac{\partial}{\partial\zeta}\int_0^{\zeta}(\zeta-x)^{-1/6}x^{-1/6}\chi_1(x,t)\,dx$$
$$=\frac{1}{\pi\sqrt{3}\,k}\,\frac{\Gamma\left(\frac{5}{6}\right)\Gamma\left(\frac{1}{3}\right)}{\Gamma\left(\frac{1}{6}\right)}\int_0^{\zeta}\frac{y^{1/6}}{\zeta^{1/3}(\zeta-y)^{5/6}}\,\frac{\partial\chi(t,y)}{\partial y}\,dy=\chi_2(t,\zeta). \tag{5.27}$$

We now estimate $\chi_2(t,\zeta)$. By Lemmas 1 and 2 of §4, it follows from (5.13) that

$$\left|\frac{\partial\chi(t,y)}{\partial y}\right|<C_1\ln\frac{1}{|t-y|}+C_2\frac{1}{(t+y-2ty)^{1/3}}. \tag{5.28}$$

Next,

$$I_1=\int_0^{\zeta}\frac{y^{1/6}}{\zeta^{1/3}(\zeta-y)^{5/6}}\ln\frac{1}{|t-y|}\,dy$$
$$=\int_0^1 v^{1/6}(1-v)^{-5/6}\ln\frac{1}{|\zeta v-t|}\,dv<C_\varepsilon\zeta^{-\varepsilon},\quad \varepsilon>0,$$

$$I_2=\int_0^{\zeta}\frac{y^{1/6}(t+y-2ty)^{-1/3}\,dy}{\zeta^{1/3}(\zeta-y)^{5/6}}$$
$$=\begin{cases}\dfrac{\Gamma\left(\frac{1}{6}\right)\Gamma\left(\frac{7}{6}\right)}{\Gamma\left(\frac{4}{3}\right)}(t+\zeta-2t\zeta)^{-1/3}F\left(\frac{1}{6},\frac{1}{3},\frac{4}{3};\frac{\zeta(1-2t)}{t+\zeta-2t\zeta}\right) & \text{for}\quad 0\leqslant t\leqslant 1/2,\\[2ex] \dfrac{\Gamma\left(\frac{1}{6}\right)\Gamma\left(\frac{7}{6}\right)}{\Gamma\left(\frac{4}{3}\right)}t^{-1/6}(t+\zeta-2t\zeta)^{-1/6}F\left(\frac{1}{6},1,\frac{4}{3};\frac{(2t-1)\zeta}{t}\right) & \text{for}\quad 1/2\leqslant t\leqslant 1.\end{cases}$$

Hence

$$I_2<\begin{cases}C_3t^{-1/6}\zeta^{-1/6} & \text{for}\quad t<1/2,\\ C_4t^{-1/6}(1-\zeta)^{-1/6} & \text{for}\quad t>1/2.\end{cases}$$

Thus

$$|\chi_2(t,\zeta)|<C_5\zeta^{-\varepsilon}+C_6t^{-1/6}\{\min(\zeta,1-\zeta)\}^{-1/6}. \tag{5.29}$$

Setting

$$\left(\frac{1-x}{1-\zeta}\right)^{1/6}\chi_3(x,\ \zeta)=\int\limits_x^1 t^{-1/6}(t-x)^{-5/6}\,\chi_2(t,\ \zeta)\,dt$$

and using (5.27) and (5.8), we get

$$\begin{aligned}\frac{d}{d\zeta}\int\limits_0^1 \nu(t)\,dt\int\limits_0^\zeta (\zeta-x)^{-1/6}x^{-1/6}\chi_1(x,\ t)\,dx\\ =\int\limits_0^1 x^{-1/6}\left(\frac{1-x}{1-\zeta}\right)^{1/6}\chi_3(x,\ \zeta)\,\mu(x)\,dx.\end{aligned}\tag{5.30}$$

Set $\epsilon = 1/6$ in (5.29); then

$$|\chi_3(x,\ \zeta)|<C_7\zeta^{-1/6}.\tag{5.31}$$

Differentiating (5.18) with respect to ζ and using (5.19), (5.22)–(5.24) and (5.30), we get

$$\begin{aligned}\frac{2}{3}\frac{d}{d\zeta}\int\limits_0^\zeta (\zeta-x)^{-1/6}x^{-1/6}\tilde{\varphi}(x)\,dx+2\pi\zeta^{-1/6}\mu(\zeta)\\ +\frac{2}{\sqrt{3}}\int\limits_0^1 t^{-1/6}\left(\frac{1-t}{1-\zeta}\right)^{1/6}\left(\frac{1}{t-\zeta}+\frac{1-2t}{t+\zeta-2t\zeta}\right)\mu(t)\,dt\\ +2\pi\int\limits_0^1 t^{-1/6}\left(\frac{1-t}{1-\zeta}\right)^{1/6}K_4(t,\ \zeta)\,\mu(t)\,dt=0,\end{aligned}$$

where

$$2\pi K_4(t,\ \zeta)=K_3(t,\ \zeta)+\chi_3(t,\ \zeta).$$

Dividing by 2π and denoting

$$\frac{\zeta^{1/6}}{2\pi}\frac{d}{d\zeta}\int\limits_0^\zeta (\zeta-x)^{-1/6}x^{-1/6}\tilde{\varphi}(x)\,dx=\lambda(\zeta),\tag{5.32}$$

we can write

$$\begin{aligned}\frac{2}{3}\lambda(\zeta)+\mu(\zeta)+\frac{1}{\pi\sqrt{3}}\int\limits_0^1\left[\frac{\zeta(1-t)}{t(1-\zeta)}\right]^{1/6}\left(\frac{1}{t-\zeta}+\frac{1-2t}{t+\zeta-2t\zeta}\right)\mu(t)\,dt\\ +\int\limits_0^1\left[\frac{\zeta(1-t)}{t(1-\zeta)}\right]^{1/6}K_4(t,\ \zeta)\,\mu(t)\,dt=0,\end{aligned}\tag{5.33}$$

where the function $K_4(t, \zeta)$ satisfies the estimate

$$|K_4(t, \zeta)| < C\left(\ln \frac{1}{t+\zeta-2t\zeta} + \zeta^{-1/6}\right), \tag{5.34}$$

which follows from (5.25) and (5.3).

We have derived a functional relation between $\lambda(x)$ and $\mu(x)$, which will play a significant role later.

LEMMA 2 [4a].

$$\int_0^1 (1-x)\lambda^2(x)\,dx \leqslant A^2 \int_0^1 (1-x)\mu^2(x)\,dx, \tag{5.35}$$

where A is an absolute constant.

The proof relies on the following

THEOREM (BABENKO). *Let $p > 1$. If $\int_{-\infty}^{\infty} |x|^\alpha |f(x)|^p\,dx < \infty$, $-1 < \alpha < p - 1$, and*

$$g(x) = \frac{1}{\pi} \int_{-\infty}^{\infty} \frac{f(t)}{t-x}\,dt,$$

then this integral exists for almost all x $(-\infty < x < \infty)$, and

$$\int_{-\infty}^{\infty} |x|^\alpha |g(x)|^p\,dx \leqslant B^p \int_{-\infty}^{\infty} |x|^\alpha |f(x)|^p\,dx, \tag{*}$$

where B depends only on α and p.

Let $t^{-1/6}(1-t)^{1/6}\mu(t) = f(t)$ and

$$G(x) = x^{1/6}(1-x)^{1/3} \int_0^1 \frac{f(t)\,dt}{t-x} = x^{1/6}(1-x)^{1/3} \left\{ \int_0^{1/2} \frac{f(t)\,dt}{t-x} + \int_{1/2}^1 \frac{f(t)\,dt}{t-x} \right\}$$

$$= x^{1/6}(1-x)^{1/3}[G_1(x) + G_2(x)].$$

Then, putting $\alpha = 1/3$ in (*), we have

$$\int_0^1 x^{1/3}(1-x)^{2/3} G_1^2(x)\,dx \leqslant \int_0^1 x^{1/3} G_1^2(x)\,dx \leqslant B^2 \int_0^1 (1-x)^{1/3}\mu^2(x)\,dx$$

$$\leqslant B_1^2 \int_0^1 (1-x)\mu^2(x)\,dx.$$

Similarly, if $\alpha = 2/3$, we get

$$\int_0^1 x^{1/3}(1-x)^{2/3}\, G_2^2(x)\,dx \leqslant \int_0^1 (1-x)^{2/3}\, G_2^2(x)\,dx$$

$$\leqslant B^2 \int_{1/2}^1 (1-x)\,x^{-1/3}\mu^2(x)\,dx \leqslant B_2^2 \int_0^1 (1-x)\mu^2(x)\,dx$$

and so

$$\int_0^1 G^2(x)\,dx \leqslant B_3^2 \int_0^1 (1-x)\mu^2(x)\,dx.$$

Consider the integral

$$I = x^{1/6}(1-x)^{1/3}\int_0^1 \frac{(1-2t)\,f(t)\,dt}{t+x-2xt} = I_1 + I_2 + I_3,$$

where the integrals I_1, I_2 and I_3 are evaluated over the intervals (0, 1/4), (1/4, 3/4) and (3/4, 1), respectively. We have

$$|I_1| \leqslant x^{1/6}(1-x)^{1/3}\int_0^{1/4} \frac{|f(t)|\,dt}{t+x},$$

and, applying (*) with $\alpha = 1/3$,

$$\int_0^1 |I_1|^2\,dx \leqslant B_4^2 \int_0^1 (1-x)\mu^2(x)\,dx.$$

Furthermore,

$$|I_2| < Cx^{1/6}(1-x)^{1/3}\int_{1/4}^{3/4} |f(t)|\,dt$$

$$\leqslant C_1 x^{1/6}(1-x)^{1/3}\left\{\int_{1/4}^{3/4}\left(\frac{1-x}{x}\right)^{1/3}\mu^2(x)\,dx\right\}^{1/2}$$

$$\leqslant C_1 x^{1/6}(1-x)^{1/3}\left\{\int_0^1 (1-x)\mu^2(x)\,dx\right\}^{1/2}.$$

Hence

$$\int_0^1 |I_2|^2\,dx \leqslant C_2 \int_0^1 (1-x)\mu^2(x)\,dx.$$

The integral I_3 is estimated in the same way as I_1. Thus,

$$\int_0^1 |I|^2\,dx \leqslant B_5^2 \int_0^1 (1-x)\,\mu^2(x)\,dx.$$

For the last integral in (5.33) we have

$$\left|\int_0^1 \left[\frac{\zeta(1-t)}{t(1-\zeta)}\right]^{1/3} K_4(t,\zeta)\,\mu(t)\,dt\right|$$

$$\leqslant (1-\zeta)^{-1/6}\left(\int_0^1 \frac{\zeta^{1/3}K_4^2(t,\zeta)\,dt}{t^{1/3}(1-t)^{2/3}}\right)^{1/2}\left(\int_0^1 (1-t)\,\mu^2(t)\,dt\right)^{1/2}$$

$$\leqslant C_3(1-\zeta)^{-1/6}\left(\int_0^1 (1-t)\,\mu^2(t)\,dt\right)^{1/2}.$$

Applying the Minkowski inequality to (5.33) and utilizing the above estimates, we obtain (5.35). Q.E.D.

Recall (see (5.15) and (5.16)) that

$$\varphi_1(x) = \frac{\sqrt{3}}{2\pi}\frac{d}{dx}\int_0^x (x-t)^{-2/3}\,\tau(t)\,dt,$$

$$\tilde{\varphi}(x) = \frac{1}{k}\varphi_1(x) - \nu(x).$$

Let

$$\lambda_1(x) = \frac{x^{1/6}}{2\pi}\frac{d}{dx}\int_0^x (x-t)^{-1/6}\,t^{-1/6}\varphi_1(t)\,dt. \tag{5.36}$$

Then, by (5.32), $\lambda(x) = \lambda_1(x)/k - \mu(x)$.

COROLLARY. *The function* (5.36) *satisfies the estimate*

$$\int_0^1 (1-x)\,\lambda_1^2(x)\,dx \leqslant A_1^2 \int_0^1 (1-x)\,\mu^2(x)\,dx. \tag{5.37}$$

REMARK. We have not yet shown that it is legitimate to change the order of integration in the proof of (5.18). The following lemma, which is of independent interest and will be used later, settles the question.

LEMMA 3 [4a]. *If $f(x) \in L_2(0, 1)$, then the integral*

$$g(x) = \frac{1}{\pi}\int_0^1 \left(\frac{t}{x}\right)^{1/2}\left(\frac{1}{t-x} - \frac{1}{t+x-2xt}\right) f(t)\,dt$$

exists almost everywhere on (0, 1), *and*

$$\int_0^1 g^2(x)\,dx \leqslant \int_0^1 f^2(x)\,dx.$$

COROLLARY. *If $h(x) \in L_2(0, 1)$ and $f(x) \in L_2(0, 1)$, then*

$$\int_0^1 g(x)\,h(x)\,dx = \frac{1}{\pi}\int_0^1 f(t)\,dt \int_0^1 \left(\frac{t}{x}\right)^{1/2}\left(\frac{1}{t-x} - \frac{1}{t+x-2xt}\right) h(x)\,dx.$$

3. *Proof of a certain inequality in the hyperbolic region of D.* Let $z(x, y)$ be a solution of equation (5.5) in class $\mathfrak{R}$, vanishing on Γ. In the domain D^- this solution can be represented by (5.7):

$$\begin{aligned} z(\xi, \eta) = {}& \gamma_1 \int_\xi^\eta \frac{(\eta-\xi)^{2/3}\,\tau(t)\,dt}{(\eta-t)^{5/6}(t-\xi)^{5/6}} - \gamma_2 \int_\xi^\eta \frac{\nu(t)\,dt}{(\eta-t)^{1/6}(t-\xi)^{1/6}} \\ &+ \int_\xi^\eta g_0(\xi, \eta, t)\,\tau(t)\,dt - \int_\xi^\eta h_0(\xi, \eta, t)\,\nu(t)\,dt, \end{aligned} \tag{5.38}$$

where

$$g_0(\xi, \eta, t) = \gamma_1 \int_\xi^t \int_t^\eta \frac{(\eta'-\xi')^{2/3}\,C(\xi', \eta')}{(\eta'-t)^{5/6}(t-\xi')^{5/6}}\, v(\xi', \eta'; \xi, \eta)\,d\xi'\,d\eta', \tag{5.39}$$

$$\begin{gathered} h_0(\xi, \eta, t) = \gamma_2 \int_\xi^t \int_t^\eta \frac{C(\xi', \eta')}{(\eta'-t)^{1/6}(t-\xi')^{1/6}}\, v(\xi', \eta'; \xi, \eta)\,d\xi'\,d\eta', \\ \gamma_1 = \frac{\Gamma\left(\frac{1}{3}\right)}{\Gamma^2\left(\frac{1}{6}\right)}, \qquad \gamma_2 = \frac{\left(\frac{4}{3}\right)^{1/3}\Gamma\left(\frac{2}{3}\right)}{2\Gamma^2\left(\frac{5}{6}\right)}. \end{gathered} \tag{5.40}$$

Setting $\xi = 0$ in (5.38), we obtain

$$\begin{aligned} z(0, \eta) = {}& \gamma_1 \int_0^\eta \frac{\eta^{2/3}\,\tau(t)\,dt}{(\eta-t)^{5/6}\,t^{5/6}} - \gamma_2 \int_0^\eta t^{-1/6}(\eta-t)^{-1/6}\,\nu(t)\,dt \\ &+ \int_0^\eta g_0(0, \eta, t)\,\tau(t)\,dt - \int_0^\eta h_0(0, \eta, t)\,\nu(t)\,dt. \end{aligned} \tag{5.41}$$

Substituting $\tau(t)$ from (5.15) into (5.41) and performing a few simple manipulations, we obtain

$$z(0, \eta) = \int_0^\eta t^{-1/6}(\eta - t)^{-1/6}\left[\frac{1}{\sqrt{3}}\varphi_1(t) - \gamma_2\nu(t)\right]dt$$
$$+ \int_0^\eta g_1(\eta, t)\varphi_1(t)\,dt - \int_0^\eta h_0(0, \eta, t)\nu(t)\,dt, \tag{5.42}$$

where

$$g_1(\eta, t) = \frac{1}{\sqrt{3}}\int_t^\eta d\eta' \int_0^t \frac{C(\xi', \eta')\,v(0, \eta; \xi', \eta')}{(\eta' - t)^{1/6}(1 - \xi')^{1/6}}\,d\xi'$$
$$+ \int_t^\eta d\eta' \int_t^{\eta'} \frac{C(\xi', \eta')\,v(0, \eta'; \xi', \eta')}{(\eta' - t)^{1/6}(\xi' - t)^{1/6}}\,d\xi'. \tag{5.43}$$

Replacing $\nu(x)$ and $\varphi_1(x)$ in (5.42) by their expressions (5.8) and (5.36), we get

$$z(0, \eta) = 2\pi \int_0^\eta x^{-1/6}\left[\frac{1}{\sqrt{3}}\lambda_1(x) - \gamma_2\mu(x)\right]dx$$
$$+ \int_0^\eta x^{-1/6} g_2(\eta, x)\lambda_1(x)\,dx - \int_0^\eta x^{-1/6} h_1(\eta, x)\mu(x)\,dx, \tag{5.44}$$

where

$$g_2(\eta, x) = \int_x^\eta t^{1/6}(t - x)^{-5/6} g_1(\eta, t)\,dt,$$

$$h_1(\eta, x) = \int_x^\eta t^{1/6}(t - x)^{-5/6} h_0(0, \eta, t)\,dt.$$

Replacing $g_1(\eta, t)$ and $h_0(0, \eta, t)$ by their expressions (5.43) and (5.40), we obtain

$$g_2(\eta, x) = \int_x^\eta d\eta' \int_0^{\eta'} C(\xi', \eta')\,v(0, \eta; \xi', \eta')\,I_2(\xi', \eta', x)\,d\xi',$$

$$h_1(\eta, x) = \int_x^\eta d\eta' \int_0^{\eta'} C(\xi', \eta')\,v(0, \eta; \xi', \eta')\,I_1(\xi', \eta', x)\,d\xi',$$

where

$$I_1(\xi', \eta', x) = \int_{\max(\xi', x)}^{\eta'} \frac{t^{1/6}\,dt}{(\eta'-t)^{1/6}(t-\xi')^{1/6}(t-x)^{5/6}};$$

$$I_2(\xi', \eta', x) = \int_0^{\eta'} \frac{t^{1/6}\,dt}{(\eta'-t)^{1/6}|t-\xi'|^{1/6}(t-x)^{5/6}}.$$

It is readily seen that

$$|I_1| \leqslant C_1\eta'^{1/6}|\xi'-x|^{-1/6}, \qquad |I_2| \leqslant C_1\eta'^{1/6}|\xi'-x|^{-1/6}.$$

We now use the following estimates for the derivatives of the Riemann function [4a]:

$$\left|\frac{\partial v(\xi', \eta'; \xi, \eta)}{\partial \xi}\right| \leqslant \frac{(\eta'-\xi')^{1/3}}{(\eta'-\xi)^{7/6}(\eta-\xi')^{1/6}} C_1(\xi', \eta'; \xi, \eta),$$
$$\left|\frac{\partial v(\xi', \eta'; \xi, \eta)}{\partial \eta}\right| \leqslant \frac{(\eta'-\xi')^{1/3}}{(\eta'-\xi)^{1/6}(\eta-\xi')^{7/6}} C_2(\xi', \eta'; \xi, \eta), \tag{5.45}$$

where $C_1 < \infty$ and $C_2 < \infty$ if $\xi \leqslant \xi' \leqslant \eta' \leqslant \eta$, $0 < \eta - \xi < \infty$, and the estimates for the integrals I_1 and I_2; the result is

$$\left|\frac{\partial g_2}{\partial \eta}\right| < C_2\eta^{1/3}, \qquad \left|\frac{\partial h_1}{\partial \eta}\right| < C_2\eta^{1/3}.$$

Differentiating (5.44) with respect to η, we get

$$\frac{\partial z(0, \eta)}{\partial \eta} = 2\pi\eta^{-1/6}\left[\frac{1}{\sqrt{3}}\lambda_1(\eta) - \gamma_2\mu(\eta)\right]$$
$$+ \int_0^{\eta} x^{-1/6}\frac{\partial g_2(\eta, x)}{\partial \eta}\lambda_1(x)\,dx - \int_0^{\eta} x^{-1/6}\frac{\partial h_1(\eta, x)}{\partial \eta}\mu(x)\,dx. \tag{5.46}$$

Hence we derive the important inequality

$$\int_0^1 (1-\eta)\,\eta^{1/3}\left[\frac{\partial z(0, \eta)}{\partial \eta}\right]^2 d\eta$$
$$\leqslant C_3^2\left[\int_0^1 (1-\eta)\,\lambda_1^2(\eta)\,d\eta + \int_0^1 (1-\eta)\,\mu^2(\eta)\,d\eta\right]. \tag{5.47}$$

4. *Properties of generalized solutions in class* $\mathfrak{R}$. Let $z(x, y)$ be a solution in class $\mathfrak{R}$, vanishing on Γ.

Let $\{\mu_n(x)\}$, $0 \leqslant x < 1$, be a sequence of functions such that

1) $\lim_{n\to\infty}\int_0^1(1-x)|\mu(x)-\mu_n(x)|^2\,dx = 0$;

2) the functions $\mu_n(x)$ are twice continuously differentiable; $\mu_n(x)$ and $\mu_n'(x)$ vanish in a sufficiently small neighborhood of the point $x = 0$.

The corresponding functions $\nu_n(x)$ defined by (5.8) are also twice continuously differentiable, and $\nu_n(x)$ and $\nu_n'(x)$ vanish in a sufficiently small neighborhood of $x = 0$. The functions $\nu_n(x)$ define a corresponding sequence of solutions $\{z_n(x, y)\}$ vanishing on Γ and continuous in $\bar{D}$. By Lemma 1,

$$\lim_{n\to\infty} z_n(x, y) = z(x, y), \quad \lim_{n\to\infty} \frac{\partial z_n}{\partial x} = \frac{\partial z}{\partial x}, \quad \lim_{n\to\infty} \frac{\partial z_n}{\partial y} = \frac{\partial z}{\partial y}$$

uniformly in $y \geqslant \delta > 0$.

LEMMA 4. *If $z(x, y)$ is a solution of equation* (5.5) *satisfying the boundary condition*

$$\lim_{y\to 0} \frac{\partial z}{\partial y} = \nu(x), \qquad z|_\Gamma = 0,$$

where $\nu(x)$ is twice continuously differentiable for $0 \leqslant x \leqslant 1$ and $\nu(0) = \nu'(0) = 0$, then the following assertions are true:

1) *$\tau'(x)$ is bounded for $0 \leqslant x \leqslant 1$ and $\tau''(x)$ is continuous for $0 < x < 1$.*

2) *$\lim_{y\to 0} \partial z/\partial x = \tau'(x)$ uniformly in each interval $(\epsilon, 1-\epsilon)$, and for small $y > 0$*

$$\left|\frac{\partial z}{\partial x}\right| < C_1[\hat{x}(y)]^{-\frac{1}{3}-\delta} + C_2 q(1, 0; x, y),$$

where $\hat{x}(y) = \min(x - x_1, x_2 - x)$, and x_1 and x_2 are the abscissas of the points of Γ with ordinate y; $\delta > 0$ is an arbitrary small number.

3) *$\lim_{y\to 0} \partial z/\partial y = \nu(x)$ uniformly in each interval $(\epsilon, 1-\epsilon)$, and for small $y > 0$*

$$\left|\frac{\partial z}{\partial y}\right| < C_3\left|\ln\left(1 - x + \sqrt{(1-x)^2 + \frac{4}{9}y^3}\right)\right| + C_4\frac{\sqrt{y}}{[\hat{x}(y)]^{1/3+\delta}}.$$

PROOF. We begin with the proof of 2). A solution $z(x, y)$ of equation (5.5) vanishing on Γ can be expressed (see (4.14) and (4.3)) as

$$z(x, y) = z_0(x, y) + \iint_{D^+} C(\xi, \eta)\, G(x, y; \xi, \eta)\, z(\xi, \eta)\, d\xi\, d\eta, \quad (5.48)$$

where

$$z_0(x, y) = \int_0^1 \nu(t)\, G(t, 0; x, y)\, dt$$

$$= \int_0^1 \nu(t)\,[q(t, 0; x, y) + p(t, 0; x, y)]\, dt = I_1 + I_2,$$

since $z_0(x, y)$ vanishes on Γ.

Consider $z_0(x, y)$. From (4.4) we have

$$\left|\frac{\partial q}{\partial x}\right| < \frac{C}{r r_1^{1/3}}, \qquad \left|\frac{\partial q}{\partial y}\right| < \frac{C\sqrt{y}}{r r_1^{1/3}},$$

$$\begin{aligned}\frac{\partial I_1}{\partial x} &= \frac{\partial}{\partial x}\int_0^1 \nu(t)\, q(t, 0; x, y)\, dt \\ &= \frac{\partial}{\partial x}\int_0^1 \nu'(t)\, dt \int_t^1 q(\zeta, 0; x, y)\, d\zeta \\ &= -\nu(1)\, q(1, 0; x, y) + \int_0^1 \nu'(t)\, q(t, 0; x, y)\, dt.\end{aligned}$$

Consequently,

$$\left|\frac{\partial I_1}{\partial x}\right| < Cq(1, 0; x, y)$$

and

$$\lim_{y\to 0}\frac{\partial I_1(x, y)}{\partial x} = \frac{\partial I_1(x, 0)}{\partial x}$$

uniformly for $0 \leqslant x < 1 - \epsilon$. The derivative $\partial^2 I_1(x, 0)/\partial x^2$ is obviously continuous for $0 < x < 1$.

The function $p(t, 0; x, y)$ can be expressed [66g] as

$$p(t, 0; x, y) = \int_0^l \rho(s, t, 0)\, q(\xi, \eta; x, y)\, ds, \qquad (\xi, \eta) \in \Gamma,$$

and

$$I_2 = \int_0^1 \nu(t)\, p(t, 0; x, y)\, dt = \int_0^l q(\xi, \eta; x, y)\, ds \int_0^1 \nu(t)\rho(s, t, 0)\, dt.$$

Set

$$\varphi(s) = \int_0^1 \nu(t)\rho(s, t, 0)\, dt, \qquad \psi(s) = \int_0^1 \nu(t)\, A(s, t, 0)\, dt.$$

Then, by (4.5),

$$\varphi(s) - 2\int_0^l K(t, s)\,\varphi(t)\, dt = 2\psi(s). \tag{5.49}$$

We have

$$\psi(s) = \int_0^1 \nu(t)\, A(s, t, 0)\, dt = \int_0^1 \nu'(t)\, dt \int_t^1 A(s, x, 0)\, dx,$$

whence

$$|\psi(s)| \leqslant \frac{k}{3} \int_0^1 |\nu'(t)|\, dt \int_t^1 \frac{\left|\eta \frac{d\eta}{ds}(\xi - x) - \frac{2}{3}\eta^2 \frac{d\xi}{ds}\right|}{\left[(\xi - x)^2 + \frac{4}{9}\eta^3\right]^{7/6}} dx$$

$$< C_5 \eta^{1/2} \int_0^1 dt \int_t^1 \frac{dx}{\left[(\xi - x)^2 + \frac{4}{9}\eta^3\right]^{1/2}} < C_6 \eta^{\frac{1}{2} - \delta_1},$$

where $\delta_1 > 0$ is an arbitrary small number. Noting that

$$|K(t, s)| < C \frac{\eta^{1/2}}{r_1^{1/3}} \left(\ln \frac{1}{\sigma} + 1\right),$$

we see via (5.49) that $|\varphi(s)| < C_7 \eta^{1/2 - \delta_1}$. Thus,

$$\left|\frac{\partial I_2(x, y)}{\partial x}\right|$$

$$= \left|\int_0^l \varphi(s) \frac{\partial q(\xi, \eta; x, y)}{\partial x} ds\right| < C_8 \int_0^l \frac{\eta^{\frac{1}{2} - \delta_1} ds}{r r_1^{1/3}} < C_9 \int_0^l \frac{\eta^{1/2} ds}{r^{\frac{4}{3} + \frac{2\delta_1}{3}}}.$$

Set $2\delta_1/3 = \delta$. To fix ideas, let $\hat{x}(y) = x_2 - x$. Then

$$\int_0^l \frac{\eta^{1/2} ds}{r^{\frac{4}{3} + \delta}} = \int_0^\varepsilon \frac{\eta^{1/2} ds}{r^{\frac{4}{3} + \delta}} + O(1).$$

But for small y and ϵ,

$$\frac{1}{r} < \frac{1}{\left[\hat{x}^2 + \frac{4}{9}(\eta^{3/2}(s) - y^{3/2})^2\right]^{1/2}}$$

and so

$$\int_0^l \frac{\eta^{1/2} ds}{r^{\frac{4}{3} + \delta}} < O(1) + C_{10} \int_{-\frac{2}{3} y^{3/2}}^{\varepsilon_1} \frac{dv}{(\hat{x}^2 + v^2)^{\frac{2}{3} + \frac{\delta}{2}}} < O(1) + C_{11} \hat{x}^{-\frac{1}{3} - \delta} < \frac{C_{12}}{\hat{x}^{\frac{1}{3} + \delta}}.$$

Clearly,

$$\lim_{y \to 0} \frac{\partial I_2(x, y)}{\partial x} = \frac{\partial I_2(x, 0)}{\partial x}$$

uniformly for $\epsilon \leqslant x \leqslant 1 - \epsilon$, and $\partial^2 I_2(x, 0)/\partial x^2$ is continuous for $0 < x < 1$. Thus 2) is proved for $z_0(x, y)$.

Next,

$$\left|\frac{\partial I_1}{\partial y}\right| = \left|\int_0^1 \nu'(t)\,dt \int_t^1 \frac{\partial q(\zeta, 0; x, y)}{\partial y}\,d\zeta\right|$$

$$= \left|\frac{2k}{9}\int_0^1 \nu'(t)\,dt \int_t^1 \frac{y^2\,d\zeta}{\left[(\zeta - x)^2 + \frac{4}{9}y^3\right]^{7/6}}\right|$$

$$< \frac{2k}{9}\int_0^1 |\nu'(t)|\,dt \int_t^1 \frac{d\zeta}{\left[(\zeta - x)^2 + \frac{4}{9}y^3\right]^{1/2}}$$

$$< C_{13}\left|\ln\left(1 - x + \sqrt{(1-x)^2 + \frac{4}{9}y^3}\right)\right|.$$

An estimate for $\partial I_2/\partial y$ can be derived in the same way as for $\partial I_1/\partial y$. Obviously,

$$\lim_{y\to 0}\frac{\partial z_0(x, y)}{\partial y} = \nu(x)$$

uniformly for $\epsilon \leqslant x \leqslant 1 - \epsilon$. Thus, 3) is proved for $z_0(x, y)$.

By Lemmas 2 and 3 of §2, parts 2) and 3) are also valid for $z(x, y)$. As we have shown that the derivative $\partial^2 z_0(x, 0)/\partial x^2$ is continuous for $0 < x < 1$, it follows from the same lemmas that $\partial^2 z(x, 0)/\partial x^2$ is also continuous for $0 < x < 1$.

It remains only to consider $\tau'(x)$. According to (5.12),

$$\tau(x) = -\int_0^1 \nu(t)\,G_0(t, 0; x, 0)\,dt - \int_0^1 \chi(t, x)\,\nu(t)\,dt.$$

It is easy to see that the derivative of the first integral is bounded for $0 \leqslant x \leqslant 1$, while the derivative of the second is bounded by virtue of (5.28). Q.E.D.

All of these solutions $z_n(x, y)$ obviously possess the properties established in Lemma 4 in the domain D^+.

The properties of $z_n(x, y)$ in D^- are given by the following lemma.

LEMMA 5. *The solution $z_n(x, y)$ has continuous derivatives $\partial^2 z_n/\partial x^2$ and $\partial^2 z_n/\partial y^2$ in D^-; moreover,*

$$\lim_{y\to 0}\frac{\partial z_n(x, y)}{\partial x} = \tau_n'(x)$$

uniformly in any interval $(\epsilon, 1 - \epsilon)$, and $\partial z_n(x, y)/\partial y$ is continuous up to the parabolic curve; $\partial z_n/\partial\eta|_{\xi=0}$ and $\partial z_n/\partial\xi|_{\eta=1}$ are continuous.

PROOF. By Lemma 4, it follows from (5.38) that $\partial^2 z_n/\partial x^2$ and $\partial^2 z_n/\partial y^2$ are continuous in D^-. Since the $z_n(x, y)$ belong to class $\mathfrak{R}_1$, it follows from Lemma 2 of §2 that the derivative $\partial z_n(x, y)/\partial y$ is continuous up to the parabolic curve, while $\partial z_n/\partial\eta|_{\xi=0}$ and $\partial z_n/\partial\xi|_{\eta=1}$ are continuous. Formula (5.38) can be written [66g] as

$$z_n(\xi, \eta) = z_{n0}(\xi, \eta) + \int_\xi^\eta d\xi' \int_{\xi'}^\eta C(\xi', \eta') z_{n0}(\xi', \eta') v(\xi', \eta'; \xi, \eta)\, d\eta' = z_{n0}(\xi, \eta) + I_{n0}(\xi, \eta), \tag{5.50}$$

where

$$z_{n0}(\xi, \eta) = \gamma_1 \int_0^1 \tau_n[\xi + (\eta - \xi) t]\, t^{-5/6}(1-t)^{-5/6}\, dt - \gamma_2 (\eta - \xi)^{2/3} \int_0^1 \nu_n[\xi + (\eta - \xi) t]\, t^{-1/6}(1-t)^{-1/6}\, dt. \tag{5.50'}$$

Using the estimates (5.46) for derivatives of the Riemann function $v(\xi', \eta'; \xi, \eta)$, one readily obtains estimates for the first derivatives of $I_{n0}(\xi, \eta)$:

$$\left| \frac{\partial I_{n0}}{\partial \xi} \right| < C_1 (\eta - \xi)^{1/3}, \qquad \left| \frac{\partial I_{n0}}{\partial \eta} \right| < C_2 (\eta - \xi)^{1/3}. \tag{5.51}$$

Differentiating (5.50′) with respect to x, we get

$$\frac{\partial z_{n0}}{\partial x} = \gamma_1 \int_0^1 \tau_n'[\xi + (\eta - \xi) t]\, t^{-5/6}(1-t)^{-5/6}\, dt - \gamma_2 (\eta - \xi)^{2/3} \int_0^1 \nu_n'[\xi + (\eta - \xi) t]\, t^{-1/6}(1-t)^{-1/6}\, dt.$$

Hence,

$$\lim_{y \to 0} \frac{\partial z_{n0}}{\partial x} = \tau_n'(x)$$

uniformly for $\epsilon < x < 1 - \epsilon$. Moreover, we note that

$$\left| \frac{\partial z_{n0}}{\partial x} \right| < C_n.$$

Thus, using (5.51), we see that

$$\lim_{y \to 0} \frac{\partial z_n(x, y)}{\partial x} = \tau_n'(x)$$

uniformly for $\epsilon < x < 1 - \epsilon$. Moreover,

$$\left| \frac{\partial z_n(x, y)}{\partial x} \right| < \widetilde{C}_n. \tag{5.52}$$

This completes the proof.

5. *Condition on* Γ. *Uniqueness theorem.* Let $z(x, y)$ be a solution of equation (5.5) in class $\mathfrak{R}$, vanishing on Γ. Let $\{z_n(x, y)\}$ be the sequence of

solutions constructed in subsection 4. To each solution $z_n(x, y)$ there corresponds a twice continuously differentiable solution $u_n(\theta, \sigma)$ of the Čaplygin equation (5.1) in the domain D:

$$u_n(\theta, \sigma) = z_n e^{-\frac{1}{2}\int_0^y b(y)\,dy}.$$

Obviously, $u_n = 0$ on Γ. Omitting the subscript n, we consider a twice continuously differentiable solution of (5.1) in D such that the function

$$z = u \exp\left(-\frac{1}{2}\int_0^y b(y)\,dy\right)$$

satisfies conditions 1, 2 and 3 of Lemma 4. Since $z = 0$ on Γ, it follows that $\partial u/\partial \theta$ and $\partial u/\partial \sigma$ are continuous right up to the boundary Γ for $\sigma > 0$.

Let $p(\theta, \sigma)$ and $q(\theta, \sigma)$ be continuous functions in $\overline{D}^+$, having continuous first derivatives there.

Consider the identities

$$\begin{aligned} 2p\frac{\partial u}{\partial \theta} K(u)) \\ = \left[p\left(ku_\theta^2 - u_\sigma^2\right)\right]_\theta + \left(2pu_\theta u_\sigma\right)_\sigma - p_\theta\left(Ku_\theta^2 - u_\sigma^2\right) - 2pu_\theta u_\sigma = 0, \end{aligned} \tag{5.53}$$

$$\begin{aligned} 2q\frac{\partial u}{\partial \sigma} K(u) = \left[q\left(u_\sigma^2 - Ku_\theta^2\right)\right]_\sigma + \left(2qKu_\theta u_\sigma\right)_\theta - q_\sigma u_\sigma^2 \\ + (qK)_\sigma u_\theta^2 - 2q_\theta K u_\theta u_\sigma = 0. \end{aligned} \tag{5.54}$$

Let D_ϵ^+ and Γ_ϵ be the parts of D^+ and Γ lying in the half-plane $y \geqslant \epsilon$; let $\theta_1(\epsilon)$ and $\theta_2(\epsilon)$ be the abscissas of the points at which Γ cuts the line $\sigma = \epsilon$ ($\epsilon > 0$ is small).

Integrating (5.53) and (5.54) over D_ϵ^+ and using Green's formula, we get

$$\begin{aligned} -2\int_{\theta_1(\varepsilon)}^{\theta_2(\varepsilon)} p(\theta, \varepsilon)\, u_\theta(\theta, \varepsilon)\, u_\sigma(\theta, \varepsilon)\, d\theta + \int_{\Gamma_\varepsilon} -2pu_\theta u_\sigma\, d\theta + p\left(Ku_\theta^2 - u_\sigma^2\right) d\sigma \\ - \iint_{D_\varepsilon^+} \left[2p_\sigma u_\theta u_\sigma + p_\theta\left(Ku_\theta^2 - u_\sigma^2\right)\right] d\theta\, d\sigma = 0, \end{aligned} \tag{5.55}$$

$$\begin{aligned} \int_{\theta_1(\varepsilon)}^{\theta_2(\varepsilon)} q\left(Ku_\theta^2 - u_\sigma^2\right) d\theta + \int_{\Gamma_\varepsilon} q\left(Ku_\theta^2 - u_\sigma^2\right) d\theta + 2qKu_\theta u_\sigma\, d\sigma \\ - \iint_{D_\varepsilon^+} \left[2Kq_\theta u_\theta u_\sigma + q_\sigma u_\sigma^2 - (qK)_\sigma u_\theta^2\right] d\theta\, d\sigma = 0. \end{aligned} \tag{5.56}$$

Since $u|_\Gamma = 0$, it follows that $u_\theta d\theta + u_\sigma d\sigma = 0$ on Γ. Using this equation and subtracting (5.55) from (5.56), we see that

$$2\int_{\theta_1(\varepsilon)}^{\theta_2(\varepsilon)} p u_\theta u_\sigma \, d\theta + \int_{\theta_1(\varepsilon)}^{\theta_2(\varepsilon)} q\left(K u_\theta^2 - u_\sigma^2\right) d\theta$$
$$- \int_{\Gamma_\varepsilon} \left(K u_\theta^2 + u_\sigma^2\right)(p\, d\sigma + q\, d\theta) - \iint_{D_\varepsilon^+} \{2\left(K q_\theta - p_\sigma\right) u_\theta u_\sigma \tag{5.57}$$
$$+ \left(q_\sigma + p_\theta\right)\left(u_\sigma^2 - K u_\theta^2\right) - q K_\sigma u_\theta^2\}\, d\theta\, d\sigma = 0.$$

The functions p and q satisfy the conditions

$$K q_\theta - p_\sigma = 0, \qquad q_\sigma + p_\theta = 0, \tag{5.58}$$
$$q \geqslant 0, \quad \lim_{\sigma \to 0} q(\theta, \sigma) = 0. \tag{5.59}$$

Then (5.57) becomes

$$2\int_{\theta_1(\varepsilon)}^{\theta_2(\varepsilon)} p u_\theta u_\sigma \, d\theta + \int_{\theta_1(\varepsilon)}^{\theta_2(\varepsilon)} q\left(K u_\theta^2 - u_\sigma^2\right) d\theta$$
$$- \int_{\Gamma_\varepsilon} \left(K u_\theta^2 + u_\sigma^2\right)(p\, d\sigma + q\, d\theta) + \iint_{D_\varepsilon^+} q K_\sigma u_\theta^2 \, d\theta\, d\sigma = 0. \tag{5.60}$$

We now assume that the following condition holds along Γ:

$$p\, d\sigma + q\, d\theta \leqslant 0. \tag{5.61}$$

Since $K_\sigma \geqslant 0$ and the function $z = u \exp(1/2 \int_0^y b(y)\, dy)$ satisfies conditions 1, 2 and 3 of Lemma 4, one can let $\epsilon \longrightarrow 0$ in (5.60). Then, by (5.59),

$$2\int_0^1 p u_\theta u_\sigma \, d\theta - \int_\Gamma \left(K u_\theta^2 + u_\sigma^2\right)(p\, d\sigma + q\, d\theta)$$
$$+ \iint_{D^+} q K_\sigma u_\theta^2 \, d\theta\, d\sigma = 0. \tag{5.62}$$

Setting $\omega_\sigma = q$ and $\omega_\sigma = -p$, we can write (5.58) as

$$K\omega_{\theta\theta} + \omega_{\sigma\sigma} = 0.$$

We now go over to the hyperbolic half-plane. Let $p(\theta, 0) = p(\theta)$, and let

$$\xi = \theta - \frac{2}{3}(-y)^{3/2} = C_1, \qquad \eta = \theta + \frac{2}{3}(-y)^{3/2} = C_2$$

be the two families of characteristics of equation (5.1).

The following identity holds in the domain D^-:

$$2p(\eta)\,u_\theta K(u) = \left[p\left(Ku_\theta^2 - u_\sigma^2\right)\right]_\theta + (2pu_\theta u_\sigma)_\sigma + p'(\eta)\left(\sqrt{-K}\,u_\theta + u_\sigma\right)^2 = 0. \tag{5.63}$$

Consider the subdomain D_γ^- of D^- bounded by the segment $[0, 1]$ of the θ-axis, a smooth curve γ emanating from the point A $(0, 0)$ and intersecting the characteristic $\eta = 1$ at a point C_γ, and the segment $C_\gamma B$ of the characteristic $\eta = 1$. Let

$$D_{\gamma\varepsilon}^- = D_\gamma^- \cap (\sigma \leqslant -\varepsilon),\quad \gamma_\varepsilon = \gamma \cap (\sigma \leqslant -\varepsilon),\ \varepsilon > 0,$$

and let $\theta_1(\epsilon)$ and $\theta_2(\epsilon)$ be the abscissas of the intersection points of γ and $\eta = 1$ with the line $\sigma = -\epsilon$; let B_ϵ be the intersection point of $\eta = 1$ with the line $\sigma = -\epsilon$.

Integrating (5.63) over the domain $D_{\gamma\epsilon}^-$, we get

$$\begin{aligned} 2\int_{\theta_1(\varepsilon)}^{\theta_2(\varepsilon)} pu_\theta(\theta, -\varepsilon)\,u_\sigma(\theta, -\varepsilon)\,d\theta &+ \int_{\gamma_\varepsilon} p\left[-2u_\theta u_\sigma\,d\theta + \left(Ku_\theta^2 - u_\sigma^2\right)d\sigma\right] \\ &- \int_{C_\gamma}^{B_\varepsilon} p(1)\left(\sqrt{-K}\,u_\theta + u_\sigma\right)^2 d\sigma \\ &+ \iint_{D_{\gamma\varepsilon}^-} p'(\eta)\left(\sqrt{-K}\,u_\theta + u_\sigma\right)^2 d\theta\,d\sigma = 0. \end{aligned} \tag{5.64}$$

By (5.59), $\lim_{\sigma\to 0} q_\sigma \geqslant 0$; thus, it follows from (5.58) that $p'(\theta, 0) = p'(\theta) \leqslant 0$. Set

$$p(1) = 0. \tag{5.65}$$

By Lemma 5, we can let $\epsilon \longrightarrow 0$ in (5.64), to obtain

$$\begin{aligned} 2\int_0^1 pu_\theta u_\sigma\,d\theta &+ \int_\gamma p\left[-2u_\theta u_\sigma\,d\theta + \left(Ku_\theta^2 - u_\sigma^2\right)d\sigma\right] \\ &+ \iint_{D_\gamma^-} p'(\eta)\left(\sqrt{-K}\,u_\theta + u_\sigma\right)^2 d\theta\,d\sigma = 0. \end{aligned} \tag{5.66}$$

Subtracting (5.62) from (5.66), we get

$$\int_{\gamma} p\left[-2u_\theta u_\sigma\, d\theta + (Ku_\theta^2 - u_\sigma^2)\, d\sigma\right]$$
$$= -\iint_{D_\gamma^-} p'(\eta)\left(\sqrt{-K}\, u_\theta + u_\sigma\right)^2 d\theta\, d\sigma \tag{5.67}$$
$$+\iint_{D^+} qK_\sigma u_\theta^2\, d\theta\, d\sigma - \int_{\Gamma} (Ku_\theta^2 + u_\sigma^2)(p\, d\sigma + q\, d\theta).$$

This equality will be of great importance in what follows.

Let γ coincide with the characteristic $\xi = 0$. Along this characteristic,

$$d\theta + \sqrt{-K}\, d\sigma = 0, \qquad d\eta = -2\sqrt{-K}\, d\sigma,$$
$$-2u_\theta u_\sigma\, d\theta + (Ku_\theta^2 - u_\sigma^2)\, d\sigma = -\frac{(du)^2}{d\sigma} = 2u_\eta^2 \sqrt{-K}\, d\eta.$$

Then it follows from (5.67) that

$$2\int_0^1 p(\eta)\, u_\eta^2 \sqrt{-K}\, d\eta = -\iint_{D^-} p'(\eta)\left(\sqrt{-K}\, u_\theta + u_\sigma\right)^2 d\theta\, d\sigma$$
$$+\iint_{D^+} qK_\sigma u_\theta^2\, d\theta\, d\sigma - \int_{\Gamma} (Ku_\theta^2 + u_\sigma^2)(p\, d\sigma + q\, d\theta). \tag{5.68}$$

Hence

$$2\int_0^1 p(\eta)\, u_\eta^2 \sqrt{-K}\, d\eta \geqslant \iint_{D^+} qK_\sigma u_\theta^2\, d\theta\, d\sigma. \tag{5.69}$$

There exists an infinite set of pairs of functions p and q satisfying (5.58), (5.59) and (5.65). Correspondingly, there is a class of curves Γ on which (5.61) holds. In particular, consider the pair of functions $p = 1 - \theta$, $q = \sigma$. Condition (5.61) becomes

$$(1-\theta)\frac{d\sigma}{ds} + \sigma\frac{d\theta}{ds} \leqslant 0, \tag{5.70}$$

where s is arc-length on Γ, $\theta(0) = 1$ and $\sigma(\theta) = 0$.

In particular, any curve Γ whose union with its reflection in the σ-axis is a convex curve satisfies (5.70).

Throughout the remainder of this section, we consider only curves Γ satisfying (5.70); this will not be mentioned again.

Returning to the solutions u_n, we have

$$\int_0^1 (1-\eta)\sqrt{-K}\left(\frac{\partial u_n}{\partial \eta}\right)^2 d\eta \geqslant \frac{1}{2}\iint_{D^+} \sigma K_\sigma \left(\frac{\partial u_n}{\partial \theta}\right)^2 d\theta\, d\sigma.$$

In view of the relation between u_n and z_n, we get

$$\frac{\partial u_n}{\partial \eta} = e^{-\frac{1}{2}\int_0^y b(y)\,dy}\left(\frac{\partial z_n}{\partial \eta} - \frac{b(y)}{2\cdot 6^{1/3}}\eta^{-1/3}z_n\right).$$

Since $\sqrt{-K} < C\eta^{1/3}$ and $\exp(-1/2\int_0^y b(y)\,dy) < C_1$, it follows that

$$\int_0^1 (1-\eta)\eta^{1/3}\left(\frac{\partial z_n}{\partial \eta} - \frac{b(y)}{2\cdot 6^{1/3}}\eta^{-1/3}z_n\right)^2 d\eta \geqslant C_2 \iint_{D^+} \sigma K_\sigma\left(\frac{\partial u_n}{\partial \theta}\right)^2 d\theta\,d\sigma. \tag{5.71}$$

By (5.37), we infer from (5.47) that

$$\int_0^1 (1-\eta)\eta^{1/3}\left(\frac{\partial z_n}{\partial \eta}\right)^2 d\eta \leqslant C_3^2\left[\int_0^1 (1-\eta)\lambda_{1n}^2(\eta)\,d\eta + \int_0^1 (1-\eta)\mu_n^2(\eta)\,d\eta\right]$$
$$\leqslant C_4^2\int_0^1 (1-\eta)\mu_n^2(\eta)\,d\eta.$$

Thus

$$\int_0^1 (1-\eta)\eta^{1/3}\left(\frac{\partial z_n}{\partial \eta} - \frac{\partial z}{\partial \eta}\right)^2 d\eta \leqslant C_4^2\int_0^1 (1-\eta)[\mu_n(\eta) - \mu(\eta)]^2\,d\eta.$$

By Hardy's inequality (see p. 84),

$$\int_0^1 (1-\eta)\eta^{-1/3}(z_n - z)^2\,d\eta \leqslant C_5^2\int_0^1 (1-\eta)\eta^{1/3}\left(\frac{\partial z_n}{\partial \eta} - \frac{\partial z}{\partial \eta}\right)^2 d\eta.$$

By virtue of the last two inequalities, we can let $n \longrightarrow \infty$ in (5.71). We have proved

LEMMA 6. *If z is a solution of equation* (5.5) *in class* $\mathfrak{R}$, *vanishing on* Γ, *then*

$$\int_0^1 (1-\eta)\eta^{1/3}\left[\frac{\partial z(0,\eta)}{\partial \eta} - \frac{b(y)}{2\cdot 6^{1/3}}\eta^{-1/3}z(0,\eta)\right]^2 d\eta \geqslant C_2\iint_{D^+}\sigma K_\sigma\left(\frac{\partial u}{\partial \theta}\right)^2 d\theta\,d\sigma. \tag{5.72}$$

THEOREM 2.10. *If condition* (5.70) *holds along* Γ, *then problem* T *for equation* (5.1) *has a unique solution in class* $\mathfrak{R}$.

PROOF. Let us suppose that there are two solutions, $z_1 \in \mathfrak{R}$ and $z_2 \in \mathfrak{R}$. Then $z = z_1 - z_2 \in \mathfrak{R}$, and this difference vanishes on Γ and on the characteristic $\xi = 0$. By Lemma 6, $u = u(\sigma)$, i.e. $u = A\sigma + B$, and so $u \equiv 0$ in D^+. But then $u \equiv z \equiv 0$ in D.

6. *Solution of the Tricomi problem in the class* $\mathfrak{R}$. Let $z(x, y)$ be a solution of the Tricomi problem for equation (5.5) in $\mathfrak{R}$. Then the function $\nu(x)$ is given by (5.8), and the unknown function is now $\mu(x)$.

In the proof of (5.18), we set

$$\varphi_1(x) = \frac{\sqrt{3}}{2\pi} \frac{d}{dx} \int_0^x (x-t)^{-2/3} \tau(t)\, dt.$$

Let

$$\tilde{\varphi}(x) = \frac{1}{k} \varphi_1(x) - \nu(x) + \frac{\sqrt{3}}{2\pi k} \frac{d}{dx} \int_0^x (x-t)^{-2/3} f_1(t)\, dt, \tag{5.73}$$

where $f_1(x) = \int_0^l \varphi(s)\tilde{\rho}(s, x, 0)\, ds$, and let

$$\lambda(x) = \frac{x^{1/6}}{2\pi} \frac{d}{dx} \int_0^x (x-t)^{-1/6} t^{-1/6} \tilde{\varphi}(t)\, dt. \tag{5.74}$$

Consider the second functional equation (see (4.57)):

$$\begin{aligned} \tau(x) = k \int_0^x (x-t)^{-1/3} \nu(t)\, dt + \int_0^x K(x, t)\, \nu(t)\, dt \\ + \frac{\Gamma\left(\frac{1}{6}\right)}{\Gamma\left(\frac{5}{6}\right)\Gamma\left(\frac{1}{3}\right)} \int_0^x \psi_1(t) \frac{t^{1/3}\, dt}{x^{1/6}(x-t)^{1/6}} + \int_0^x l(x, t)\, \psi_1(t)\, dt. \end{aligned} \tag{5.75}$$

Applying to (5.75) the operator

$$\frac{d}{dy} \int_0^y (y-x)^{-2/3} \ldots dx$$

and using (5.73), we obtain (see §4.3)

$$\tilde{\varphi}(x)=\frac{\sqrt{3}}{2\pi k}\left\{\int_0^x \nu(t)\,dt\int_t^x (x-y)^{-2/3}\frac{\partial K(y,t)}{\partial y}\,dy\right.$$

$$+x^{1/6}\int_0^x (x-t)^{-5/6}\psi'(t)\,dt+\int_0^x \psi_1(t)\,dt\int_t^x (x-y)^{-2/3}\frac{\partial l(y,t)}{\partial y}\,dy$$

$$\left.+\int_0^x (x-t)^{-2/3} f_1'(t)\,dt\right\}.$$

Substituting the last expression into (5.74), we find that

$$\lambda(x)=\frac{\sqrt{3}\,\Gamma\left(\frac{1}{3}\right)}{2\pi k\Gamma^2\left(\frac{1}{6}\right)}\left\{\int_0^x \nu(t)\,dt\int_t^x \left(\frac{y}{x}\right)^{1/6}(x-y)^{-5/6}\frac{\partial K(y,t)}{\partial y}\,dy\right.$$

$$+\int_0^x \psi_1(t)\,dt\int_t^x \left(\frac{y}{x}\right)^{1/6}(x-y)^{-5/6}\frac{\partial l(y,t)}{\partial y}\,dy \tag{5.76}$$

$$\left.+\int_0^x \left(\frac{y}{x}\right)^{1/6}(x-y)^{-5/6} f_1'(y)\,dy\right\}+\frac{\sqrt{3}}{2\pi k}x^{1/6}\psi'(x).$$

Finally, substituting (5.76) in (5.33) we get

$$\mu(x)+\frac{1}{\pi\sqrt{3}}\int_0^1\left[\frac{x(1-t)}{t(1-x)}\right]^{1/6}\left(\frac{1}{t-x}+\frac{1-2t}{t+x-2xt}\right)\mu(t)\,dt$$

$$+\int_0^1\left[\frac{x(1-t)}{t(1-x)}\right]^{1/6}K_4(t,x)\mu(t)\,dt \tag{5.77}$$

$$+\int_0^x K_5(x,t)\mu(t)\,dt=g(x),$$

where

$$K_5(x,t)=\frac{\Gamma\left(\frac{1}{3}\right)}{\pi\sqrt{3}\,\Gamma^2\left(\frac{1}{6}\right)}\int_t^x\left(\frac{y}{x}\right)^{1/6}(x-y)^{-5/6}\,dy$$

$$\times\int_t^y\left(\frac{v}{t}\right)^{1/6}(v-t)^{-5/6}\frac{\partial K(y,v)}{\partial y}\,dv,$$

$$g(x) = -\frac{1}{\pi\sqrt{3}k} x^{1/6}\psi'(x)$$
$$-\frac{\Gamma\left(\frac{1}{3}\right)}{\pi\sqrt{3}k\Gamma^2\left(\frac{1}{6}\right)}\left\{\int_0^x \psi_1(t)\,dt\int_t^x \left(\frac{y}{x}\right)^{1/6}(x-y)^{-5/6}\frac{\partial l(y,t)}{\partial y}\,dy \right. \tag{5.78}$$
$$\left. + \int_0^x \left(\frac{y}{x}\right)^{1/6}(x-y)^{-5/6} f_1'(y)\,dy\right\}.$$

Note that, by Lemma 5 of §4.2,

$$|K_5(x, t)| < K_5 x^{1/3} t^{-1/6}. \tag{5.79}$$

Let $\mathfrak{L}_2(0, 1)$ denote the space of all functions $f(x)$, $0 < x < 1$, such that

$$\int_0^1 (1-x) f^2(x)\,dx < \infty.$$

LEMMA 7. *If* $f(x), h(x) \in \mathfrak{L}_2(0, 1)$, *then the formulas*

$$h(x) = f(x) + \frac{1}{\pi\sqrt{3}}\int_0^1 \left[\frac{x(1-t)}{t(1-x)}\right]^{1/6}\left(\frac{1}{t-x} + \frac{1-2t}{t+x-2xt}\right) f(t)\,dt, \tag{5.80}$$

and

$$f(x) = \frac{3}{4}\left\{h(x) - \frac{1}{\pi\sqrt{3}}\int_0^1 \left[\frac{t(1-t)}{x(1-x)}\right]^{1/2}\left(\frac{1}{t-x} - \frac{1}{t+x-2xt}\right) h(t)\,dt\right\} \tag{5.81}$$

are inverses of one another, and

$$m^2 \int_0^1 (1-x) h^2(x)\,dx \leqslant \int_0^1 (1-x) f^2(x)\,dx \leqslant M^2 \int_0^1 (1-x) h^2(x)\,dx. \tag{5.82}$$

PROOF. If $f(x) \in \mathfrak{L}_2(0, 1)$, then, by Lemma 2,

$$\int_0^1 (1-x) h^2(x)\,dx \leqslant B_1^2 \int_0^1 (1-x) f^2(x)\,dx. \tag{5.83}$$

Then, according to Lemma 3, if $h(x) \in \mathfrak{L}_2(0, 1)$, it follows from (5.81) that

$$\int_0^1 (1-x) f^2(x)\,dx \leqslant B_2^2 \int_0^1 (1-x) h^2(x)\,dx. \tag{5.84}$$

Using the identity

$$\frac{1}{t-x}+\frac{1-2t}{t+x-2xt}=\frac{t}{x}\left(\frac{1}{t-x}-\frac{1}{t+x-2xt}\right),$$

we can rewrite (5.80) as

$$h(x)=f(x)+\frac{1}{\pi\sqrt{3}}\int_0^1\frac{t^{5/6}(1-t)^{1/6}}{x^{5/6}(1-x)^{1/6}}\left(\frac{1}{t-x}-\frac{1}{t+x-2xt}\right)f(t)\,dt.$$

Therefore, if $h(x)$ satisfies a Hölder condition for $0<x<1$, it follows from Lemma 9 of §4 that

$$f(x)=\frac{3}{4}\left\{h(x)-\frac{1}{\pi\sqrt{3}}\int_0^1\left[\frac{t(1-t)}{x(1-x)}\right]^{1/2}\right.$$
$$\left.\times\left(\frac{1}{t-x}-\frac{1}{t+x-2xt}\right)h(t)\,dt\right\}+Ax^{-3/2}(1-x)^{-1/2}.$$

But, if $f(x)\in\mathfrak{L}_2(0,1)$, then $A=0$ and so (5.81) holds. Since the set of functions satisfying a Hölder condition is dense in $\mathfrak{L}_2(0,1)$, (5.81) follows from (5.80). Combining (5.83) and (5.84), we get (5.82). Q.E.D.

Applying this lemma to (5.77), we see that

$$\mu(x)+\int_0^1 M(x,t)\,\mu(t)\,dt=h(x). \tag{5.85}$$

Here

$$M(x,t)=\frac{3}{4}\left\{\left[\frac{x(1-t)}{t(1-x)}\right]^{1/6}K_4(t,x)\right.$$
$$\left.-\frac{1}{\pi\sqrt{3}}\int_0^1\left[\frac{y(1-y)}{x(1-x)}\right]^{1/2}\left(\frac{1}{y-x}-\frac{1}{y+x-2xy}\right)\left[\frac{y(1-t)}{t(1-y)}\right]^{1/6}K_4(t,y)\,dy\right\}$$
$$+\frac{3}{4}\left\{\delta K_5(x,t)-\frac{1}{\pi\sqrt{3}}\int_t^1\left[\frac{y(1-y)}{x(1-x)}\right]^{1/2}\right. \tag{5.86}$$
$$\left.\times\left(\frac{1}{y-x}-\frac{1}{y+x-2xy}\right)K_5(y,t)\,dy\right\}=M_1(x,t)+M_2(x,t),$$

where $\delta=1$ if $t<x$, and $\delta=0$ if $t>x$; and

$$h(x)=\frac{3}{4}\left\{g(x)-\frac{1}{\pi\sqrt{3}}\right. \tag{5.87}$$
$$\left.\times\int_0^1\left[\frac{t(1-t)}{x(1-x)}\right]^{1/2}\left(\frac{1}{t-x}-\frac{1}{t+x-2xt}\right)g(t)\,dt\right\}.$$

We now investigate the kernel $M(x, t)$. By (5.82) and (5.34), we have

$$\int_0^1 (1-x)\, M_1^2(x, t)\, dx$$

$$\leqslant M_1^2 \int_0^1 x^{1/3}(1-x)^{2/3}\left(\frac{1-t}{t}\right)^{1/3} K_4^2(t, x)\, dx < M_2^2\left(\frac{1-t}{t}\right)^{1/3}$$

and so

$$\int_0^1 dt \int_0^1 \frac{1-x}{1-t} M_1^2(x, t)\, dx < C_1^2 < \infty.$$

Similarly, by (5.82) and (5.79),

$$\int_0^1 (1-x)\, M_2^2(x, t)\, dx \leqslant M_3^2 \int_t^1 (1-x)\, K_5^2(x, t)\, dx < M_4^2 t^{-1/3}(1-t)^2$$

and so

$$\int_0^1 dt \int_0^1 \frac{1-x}{1-t} M_2^2(x, t)\, dx < C_2^2 < \infty.$$

Thus,

$$\int_0^1 dt \int_0^1 \frac{1-x}{1-t} M^2(x, t)\, dx < C^2 < \infty. \tag{5.88}$$

Consider the function $h(x)$. It follows easily from (5.78) that if the function $z|_\Gamma = \varphi(s)$ satisfies the conditions of Lemma 7 of §4.4, then

$$\int_0^1 (1-x)\, g^2(x)\, dx \leqslant C_1 \int_0^1 (1-x)\, x^{1/3} \psi'^2(x)\, dx + C_2,$$

where the constant C_2 depends on $\varphi(s)$, and $C_2 = 0$ if $\varphi \equiv 0$. By the last inequality and Lemma 7, it follows from (5.87) that

$$\int_0^1 (1-x)\, h^2(x)\, dx \leqslant C_3 \int_0^1 (1-x)\, x^{1/3} \psi'^2(x)\, dx + C_4, \tag{5.89}$$

where $C_4 = 0$ if $\varphi(s) \equiv 0$.

Set

$$\tilde{\mu}(x) = (1-x)^{1/2} \mu(x), \quad \tilde{h}(x) = (1-x)^{1/2} h(x),$$
$$\tilde{M}(x, t) = \left(\frac{1-x}{1-t}\right)^{1/2} M(x, t).$$

Then (5.85) can be rewritten as

$$\tilde{\mu}(x) + \int_0^1 \tilde{M}(x, t)\tilde{\mu}(t)\,dt = \tilde{h}(x). \tag{5.90}$$

Inequality (5.88) implies that Fredholm theory can be applied to (5.90).

We now prove that the homogeneous equation

$$\tilde{\mu}(x) + \int_0^1 \tilde{M}(x, t)\tilde{\mu}(t)\,dt = 0$$

has no nontrivial solutions in $L_2(0, 1)$. Suppose the contrary. Then the function

$$\nu(x) = x^{1/6}\int_0^x (x-t)^{-5/6}t^{-1/6}\mu(t)\,dt$$

satisfies the homogeneous equation (see §4.3)

$$\nu(x) + \frac{1}{\pi\sqrt{3}}\int_0^1 \left(\frac{t}{x}\right)^{2/3}\left(\frac{1}{t-x} - \frac{1}{t+x-2xt}\right)\nu(t)\,dt + \int_0^1 L(x, t)\nu(t)\,dt = 0.$$

By virtue of Lemma 1,

$$\int_0^1 (1-x)^{3/2}|\nu(x)|^3\,dx < \infty.$$

It follows that $\int_0^1 |\nu(x)|^p\,dx < \infty$ if $1 < p < 6/5$. Then, by the results of §4.5, $\nu(x) \equiv 0$, i.e. $\mu(x) \equiv 0$.

Thus (5.90) has a unique solution $\tilde{\mu}(x)$, and

$$\int_0^1 \tilde{\mu}^2(x)\,dx \leqslant A^2\int_0^1 (1-x)\,x^{1/3}\psi'^2(x)\,dx + B,$$

as follows from (5.89). If $\varphi \equiv 0$, then $B = 0$.

We have thus proved

THEOREM 2.11. *If $\varphi(s)$ satisfies the conditions of Lemma 7 of §4.4, if $\psi(\eta)$ is such that*

$$\int_0^1 (1-x)\,x^{1/3}\psi'^2(x)\,dx < \infty,$$

with $\varphi(l) = \psi(0)$, and if Γ satisfies (5.70), then there exists a solution of equation (5.5) in D, belonging to the class $\mathfrak{R}$, such that $z|_\Gamma = \varphi(s)$ and $z|_{\xi=0} = \psi(\eta)$. Moreover,

$$\int_0^1 (1-x)\mu^2(x)\,dx \leqslant A^2 \int_0^1 (1-x)\,x^{1/3}\psi'^2(x)\,dx + B. \tag{5.91}$$

§6. The Neumann-Tricomi problem[8]

Consider the equation

$$y^m u_{xx} + u_{yy} = 0 \qquad (m > 0). \tag{6.1}$$

Let D be the domain bounded by a simple Jordan curve Γ with end points A (0, 0), B (1, 0), lying in the upper half-plane $y > 0$, and by characteristics

$$AC\colon\ \xi = x - \frac{2}{m+2}(-y)^{\frac{m+2}{2}} = 0$$

and

$$BC\colon\ \eta = x + \frac{2}{m+2}(-y)^{\frac{m+2}{2}} = 1$$

of equation (6.1).

PROBLEM T_N. *Find a solution of equation* (6.1) *in the domain D, continuous in the closed domain $\overline{D}$, having continuous partial derivatives u_x and u_y in $D + \Gamma$ and satisfying boundary conditions*

$$A_s[u] \equiv y^m \frac{dy}{ds}\frac{\partial u}{\partial x} - \frac{dx}{ds}\frac{\partial u}{\partial y}\bigg|_{\Gamma} = \varphi(s) \qquad (0 < s < l), \tag{6.2}$$

$$u\,|_{AC} = \psi(x) \qquad \left(0 \leqslant x \leqslant \tfrac{1}{2}\right), \tag{6.3}$$

where $\varphi(s)$ and $\psi(x)$ are given functions.

The assumptions of problem T_N imply that the function $u(x, 0) = \tau(x)$ must be continuous on [0, 1]. The functions $\nu(x) = \partial u(x, 0)/\partial y$ and $\tau'(x)$ must be continuous and differentiable on $0 < x < 1$. In addition, we assume that $\nu(x)$ may have poles of order less than $2/(m + 2)$ at the points $x = 0$ and $x = 1$ (with respect to x and $(1 - x)$).

1. *Extremum principle and uniqueness theorem for problem T_N.* A solution of (6.1) in the hyperbolic half-plane $y < 0$, satisfying the Cauchy data

$$u(x, 0) = \tau(x), \qquad \lim_{y \to 0} \frac{\partial u(x, y)}{\partial y} = \nu(x), \tag{6.4}$$

is given by

(8) See [66e] and [66g].

$$u(x, y)=\gamma_1 \int_0^1 \tau\left[x+\frac{2}{m+2}(-y)^{\frac{m+2}{2}}(2t-1)\right] t^{\beta-1}(1-t)^{\beta-1}dt$$

$$+\left(\frac{4}{m+2}\right)^{\frac{2}{m+2}} \gamma_2 y \int_0^1 \nu\left[x+\frac{2}{m+2}(-y)^{\frac{m+2}{2}}(2t-1)\right] t^{-\beta}(1-t)^{-\beta}\,dt, \tag{6.5}$$

where

$$\gamma_1=\frac{\Gamma(2\beta)}{\Gamma^2(\beta)}, \quad \gamma_2=\left(\frac{m+2}{4}\right)^{\frac{2}{m+2}} \frac{\Gamma(2-2\beta)}{\Gamma^2(1-\beta)}, \quad \beta=\frac{m}{2(m+2)}.$$

This solution of the Cauchy problem for equation (6.1) is unique.

DEFINITION. A generalized solution (6.5) of equation (6.1) is said to belong to *class* R_1 if the function $\tau(x)$ satisfies a Hölder condition with exponent $\alpha_1 > 1-\beta$ for $0 \leqslant x < 1$ and $\nu(x)$ satisfies a Hölder condition with exponent $\alpha_2 > \beta$ for $0 \leqslant x < 1$.

We have the following

LEMMA. *If $u \in R_1$, then the derivatives u_x and u_y are continuous in ΔABC, u_y is continuous right up to the parabolic curve, and*

$$\lim_{y\to 0}\frac{\partial u}{\partial y}=\nu(x) \qquad (0<x<1).$$

If the solution (6.5) of the Cauchy problem satisfies the boundary condition (6.3), it follows from (6.5) that

$$\psi(x)=\gamma_1 \int_0^1 \tau(2xt)\, t^{\beta-1}(1-t)^{\beta-1}\,dt-\gamma_2(2x)^{\frac{2}{m+2}} \int_0^1 \nu(2xt)\, t^{-\beta}(1-t)^{-\beta}\,dt$$

or, replacing x by $x/2$ and putting $\zeta = xt$,

$$\psi\left(\frac{x}{2}\right)=\gamma_1 x^{1-2\beta} \int_0^x \tau(\zeta)\, \zeta^{\beta-1}(x-\zeta)^{\beta-1}\,d\zeta-\gamma_2 \int_0^x \nu(\zeta)\, \zeta^{-\beta}(x-\zeta)^{-\beta}\,d\zeta \tag{6.6}$$

$$(0 \leqslant x \leqslant 1).$$

Let us solve (6.6) for $\tau(x)$, using the inversion formula for the Abel integral equation. This gives

$$\tau(x)=\psi_1(x)+\gamma \int_0^x (x-t)^{-2\beta}\, \nu(t)\,dt, \tag{6.7}$$

where

$$\psi_1(x) = \frac{\sin\beta\pi}{\pi\gamma_1} x^{1-\beta} \frac{d}{dx} \int_0^x t^{2\beta-1}(x-t)^{-\beta}\psi\left(\frac{t}{2}\right)dt,$$
$$\gamma = \frac{\gamma_2}{\gamma_1} \frac{\Gamma(1-\beta)}{\Gamma(\beta)\Gamma(1-2\beta)}. \tag{6.8}$$

(6.7) gives the first functional relation between $\tau(x)$ and $\nu(x)$, determined by the condition that the solution $u(x, y)$ of (6.1) in the domain D^- must take prescribed values $\psi(x)$ on $\xi = 0$.

If $\psi(x) \equiv 0$, (6.7) gives

$$\tau(x) = \gamma \int_0^x (x-t)^{-2\beta}\nu(t)\,dt. \tag{6.9}$$

Again using the inversion formula for the Abel equation, we can write (6.9) as

$$\nu(x) = \frac{\sin 2\beta\pi}{\gamma\pi} \frac{d}{dx} \int_0^x \frac{\tau(t)\,dt}{(x-t)^{1-2\beta}}. \tag{6.10}$$

This immediately implies the *extremum principle* for problem T_N: *a solution $u(x, y)$ of problem T_N vanishing on the characteristic AC achieves its positive maximum and negative minimum in the closed domain $\overline{D}^+$ on the curve* Γ.

Indeed, $u(x, y)$ obviously cannot have an extremum in the interior of D^+. Suppose that its positive maximum in the closed domain $\overline{D}^+$ is achieved at an interior point $P(\xi, 0)$ of the interval $0 < x < 1$. At the point ξ, $0 < \xi < 1$, we have

$$\tau'(\xi) = 0, \tag{6.11}$$

and so (6.10) can be rewritten as

$$\frac{\gamma\pi}{\sin 2\beta\pi}\nu(\xi) = (1-2\beta)\int_0^\xi \frac{\tau(\xi)-\tau(t)}{(\xi-t)^{2-2\beta}}dt + \tau(\xi)\,\xi^{2\beta-1}.$$

Hence, in view of the fact that $\tau(\xi) > 0$ and $\tau(\xi) - \tau(t) > 0$, we get $\nu(\xi) > 0$. This contradicts Lemma 5 of §2.

The extremum principle immediately implies the uniqueness of the solution of problem T_N.

2. *Problem* T_N^+. *Find a regular solution $u(x, y)$ of equation* (6.1) *in* D^+, *continuous in* $\overline{D}^+$, *having continuous derivatives u_x and u_y in $D^+ \cup \Gamma$ which may have poles of order less than $2/(m+2)$ at the points A and B, and satisfying the boundary conditions*

$$A_s[u] \equiv y^m \frac{dy}{ds}\frac{\partial u}{\partial x} - \frac{dx}{ds}\frac{\partial u}{\partial y}\Big|_\Gamma = \varphi(s) \qquad (0 < s < l), \tag{6.12}$$

$$u\,|_{y=0} = \tau(x) \qquad (0 \leqslant x \leqslant 1). \tag{6.13}$$

The curve Γ is assumed to satisfy two conditions: 1) condition 1 of §4; 2) the inequality $|dx/ds| \leqslant C^2 y^{m+1}(s)$, where C is a constant, holds in a neighborhood of the points A and B.

The uniqueness of the solution of problem T_N^+ follows easily from the formula

$$\iint_{D^+} (y^m u_x^2 + u_y^2)\, dx\, dy = \int_{\Gamma + AB} u A_s[u]\, ds, \tag{6.14}$$

where u is a solution of (6.1).

In the elliptic half-plane, (6.1) has a fundamental solution

$$\begin{aligned} &q_2(x, y; x_0, y_0) \\ &= k_2 \left(\frac{4}{m+2}\right)^{4\beta-2} \left(r_1^2\right)^{-\beta} (1-\sigma)^{1-2\beta} F(1-\beta, 1-\beta, 2-2\beta; 1-\sigma), \end{aligned} \tag{6.15}$$

where

$$\left.\begin{matrix} r^2 \\ r_1^2 \end{matrix}\right\} = (x - x_0)^2 + \frac{4}{(m+2)^2}\left(y^{\frac{m+2}{2}} \mp y_0^{\frac{m+2}{2}}\right)^2,$$

$$\sigma = \frac{r^2}{r_1^2}, \qquad k_2 = \frac{1}{4\pi}\left(\frac{4}{m+2}\right)^{2-2\beta} \frac{\Gamma^2(1-\beta)}{\Gamma(2-2\beta)}. \tag{6.15'}$$

Denoting the coordinates of a variable point on the curve Γ by (ξ, η), consider the function

$$A(s, x, y) = \eta^m \frac{d\eta}{ds} \frac{\partial q_2(\xi, \eta; x, y)}{\partial \xi} - \frac{d\xi}{ds} \frac{\partial q_2(\xi, \eta; x, y)}{\partial \eta}. \tag{6.16}$$

We shall need several lemmas.

LEMMA 1 (GELLERSTEDT).

$$\int_0^l A(s, x, y)\, ds = \begin{cases} i(x, y) - 1, & \text{if } (x, y) \text{ is inside } D \text{ or on the } x\text{-axis for } 0 < x < 1, \\ i(x, y) - \tfrac{1}{2}, & \text{if } (x, y) \in \Gamma, \\ i(x, y), & \text{if } (x, y) \text{ lies outside } D, \end{cases}$$

where

$$i(x, y) = \int_0^1 \frac{\partial q_2(\xi, 0; x, y)}{\partial \eta}\, d\xi.$$

LEMMA 2. *If the point (x, y) lies anywhere in the upper half-plane, then*

$$\int_0^l |A(s; x, y)|\, ds \leqslant A, \tag{6.17}$$

where A is a constant.

LEMMA 3. *If the point* (x, y) *lies on* Γ, *then*

$$|A(s, x, y)| \leqslant A_1 \frac{\eta^{m/2}}{(r_1^2)^\beta}\left(\ln\frac{1}{\sigma}+1\right). \tag{6.18}$$

Consider the potential of a simple layer with density $\mu(s)$:

$$v(x, y) = \int_0^l \mu(t)\, q_2(\xi, \eta;\ x, y)\, dt. \tag{6.19}$$

Consider the conormal[9] to Γ at an arbitrary point $N(s)$ on Γ. Let $M(x, y)$ be some point of the conormal, not on Γ, and find the conormal derivative of the simple-layer potential (6.19):

$$A_s[v(x, y)] = \int_0^l \mu(t)\, A_s[q_2(\xi, \eta;\ x, y)]\, dt. \tag{6.20}$$

The integral (6.20) also exists if $M(x, y)$ is the point $N(s)$.

We have

THEOREM 2.12. *If the density* $\mu(t)$ *is continuous, then*

$$\begin{aligned} A_s[v(x, y)]_i &= \frac{1}{2}\mu(s) + A_s[v(x, y)]_0, \\ A_s[v(x, y)]_e &= -\frac{1}{2}\mu(s) + A_s[v(x, y]_0, \end{aligned} \tag{6.21}$$

where

$$A_s[v(x, y)]_0 = \int_0^l \mu(t)\, A_s[q_2(\xi(t), \eta(t);\ x(s), y(s))]\, dt = \int_0^l K_2(t, s)\, \mu(t)\, dt. \tag{6.22}$$

The proof follows from Lemmas 1 and 2.

Obviously, the potential $v(x, y)$ is a regular solution of (6.1) in any domain contained in the upper half-plane which has no points in common with either Γ or the x-axis.

It is easy to see that if (x, y) goes to infinity, then $v(x, y)$ tends to zero. Indeed, let (x, y) be a point on the "normal" curve C_R:

$$x^2 + \frac{4}{(m+2)^2}\, y^{m+2} = R^2.$$

---

(9) The direction cosines of the conormal are

$$\frac{y^m \cos(nx)}{\sqrt{y^{2m}\cos^2(nx) + \cos^2(ny)}}, \quad \frac{\cos(ny)}{\sqrt{y^{2m}\cos^2(nx) + \cos^2(ny)}},$$

where n is the exterior normal to Γ.

Then, by (6.15),

$$|v(x, y)| \leqslant \int_0^l |\mu(t)||q_2(\xi, \eta; x, y)| dt \leqslant \frac{M}{R} \qquad (R \geqslant R_0). \tag{6.23}$$

Similarly

$$|A_s[v(x, y)]| \leqslant MR^{2(\beta-1)} \qquad (R \geqslant R_0), \tag{6.24}$$

where M is a constant.

As in potential theory for the Laplace equation, it can be shown that (6.14) is applicable to the potential $v(x, y)$ in D^+. We have

$$\iint_{D^+} (y^m v_x^2 + v_y^2)\, dx\, dy = \int_0^l v A_s[v]_i\, ds. \tag{6.25}$$

We now apply (6.14) to the domain D' bounded by the curve Γ, part of the x-axis and by the "normal" curve C_R, which contains D^+. Letting $R \longrightarrow \infty$ and using (6.23) and (6.24), we see that

$$\iint_{D_e^+} (y^m v_x^2 + v_y^2)\, dx\, dy = -\int_0^l v A_s[v]_e\, ds. \tag{6.26}$$

Definition. A function $G_2(x, y; x_0, y_0)$ is a *Green's function for problem* T_N^+ if it satisfies the following conditions:

1) It is a solution of equation (6.1) in the interior of D^+, with the exception of the point (x_0, y_0).

2) It satisfies the boundary conditions

$$\begin{aligned} A_s[G_2(\xi, \eta; x_0\ y_0)]|_\Gamma &= 0, \qquad (x_0, y_0) \in D^+, \\ G_2(x, 0; x_0, y_0) &= 0 \qquad (y_0 > 0); \end{aligned} \tag{6.27}$$

3) It can be expressed as

$$G_2(x, y; x_0, y_0) = q_2(x, y; x_0, y_0) + v(x, y; x_0, y_0), \tag{6.28}$$

where $v(x, y; x_0, y_0)$ is a regular solution of (6.1) in D^+.

By virtue of (6.27), $v(x, y; x_0, y_0)$ satisfies the boundary conditions

$$A_s[v(\xi, \eta; x_0, y_0)]|_\Gamma = -A(s; x_0, y_0), \tag{6.29}$$

$$v(x, 0; x_0, y_0) = 0 \qquad (y_0 > 0). \tag{6.30}$$

We shall assume that $v(x, y; x_0, y_0)$ is a simple-layer potential:

$$v(x, y; x_0, y_0) = \int_0^l \mu(t; x_0, y_0)\, q_2(\xi, \eta; x, y)\, dt. \tag{6.31}$$

Condition (6.30) is satisfied. The boundary condition (6.29) and the first formula of (6.21) imply an integral equation for the density $\mu(s; x_0, y_0)$:

$$\mu(s; x_0, y_0) + 2\int_0^l K_2(t, s)\mu(t; x_0, y_0)\,dt = -2A(s, x_0, y_0). \qquad (6.32)$$

By (6.18), the kernel $K_2(t, s)$ has a weak singularity. We claim that $\lambda = -2$ is not an eigenvalue of $K_2(t, s)$. This is true if and only if the homogeneous equation

$$\mu(s) + 2\int_0^l K_2(t, s)\mu(t)\,dt = 0 \qquad (6.33)$$

has no nontrivial solution.

Suppose the contrary. Let $\mu_0(t)$ be a bounded solution of (6.33). Then the simple-layer potential with density $\mu_0(t)$ gives a function $v_0(x, y)$ which is a solution of (6.1) in D^+ and D_e^+ and is continuous throughout the upper half-plane $y > 0$. Since $\mu_0(t)$ satisfies the homogeneous equation (6.33) by assumption, it follows that $A_s[v_0]_i = 0$ on Γ. It then follows from (6.25) that $v_0(x, y) \equiv 0$ in D^+ and, in particular, on Γ. But then, by (6.26), $v_0(x, y) \equiv 0$ in D_e^+, i.e. $v_0(x, y) \equiv 0$ in the half-plane $y > 0$. Consequently, $A_s[v_0]_i = A_s[v_0]_e = 0$ on Γ. Thus, by (6.21), $\mu_0(t) \equiv 0$.

Thus, $\lambda = -2$ is not an eigenvalue of the kernel $K_2(t, s)$.

Let $R_2(t, s; \lambda)$ be the resolvent of the kernel $K_2(t, s)$. Then the solution of (6.32) can be written as

$$\begin{aligned}\mu(s; x_0, y_0) &\\ = -2A(s; x_0, y_0) &+ 4\int_0^l A(t; x_0, y_0) R_2(t, s; -2)\,dt.\end{aligned} \qquad (6.34)$$

Substituting (6.34) into (6.31), we get

$$\begin{aligned} v(x, y; x_0, y_0) &= -2\int_0^l A(s; x_0, y_0) q_2(\xi, \eta; x, y)\,ds \\ &+ 4\int_0^l\int_0^l R_2(t, s; -2) A(t; x_0, y_0) q_2(\xi(s), \eta(s); x, y)\,dt\,ds. \end{aligned} \qquad (6.35)$$

Let $\lambda(s; x, y)$ be a solution of the integral equation

$$\begin{aligned}\lambda(s; x, y) + 2\int_0^l K_2(s, t)\lambda(t; x, y)\,dt &\\ = -2q_2(\xi(s), \eta(s); x, y).&\end{aligned} \qquad (6.36)$$

Then

$$\lambda(s;\ x,\ y) = -2q_2(\xi(s),\ \eta(s);\ x,\ y) + 4\int_0^l R_2(s,\ t;\ -2)\, q_2(\xi(t),\ \eta(t);\ x,\ y)\, dt. \tag{6.37}$$

Multiplying (6.37) by $A(s; x_0, y_0)$ and integrating with respect to s from 0 to l, we get

$$\int_0^l \lambda(s;\ x,\ y)\, A(s;\ x_0,\ y_0)\, ds = -2\int_0^l q_2(\xi(s),\ \eta(s);\ x,\ y)\, A(s;\ x_0,\ y_0)\, ds$$

$$+4\int_0^l\int_0^l R_2(s,\ t;\ -2)\, A(s;\ x_0,\ y_0)\, q_2(\xi(t),\ \eta(t);\ x,\ y)\, dt\, ds$$

or, by (6.35),

$$v(x,\ y;\ x_0,\ y_0) = \int_0^l \lambda(s;\ x,\ y)\, A(s;\ x_0,\ y_0)\, ds. \tag{6.38}$$

As in the case of the Laplace equation, we have

LEMMA 4. *The Green's function $G_2(x, y; x_0, y_0)$ is symmetric with respect to the points (x, y) and (x_0, y_0), provided both of them lie in the interior of D^+.*

By virtue of this lemma, it follows from (6.28) that $v(x, y; x_0, y_0)$ is symmetric with respect to the points (x, y) and (x_0, y_0), and (6.38) can also be written as

$$v(x,\ y;\ x_0,\ y_0) = \int_0^l \lambda(t;\ x_0,\ y_0)\, A(t;\ x,\ y)\, dt. \tag{6.39}$$

Hence $v(x, y; x_0, y_0)$, regarded as a function of (x, y), is a double-layer potential with density $\lambda(t, x_0, y_0)$. Therefore

$$v_i(\xi,\ \eta;\ x_0,\ y_0) = -\frac{1}{2}\lambda(s;\ x_0,\ y_0) + \int_0^l K_2(s,\ t)\, \lambda(t;\ x_0,\ y_0)\, dt$$

or, by (6.36),

$$v_i(\xi,\ \eta;\ x_0,\ y_0) = -\lambda(s;\ x_0,\ y_0) - q_2(\xi,\ \eta;\ x_0,\ y_0).$$

Hence

$$G_2(\xi,\ \eta;\ x_0,\ y_0) = -\lambda(s;\ x_0,\ y_0), \qquad (\xi,\ \eta) \in \Gamma. \tag{6.40}$$

It is easily checked that

$$G_{02}(x, y; x_0, y_0) = q_2(x, y; x_0, y_0) - v_0(x, y; x_0, y_0), \tag{6.41}$$

where

$$v_0(x, y; x_0, y_0) = (4R^2)^{-\beta} q_2\left(x - \frac{1}{2}, y; \bar{x}_0, \bar{y}_0\right), \tag{6.42}$$

$$R^2 = \left(x_0 - \frac{1}{2}\right)^2 + \frac{4}{(m+2)^2} y_0^{m+2}; \qquad \bar{x}_0 = \frac{x_0 - \frac{1}{2}}{4R^2},$$
$$\bar{y}_0^{\frac{m+2}{2}} = \frac{y_0^{\frac{m+2}{2}}}{4R^2}, \tag{6.43}$$

is the Green's function of the Dirichlet problem for the normal domain D_0^+ bounded by the segment $[0, 1]$ of the x-axis and the normal curve Γ_0:

$$\left(x - \frac{1}{2}\right)^2 + \frac{4}{(m+2)^2} y^{m+2} = \frac{1}{4}.$$

We shall show that $v(x, y; x_0, y_0)$ can be expressed as

$$v_0(x, y; x_0, y_0) = -\int_0^l \lambda(s; x, y) A_s[v_0(\xi, \eta; x_0, y_0)]\, ds. \tag{6.44}$$

To this end, we consider the function

$$u(x, y; x_0, y_0) = -\int_0^l \lambda(s; x, y) A_s[v_0(\xi, \eta; x_0, y_0)]\, ds. \tag{6.45}$$

Since $\lambda(s; x, y)$ satisfies (6.1), the above function also satisfies it, as a function of (x, y).

Substituting expression (6.37) for $\lambda(s; x, y)$ into (6.45), we get

$$\begin{aligned} u(x, y; x_0, y_0) &= \int_0^l q_2(\xi(s), \eta(s); x, y) \Big\{ 2A_s[v_0(\xi(s), \eta(s); x_0, y_0)] \\ &\quad - 4\int_0^l R(t, s, -2) A_t[v_0(\xi(t), \eta(t); x_0, y_0)]\, dt \Big\}\, ds \end{aligned}$$

or

$$u(x, y; x_0, y_0) = -\int_0^l \chi(s; x_0, y_0) q_2(\xi, \eta; x, y)\, ds, \tag{6.46}$$

where

$$\chi(s; x_0, y_0) = -2A_s[v_0(\xi, \eta; x_0, y_0)] + 4\int_0^l R_2(t, s; -2)\, A_t[v_0(\xi(t), \eta(t); x_0, y_0)]\, dt,$$

i.e. $\chi(s; x_0, y_0)$ is a solution of the integral equation

$$\chi(s; x_0, y_0) + 2\int_0^l K_2(t, s)\,\chi(t; x_0, y_0)\, dt = -2A_s[v_0(\xi(s), \eta(s); x_0, y_0)]. \tag{6.47}$$

Using the first formula of (6.21), we infer from (6.46) that

$$A_s[u(x, y; x_0, y_0)]_l = -\frac{1}{2}\chi(s; x_0, y_0) - \int_0^l K_2(t, s)\,\chi(t; x_0, y_0)\, dt.$$

Hence, by (6.47)

$$A_s[u(x, y; x_0, y_0)]_l = A_s[v_0(\xi(s), \eta(s); x_0, y_0)].$$

It is easy to verify that

$$u(x, 0; x_0, y_0) = 0, \qquad v_0(x, 0; x_0, y_0) = 0.$$

Consequently, $u(x, y; x_0, y_0)$ and $v_0(x, y; x_0, y_0)$ satisfy the same equation (6.1) and the same boundary conditions (6.10) and (6.11), and thus, by the uniqueness theorem for problem T_N^+,

$$u(x, y; x_0, y_0) \equiv v_0(x, y; x_0, y_0).$$

Subtracting (6.41) from (6.28), we get

$$H(x, y; x_0, y_0) = G_2(x, y; x_0, y_0) - G_{02}(x, y; x_0, y_0) = v(x, y; x_0, y_0) + v_0(x, y; x_0, y_0)$$

or, by (6.38), (6.41) and (6.44),

$$H(x, y; x_0, y_0) = \int_0^l \lambda(s; x, y)\, A_s[G_{02}(\xi, \eta; x_0, y_0)]\, ds. \tag{6.48}$$

Thus

$$G_2(x, y; x_0, y_0) = G_{02}(x, y; x_0, y_0) + H(x, y; x_0, y_0). \tag{6.49}$$

This representation of the Green's function is more convenient for our subsequent treatment.

THEOREM 2.13. *The function*

$$u(x_0, y_0) = \int_0^1 \tau(x) \frac{\partial G_2(x, 0; x_0, y_0)}{\partial y} dx + \int_0^l \varphi(s) G_2(\xi, \eta; x_0, y_0) ds, \tag{6.50}$$

where $\tau(x)$ *and* $\varphi(s)$ *are continuous on* $[0, 1]$, *is a solution of problem* T_N^+ *in the domain* D^+.

PROOF. The first integral in (6.50), call it $I_1(x_0, y_0)$, is a solution of (6.1) continuous in $\overline{D}^+$. By (6.28) and (6.39), it can be written as

$$I_1(x_0, y_0) = \int_0^1 \tau(x) \frac{\partial q_2(x, 0; x_0, y_0)}{\partial y} dx + \int_0^1\int_0^l \tau(x) \lambda(t; x_0, y_0) \frac{\partial A(t; x, 0)}{\partial y} dx\, dt. \tag{6.51}$$

We claim that

$$\lim_{y_0 \to 0} I_1(x_0, y_0) = \tau(x_0) \qquad (0 < x_0 < 1). \tag{6.52}$$

It is easy to see that the second integral in (6.51) vanishes for $y_0 = 0$, $0 < x_0 < 1$. Thus we must prove that the first integral, $i_1(x_0, y_0)$ say, tends to $\tau(x_0)$ as $y_0 \to 0$ $(0 < x_0 < 1)$.

We have

$$i_1(x_0, y_0) = \int_0^1 \tau(x) \frac{\partial q_2(x, 0; x_0, y_0)}{\partial y} dx = k_2 y_0 \int_0^1 \left[(x - x_0)^2 + \frac{4}{(m+2)^2} y_0^{m+2}\right]^{\beta-1} \tau(x) dx.$$

Setting

$$x - x_0 = \frac{2}{m+2} y_0^{\frac{m+2}{2}} t,$$

we obtain

$$i_1(x_0, y_0) = k_2(1 - 2\beta)\left[\frac{4}{(m+2)^2}\right]^{\beta-1} \int_{A_1}^{B_1} \tau\left(x_0 + \frac{2}{m+2} y_0^{\frac{m+2}{2}} t\right)(1 + t^2)^{\beta-1} dt,$$

where

$$A_1 = -\frac{(m+2)\,x_0}{2y_0^{\frac{m+2}{2}}}, \qquad B_1 = \frac{(m+2)\,(1-x_0)}{2y_0^{\frac{m+2}{2}}}.$$

Letting $y_0 \longrightarrow 0$ $(0 < x_0 < 1)$, we get

$$\lim_{y_0 \to 0} i_1(x_0, y_0) = k_2(1-2\beta)\left[\frac{4}{(m+2)^2}\right]^{\beta-1} \tau(x_0) \int_{-\infty}^{\infty} (1+t^2)^{\beta-1}\,dt. \qquad (6.53)$$

By a standard formula,

$$\int_{-\infty}^{\infty} (1+t^2)^{\beta-1}\,dt = 2^{2\beta}\pi \frac{\Gamma(1-2\beta)}{\Gamma^2(1-\beta)}.$$

Therefore, by (6.15), it follows from (6.53) that

$$\lim_{y_0 \to 0} i_1(x_0, y_0) = \tau(x_0), \quad 0 < x_0 < 1.$$

If (x_0, y_0) lies on Γ and coincides with neither A nor B, then

$$\frac{\partial}{\partial y} A_s[G_2(x, 0;\, x_0(s), y_0(s))] = 0, \qquad 0 \leqslant x \leqslant 1.$$

Hence

$$A_s[I_1(x_0, y_0)]_\Gamma = 0. \qquad (6.54)$$

Consider the second integral in (6.50), $I_2(x_0, y_0)$ say. By (6.40) and (6.37), it may be written as

$$I_2(x_0, y_0) = -\int_0^l \varphi(s)\,\lambda(s;\, x_0, y_0)\,ds = \int_0^l \varkappa(s)\,q_2(\xi, \eta;\, x_0, y_0)\,ds; \qquad (6.55)$$

where

$$\varkappa(s) = 2\varphi(s) - 4\int_0^l R_2(t, s;\, -2)\,\varphi(t)\,dt,$$

i.e., $\kappa(s)$ is a solution of the integral equation

$$\varkappa(s) + 2\int_0^l K_2(t, s)\,\varkappa(t)\,dt = 2\varphi(s). \qquad (6.56)$$

Since $\kappa(s)$ is a continuous function, it follows immediately that $I_2(x_0, y_0)$ is a solution of (6.1), continuous in $\overline{D}^+$, which, thanks to (6.21) and (6.56), satisfies the condition

$$A_s[I_2(x_0, y_0)]_\Gamma = \varphi(s) \qquad (0 < s < l).$$

It is readily seen that

$$I_2(x_0, 0) = 0 \qquad (0 \leqslant x_0 \leqslant 1). \tag{6.57}$$

Q.E.D.

By virtue of (6.49) and (6.55), the solution (6.50) of problem T_N^+ can be written as

$$u(x_0, y_0) = \int_0^1 \tau(x) \left[\frac{\partial G_{02}(x, 0; x_0, y_0)}{\partial y} + \frac{\partial H(x, 0; x_0, y_0)}{\partial y} \right] dx$$
$$+ \int_0^l \varkappa(s)\, q_2(\xi, \eta; x_0, y_0)\, ds. \tag{6.58}$$

Without loss of generality, we may assume that $\tau(0) = 0$. Then, by (6.58),

$$\nu(x_0) = \frac{k_2}{1-2\beta} \int_0^1 \frac{(x - x_0)\, \tau'(x)\, dx}{|x - x_0|^{2-2\beta}} - k \int_0^1 \frac{\tau(x)\, dx}{(x + x_0 - 2xx_0)^{2-2\beta}}$$
$$- \frac{k_2 \tau(1)}{(1-2\beta)(1-x_0)^{1-2\beta}} + \int_0^1 \frac{\partial^2 H(x, 0; x_0, 0)}{\partial y\, \partial y_0} \tau(x)\, dx \tag{6.59}$$
$$+ \int_0^l \varkappa(s) \frac{\partial q_2(\xi, \eta; x_0, 0)}{\partial y_0}\, ds.$$

This is the second functional relation between $\tau(x)$ and $\nu(x)$, representing the fact that the solution $u(x, y)$ of (6.1) in D^+ must satisfy condition (6.2) on Γ.

3. *Reduction of problem T_N to a singular equation.* According to the results of subsections 1 and 2, the existence problem for a solution of problem T_N for equation (6.1) is equivalent to the solvability of equations (6.7) and (6.59). Elimination of $\nu(x_0)$ from these equations gives

$$\tau(x) = \frac{k_2 \gamma}{1-2\beta} \int_0^x (x - \zeta)^{-2\beta}\, d\zeta \int_0^1 \frac{(t - \zeta)\, \tau'(t)}{|t - \zeta|^{2-2\beta}}\, dt$$
$$- k_2 \gamma \int_0^x (x - \zeta)^{-2\beta}\, d\zeta \int_0^1 \frac{\tau(t)\, dt}{(t + \zeta - 2t\zeta)^{2-2\beta}} - \frac{k_2 \tau(1)}{1-2\beta} \int_0^x \frac{d\zeta}{(x-\zeta)^{2\beta}(1-\zeta)^{1-2\beta}}$$
$$+ \gamma \int_0^x (x - \zeta)^{-2\beta}\, d\zeta \int_0^1 \frac{\partial^2 H(t, 0; \zeta, 0)}{\partial y\, \partial y_0} \tau(t)\, dt$$
$$+ \gamma \int_0^x (x - \zeta)^{-2\beta}\, d\zeta \int_0^l \varkappa(s) \frac{\partial q_2(\xi(s), \eta(s); \zeta, 0)}{\partial y_0}\, ds + \psi_1(s).$$

After the necessary manipulations, we obtain a singular integral equation for $\tau(x)$:

$$\tau(x) - \lambda \int_0^1 \left(\frac{x}{t}\right)^{1-2\beta} \left(\frac{1}{t-x} - \frac{1}{t+x-2xt}\right) \tau(t)\, dt \\ - \lambda \int_0^1 H_1(x, t)\, \tau(t)\, dt = F(x), \tag{6.60}$$

where

$$\lambda = \frac{\cos \beta\pi}{\pi(1 + \sin \beta\pi)}, \tag{6.61}$$

$$H_1(x, t) = \frac{1-2\beta}{k_2} \int_0^x (x-\zeta)^{-2\beta} \frac{\partial^2 H(t, 0; \zeta, 0)}{\partial y\, \partial y_0}\, d\zeta, \tag{6.62}$$

$$F(x) = \frac{\gamma}{1 + \sin \beta\pi} \left[\psi_1(x) + \int_0^x (x-\zeta)^{-2\beta}\, d\zeta \int_0^l \varkappa(s) \frac{\partial q_2(\xi, \eta; \zeta, 0)}{\partial y_0}\, ds \right]. \tag{6.63}$$

It can be shown that the function $H_1(x, t)$ is continuous, except possibly at the point (1, 1), where it may have a pole of order less than 1.

We shall impose the following restrictions on $\varphi(s)$ and $\psi(x)$: the function $\varphi(s)$ has a continuous first derivative on $(0, l)$, and $\psi(x)$ has a continuous first derivative for $0 \leqslant x \leqslant 1/2$, having a zero of order 2β at $x = 0$; moreover, $\psi(x)$ has a continuous second derivative for $0 < x < 1/2$, which may have a pole of order at most $1 - 2\beta$ at $x = 0$. Then the function $F(x)$ is continuous for $0 \leqslant x \leqslant 1$ and for $0 < x < 1$ has a continuous first derivative which may have a pole of order at most 2β at $x = 0$.

4. *Regularization of the singular equation* (6.60). Consider the equation

$$\rho(x) - \lambda \int_0^1 \left(\frac{x}{t}\right)^{1-2\beta} \left(\frac{1}{t-x} - \frac{1}{t+x-2xt}\right) \rho(t)\, dt = f(x), \tag{6.64}$$

where $f(x)$ is a given function, satisfying a Hölder condition. We shall look for a solution in the class of functions satisfying a Hölder condition at interior points of the interval (0, 1) and continuous at its endpoints.

The solution of (6.64) has the form (see [66a]).

$$\rho(x) = \frac{1 + \sin \beta\pi}{2} \\ \times \left\{ f(x) + \lambda \int_0^1 \left[\frac{x(1-x)}{t(1-t)}\right]^{\frac{1}{2}-\beta} \left(\frac{1}{t-x} - \frac{1}{t+x-2xt}\right) f(t)\, dt \right\}. \tag{6.65}$$

We rewrite (6.60) as

$$\tau(x) - \lambda \int_0^1 \left(\frac{x}{t}\right)^{1-2\beta} \left(\frac{1}{t-x} - \frac{1}{t+x-2xt}\right) \tau(t)\, dt$$

$$= F(x) + \lambda \int_0^1 H_1(x, t)\, \tau(t)\, dt.$$

Treating the right-hand side as a known function, and using (6.64) and (6.65), we obtain

$$\tau(x) - \frac{\cos \beta\pi}{2\pi} \int_0^1 K(x, t)\, \tau(t)\, dt = g(x), \tag{6.66}$$

where

$$K(x, t) = H_1(x, t) + \lambda \int_0^1 \left[\frac{x(1-x)}{\xi(1-\xi)}\right]^{\frac{1}{2}-\beta} \left(\frac{1}{\xi - x} - \frac{1}{\xi + x - 2x\xi}\right) H_1(\xi, t)\, d\xi,$$

$$g(x) = \frac{1 + \sin \beta\pi}{2} \left\{ F(x) + \lambda \int_0^1 \left[\frac{x(1-x)}{t(1-t)}\right]^{\frac{1}{2}-\beta} \left(\frac{1}{t-x} - \frac{1}{t+x-2xt}\right) F(t)\, dt \right\}.$$

(6.66) is a Fredholm equation. Its solvability follows from the uniqueness theorem for problem T_N. Defining $\tau(x)$ from (6.66), we determine $\nu(x)$ from (6.7) and then, using (6.5) and (6.50), find the solution $u(x, y)$ in the domains D^- and D^+, respectively. This proves the existence of a solution to problem T_N.

CHAPTER III

THE GENERALIZED TRICOMI PROBLEM

§1. Statement of the generalized Tricomi problem

Consider the equation

$$K(y)u_{xx}+u_{yy}+a(x,y)u_x+b(x,y)u_y+c(x,y)u=f(x,y), \tag{1.1}$$

where $K(y)>0$ for $y>0$, $K(y)<0$ for $y<0$, and $K(0)=0$.

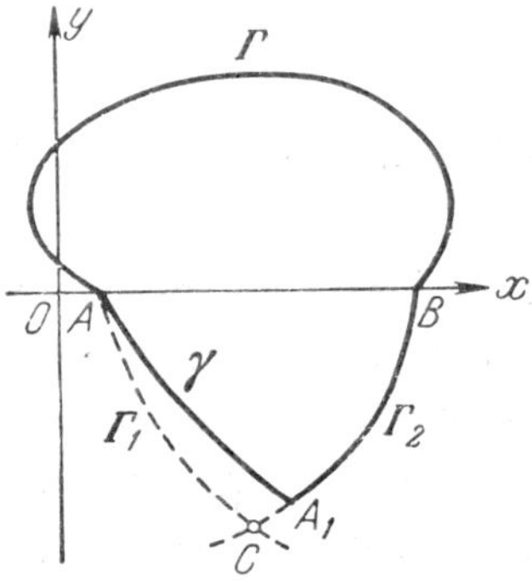

FIGURE 7.

Let Γ be a rectifiable Jordan curve in the upper half-plane, with endpoints $A\ (a, 0)$ and $B\ (b, 0)$. In the half-plane $y<0$, equation (1.1) has two different families of real characteristics defined by the equations

$$\frac{dy}{dx}=-\frac{1}{\sqrt{-K(y)}}, \tag{1.2a}$$

$$\frac{dy}{dx}=\frac{1}{\sqrt{-K(y)}}. \tag{1.2b}$$

Let Γ_1 and Γ_2 be characteristics of the families (1.2a) and (1.2b), emanating from the points A and B, respectively, and intersecting at a point C. Let D denote the domain bounded by the curve Γ and the characteristic arcs AC and BC of equation (1.1). The part of D lying in the half-plane $y>0$ $(y<0)$ is denoted by D^+ (D^-).

Let γ be a smooth curve in the interior of D^-, emanating from A and cutting the characteristic Γ_2 at a point A_1 (Figure 7), such that, at all points of γ,

$$0 \geqslant \frac{dy}{dx} \geqslant -\frac{1}{\sqrt{-K(y)}}, \tag{1.3}$$

i.e. the slope of γ at any point does not exceed that of the characteristics passing through that point.

Let D_γ denote the domain bounded by the curve Γ, the arc AA_1, and the characteristic arc A_1B of equation (1.1).

The *generalized Tricomi problem* can be stated as follows: *Find a solution of equation* (1.1) *in the domain* D_γ, *assuming prescribed values on the curves* Γ *and* γ:

$$u|_\Gamma = \varphi(s), \qquad u|_\gamma = \psi(x), \tag{1.4}$$

where $\varphi(l) = \psi(a)$.

On the parabolic curve $y = 0$ we impose "matching" conditions:

$$\begin{aligned} \lim_{y \to +0} u(x, y) &= \lim_{y \to -0} u(x, y) \qquad (a \leqslant x \leqslant b), \\ \lim_{y \to +0} \frac{\partial u(x, y)}{\partial y} &= \lim_{y \to -0} \frac{\partial u(x, y)}{\partial y} \qquad (a < x < b). \end{aligned} \tag{1.5}$$

The generalized Tricomi problem can also be considered in more complicated domains. Let γ_1 and γ_2 be characteristics of equation (1.1) emanating from the point O (0, 0), and let C_1 (C_2) be a smooth curve emanating from A (B) and intersecting the characteristic γ_1 (γ_2) at A_1 (B_1) (see Figure 2, p. 8). We assume that the following inequalities hold along the curves C_1 and C_2:

$$0 \geqslant \frac{dy}{dx} \geqslant -\frac{1}{\sqrt{-K(y)}} \ (\text{on } C_1) \qquad 0 \leqslant \frac{dy}{dx} \leqslant \frac{1}{\sqrt{-K(y)}} \ (\text{on } C_2).$$

Let D_1 denote the domain bounded by the curves Γ, C_1, γ_1, γ_2, and C_2.

The generalized Tricomi problem consists in finding a solution of equation (1.1) in the domain D_1 which takes prescribed values on the curves Γ, C_1 and C_2:

$$u|_\Gamma = \varphi(s), \quad u|_{C_1} = \psi_1(x), \quad u|_{C_2} = \psi_2(x),$$

with the "matching" condition (1.5) satisfied on the line $y = 0$.

§2. Uniqueness of the solution of the generalized Tricomi problem for the Čaplygin equation

1. *Weak extremum principle.*(1) Consider the Čaplygin equation

(1) See [56b].

$$K(y)u_{xx} + u_{yy} = 0, \tag{2.1}$$

where $K(y) > 0$ for $y > 0$, $K(y) < 0$ for $y < 0$, $K(0) = 0$ and $K(y)$ has a bounded derivative.

Let $u(x, y)$ be a solution of (2.1) in a simply connected domain D with boundary S. Consider the function

$$\Psi(x, y) = \int_{(x_0, y_0)}^{(x, y)} -2u_x u_y\, dx + (Ku_x^2 - u_y^2)\, dy, \tag{2.2}$$

defined in D, where (x_0, y_0) is a point of D. By (2.1), the value of the integral (2.2) does not depend on the integration path. We shall assume, in addition, that the derivatives of $u(x, y)$ satisfy conditions guaranteeing the continuity of the function $\Psi(x, y)$ in the closed domain $\bar{D}$. A suitable set of conditions is that u_x and u_y are continuous in $\bar{D}$, with the possible exception of isolated points of the boundary S, where the derivatives are $O(r^{-1/2+\epsilon})$, r being the distance from the boundary, $\epsilon > 0$.

LEMMA 1. *The function $\Psi(x, y)$ achieves its maximum in the closed domain $\bar{D}$ on the boundary S.*

Let D^+ be the region of D where $y > 0$ and D^- the region of D where $y < 0$. It follows from (2.1) and (2.2) that $\Psi(x, y)$ satisfies the equation

$$K\Psi_{xx} + \Psi_{yy} = K' u_x^2. \tag{2.3}$$

Since $K(y) > 0$ in D^+, we can express u_x^2 in terms of Ψ_x and Ψ_y using (2.2):

$$\begin{aligned} u_x^2 &= \frac{1}{2K}\left(\Psi_y + \sqrt{K\Psi_x^2 + \Psi_y^2}\right) \\ &= \frac{1}{2K}\left[\frac{K\Psi_x}{\sqrt{K\Psi_x^2 + \Psi_y^2}}\Psi_x + \left(1 + \frac{\Psi_y}{\sqrt{K\Psi_x^2 + \Psi_y^2}}\right)\Psi_y\right]. \end{aligned} \tag{2.4}$$

Substituting (2.4) into (2.3), we get

$$K(y)\Psi_{xx} + \Psi_{yy} = \alpha\Psi_x + \beta\Psi_y, \tag{2.5}$$

where

$$\alpha = \frac{K'\Psi_x}{2\sqrt{K\Psi_x^2 + \Psi_y^2}}, \qquad \beta = \frac{K'}{2K}\left(1 + \frac{\Psi_y}{\sqrt{K\Psi_x^2 + \Psi_y^2}}\right).$$

The coefficients α and β are bounded in any closed subdomain of D^+, and equation (2.5) is elliptic in D^+. Consequently, $\Psi(x, y)$ satisfies the maximum principle in D^+.

$\Psi(x, y)$ is not an increasing function of y in D^-, since $\Psi_y = Ku_x^2 - u_y^2 \leqslant 0$ for $y \leqslant 0$, and therefore, if $\Psi(x, y)$ achieves a maximum value for $y \leqslant 0$, it must do so at points of the boundary S for $y \leqslant 0$. Q.E.D.

2. *Uniqueness theorem.* Let the boundary S of the domain D be the union of the five arcs Γ, C_1, γ_1, γ_2 and C_2, all having continuous tangents. The curve T, with endpoints A $(a, 0)$ and B $(b, 0)$, $a < 0 < b$, lies in the upper half-plane. The curves C_1 and C_2 lie in the lower half-plane, emanate from points A and B, respectively, and intersect characteristics γ_1 and γ_2 of equation (2.1) (see Figure 2, p. 8) emanating from the origin; along C_1 and C_2 we have

$$\begin{aligned} 0 \geqslant \frac{dy}{dx} = \tan \alpha \geqslant -\frac{1}{\sqrt{-K(y)}} \quad &\text{on} \quad C_1, \\ 0 \leqslant \frac{dy}{dx} = \tan \alpha \leqslant \frac{1}{\sqrt{-K(y)}} \quad &\text{on} \quad C_2. \end{aligned} \tag{2.6}$$

Here α is the angle between the positive x-axis and the tangent to S, in the sense of increasing of arc-length s on going counterclockwise around S.

We assume that

$$0 \leqslant \alpha \leqslant 2\pi \quad \text{on} \quad \Gamma. \tag{2.7}$$

THEOREM 3.1. *Assume that $u(x, y)$ satisfies equation* (2.1) *in the domain D, is continuous in $\bar{D}$ and has partial derivatives u_x and u_y such that the function $\Psi(x, y)$ defined by* (2.2) *is continuous in $\bar{D}$. Let* (2.6) *hold on C_1 and C_2, and* (2.7) *on Γ. If*

$$u = 0 \quad on \quad \Gamma + C_1 + C_2, \tag{2.8}$$

then $u(x, y) \equiv 0$ in D.

PROOF. Suppose that $u(x, y) \not\equiv 0$ in D. Then, according to Lemma 1, the function $\Psi(x, y)$ achieves its maximum on the boundary S of D. By (2.8) we have $u_x dx + u_y dy = 0$ on $\Gamma + C_1 + C_2$, and

$$\frac{\partial \Psi}{\partial s} = (Ku_x^2 + u_y^2) \frac{dy}{ds} = (K + \cot^2 \alpha)\, u_x^2 \sin \alpha. \tag{2.9}$$

Along the characteristics γ_1 and γ_2 we have

$$\frac{dy}{dx} = \pm \frac{1}{\sqrt{-K(y)}},$$

and it follows from (2.2) that

$$\frac{\partial \Psi}{\partial s} = -\left(\sqrt{-K}\, u_x \pm u_y\right)^2 \frac{dy}{ds}. \tag{2.10}$$

Let $\Psi(x, y)$ achieve a maximum on γ_1. Then by (2.10) this occurs at the intersection of γ_1 and C_1, i.e. on C_1. The same is true for γ_2. Consequently, according to Lemma 1, $\Psi(x, y)$ achieves its maximum on $\Gamma + C_1 + C_2$. But it follows from (2.6) and (2.9) that $\Psi(x, y)$ is a nondecreasing function of y on C_1 and C_2. Hence, if the maximum is achieved either on C_1 or on C_2, it must in fact be achieved either at A or at B.

Let $\Psi(x, y)$ achieve its maximum at B. By (2.7), $\sin\alpha$ is positive for increasing s and y. Then it follows from (2.9) that $\Psi(x, y)$ does not decrease when s increases near the point B. Similarly, $\Psi(x, y)$ does not decrease when s increases near A. Thus $\Psi(x, y)$ achieves its maximum somewhere on Γ, where $y > 0$. By (2.9), $\Psi(x, y)$ is a monotone function of s on any section of the curve Γ where $\sin\alpha$ does not change sign. Hence $\Psi(x, y)$ must have its maximum at a point P_1 where $\sin\alpha = 0$ and $\sin\alpha(P) < 0$ for $s(P) > s(P_1)$, $|P - P_1| < \delta$. At P_1 we have $\alpha = (2n + 1)\pi$, or, by (2.7), $\alpha = \pi$. Thus, at P_1 we have

$$\frac{\partial\Psi}{\partial n} = \frac{\partial\Psi}{\partial y}\cos\alpha - \frac{\partial\Psi}{\partial x}\sin\alpha = -\Psi_y,$$

where n is the direction of the interior normal. However, in view of (2.8) it follows from (2.2) that $\Psi_y = -u_y^2$ at P_1, so that $\partial\Psi/\partial n \geqslant 0$. On the other hand, $\Psi(x, y)$ satisfies the elliptic equation (2.5), and, by Giraud's theorem, $\partial\Psi/\partial n < 0$ at any boundary point where $\Psi(x, y)$ has a maximum. This contradiction implies that $\Psi = \text{const}$ in D^+, and then it follows from (2.2) that $u_x = u_y = 0$ in D^+. Since $u = 0$ on Γ, we have $u(x, y) \equiv 0$ in D^+. In particular, $\Psi(x, y) = 0$ on the interval AB. We claim that $u(x, y) \equiv 0$ in D^-. Let Q be any point of AB at which $x(P) < 0$, and let Q_1 be the point at which the characteristic $(\partial y/\partial x = 1/\sqrt{-K(y)})$, emanating from Q, cuts the curve C_1. Then, by (2.10), if $\sqrt{-K}u_x + u_y \neq 0$ on QQ_1, we have $\Psi(Q) < \Psi(Q_1)$. But it follows from (2.6) and (2.9) that $\Psi(Q_1) < \Psi(A)$, and so $\Psi(Q) < \Psi(A)$. On the other hand, $\Psi(Q) = \Psi(A)$, a contradiction. Therefore, $\sqrt{-K}u_x + u_y \equiv 0$ on QQ_1, and thus $u(x, y) \equiv 0$ in D^- for $x \leqslant 0$. Using similar reasoning, we can prove that $u(x, y) \equiv 0$ in D^- for $x \geqslant 0$. Thus $u(x, y) \equiv 0$ in D. Q.E.D.

3. *a, b, c-method.*(2) If the uniqueness theorem is proved using the a, b, c-method, one can weaken the restrictions on the solution $u(x, y)$ of equation (2.1), at the cost of stronger restrictions on the curve Γ, which must be star-shaped with respect to the point O; in analytical terms, this means that on going counterclockwise around Γ we have

$$x\,dy - y\,dx > 0. \tag{2.11}$$

---

(2) See [56a].

Let $u(x, y)$ be a solution of (2.1) which is continuous in $\bar{D}$ and has continuous derivatives u_x and u_y in $\bar{D}$, with the possible exception of points A $(a, 0)$, O $(0, 0)$ and B $(b, 0)$, at which they may have poles of order less than 1:

$$u_x = O(|OP|^{-\omega}), \qquad u_y = O(|OP|^{-\omega}) \quad (0 \leqslant \omega < 1), \tag{2.12}$$

$$u_x = O(|AP|^{-\alpha}), \qquad u_y = O(|AP|^{-\alpha}) \quad (0 \leqslant \alpha < 1), \tag{2.13}$$

$$u_x = O(|BP|^{-\beta}), \qquad u_y = O(|BP|^{-\beta}) \quad (0 \leqslant \beta < 1). \tag{2.14}$$

The point P is in D.

If $1/2 \leqslant \alpha < 1$, we demand additionally that

$$u_{xx} = O(|AP|^{-\alpha-1}), \qquad u_{xy} = O(|AP|^{-\alpha-1}) \tag{2.15}$$

and

$$x(A) - x = O(y^{2\alpha_1}) \tag{2.16}$$

on the curve Γ in a neighborhood of the point A, $y > 0$, where $\alpha_1 > \alpha$.

If $1/2 \leqslant \beta < 1$, we demand that

$$u_{xx} = O(|BP|^{-\beta-1}), \qquad u_{xy} = O(|BP|^{-\beta-1}) \tag{2.17}$$

and

$$x - x(B) = O(y^{2\beta_1}) \tag{2.18}$$

on Γ in a neighborhood of B, $y \geqslant 0$, where $\beta_1 > \beta$.

Conditions (2.16) and (2.18) mean that Γ is orthogonal to the x-axis at the points A and B.

THEOREM 3.2. *A solution $u(x, y)$ of equation* (2.1) *in the domain D vanishes identically if the following conditions are satisfied*:

1) *$u(x, y) = 0$ on $\Gamma + C_1 + C_2$.*

2) *$u(x, y)$ is continuous in $\bar{D}$ and has continuous derivatives u_x and u_y everywhere in $\bar{D}$ with the possible exception of the points A, O and B, in whose neighborhoods they satisfy* (2.12)–(2.15) *and* (2.17).

3) *The boundary $S = \Gamma + C_1 + \gamma_1 + \gamma_2 + C_2$ satisfies conditions* (2.6), (2.11), (2.16) *and* (2.18).

4) *$K(y)$ is continuously differentiable, $K(0) = 0$, and $K'(y) \geqslant 0$ for $y \geqslant 0$. If $0 \leqslant \alpha < 1/2$ and $0 \leqslant \beta < 1/2$, conditions* (2.15)–(2.18) *do not apply.*

PROOF. To fix ideas, we give the uniqueness proof for the case in which conditions (2.12) hold in the neighborhood of O, conditions (2.13) in the neighborhood of A $(0 \leqslant \alpha \leqslant 1/2)$, and conditions (2.14), (2.17) and (2.18) in the neighborhood of B. Let D_δ be the subdomain of D bounded near O by an arc

$\gamma_{O\delta}$ of the circle of radius δ about O, near A by an arc $C_{A\delta}$ of the circle of the same radius δ about A, and near B by the interval $C_{B\delta}$ of the straight line

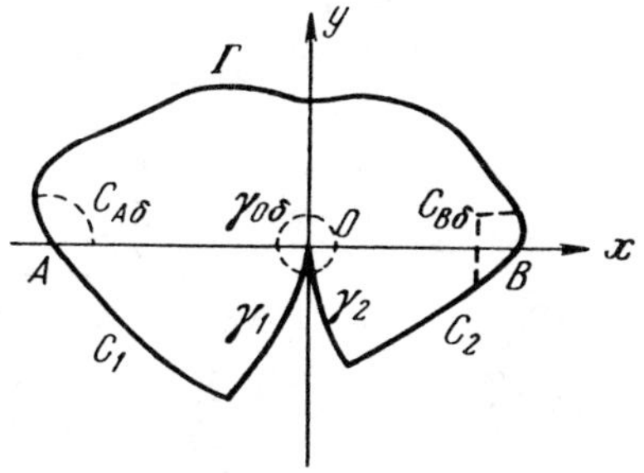

FIGURE 8

$$x = x(B) - \delta, \tag{2.19}$$

if $x < x(B)$ on Γ near B. If $x \geqslant x(B)$ on Γ near B, then $C_{B\delta}$ consists of two intervals of the line (2.19), and of the line

$$y = \delta^{1/2}, \tag{2.20}$$

and the rest of the boundary S_δ of D_δ coincides with the boundary of D (Figure 8).

Consider the identity

$$\begin{aligned} 2 \iint_{D_\delta} & (bu_x + cu_y)(Ku_{xx} + u_{yy})\, dx\, dy \\ = \iint_{D_\delta} & \left[\frac{\partial}{\partial x}(Kbu_x^2 - bu_y^2 + 2Kcu_xu_y) - \frac{\partial}{\partial y}(Kcu_x^2 - cu_y^2 - 2bu_xu_y)\right] dx\, dy \\ + \iint_{D_\delta} & [u_x^2(-Kb_x + (Kc)_y) + u_y^2(b_x - c_y) - 2u_xu_y(Kc_x + b_y)]\, dx\, dy, \end{aligned} \tag{2.21}$$

where $b(x, y)$ and $c(x, y)$ are continuous functions with piecewise continuous first derivatives, and $u(x, y)$ is a solution of (2.1) such that $u = 0$ on $\Gamma + C_1 + C_2$.

Applying Green's formula to the integral (2.21), we obtain

$$\begin{aligned} & \int_{S_\delta} (Kcu_x^2 - cu_y^2 - 2bu_xu_y)dx + (Kbu_x^2 - bu_y^2 + 2Kcu_xu_y)\, dy \\ & + \iint_{D_\delta} [u_x^2(-Kb_x + (Kc)_y) - 2u_xu_y(Kc_x + b_y) + u_y^2(b_x - c_y)]\, dx\, dy = 0. \end{aligned} \tag{2.22}$$

By condition 1, we have $u_x dx + u_y dy = 0$ on $\Gamma + C_1 + C_2$. Along the characteristics γ_1 and γ_2, we have $dx = \sqrt{-K}\, dy$ and $dx = -\sqrt{-K} dy$.

Then (2.22) can be written as

$$\int\limits_{\Gamma_\delta + C_{1\delta} + C_{2\delta}} \left(K + \left(\frac{dx}{dy}\right)^2\right) u_x^2 (b\,dy - c\,dx) + \int\limits_{\gamma_{1\delta}+\gamma_{2\delta}} \left(\frac{du}{dy}\right)^2 (-b\,dy - c\,dx)$$

$$+ \iint\limits_{D_\delta} [u_x^2(-Kb_x + (Kc)_y) - 2u_x u_y (Kc_x + b_y) + u_y^2 (b_x - c_y)]\,dx\,dy \tag{2.23}$$

$$+ \int\limits_{\gamma_{O\delta}+C_{A\delta}+C_{B\delta}} (Kcu_x^2 - cu_y^2 - 2bu_x u_y)\,dx$$

$$+ (Kbu_x^2 - bu_y^2 + 2Kcu_x u_y)\,dy = 0.$$

We now impose conditions on the functions b and c ensuring that the first three integrals in (2.23) are nonnegative. It suffices to assume that

$$b\,dy - c\,dx \geqslant 0 \quad \text{on} \quad \Gamma + C_1 + C_2, \tag{2.24}$$

$$-b\,dy - c\,dx \geqslant 0 \quad \text{on} \quad \gamma_1 + \gamma_2, \tag{2.25}$$

$$-Kb_x + (Kc)_y \geqslant 0, \tag{2.26}$$

$$b_x - c_y \geqslant 0, \tag{2.27}$$

$$(Kc_x + b_y)^2 \leqslant (b_x - c_y)(-Kb_x + (Kc)_y). \tag{2.28}$$

For example, for $y \geqslant 0$ we can take

$$b = x, \qquad c = y, \tag{2.29}$$

and for $y \leqslant 0$

$$b = x, \qquad c = 0. \tag{2.30}$$

Then conditions (2.25)–(2.28) hold. Condition (2.24) holds for $y \leqslant 0$ by virtue of (2.6), and for $y \geqslant 0$ it becomes condition (2.11).

With this choice of b and c, (2.23) can be written as

$$\iint\limits_{D_\delta^+} yK' u_x^2\,dx\,dy + \iint\limits_{D_\delta^-} (-Ku_x^2 + u_y^2)\,dx\,dy. \tag{2.31}$$

We claim that the integrals along $\gamma_{O\delta}$, $C_{A\delta}$ and $C_{B\delta}$ either are nonnegative or tend to zero as $\delta \to 0$. In fact, by virtue of (2.12), (2.29) and (2.30), the integral is $O(r^{1-2\omega})$ along $\gamma_{O\delta}$, where $r^2 = x^2 + y^2$, and hence this integral tends to zero as $\delta \to 0$. The same is true of the integral along $C_{A\delta}$, if $0 \leqslant \alpha < 1/2$. The integral along $C_{B\delta}$ can be rewritten as

$$\int\limits_{C_{B\delta}} c(Ku_x^2 - u_y^2)\,dx + Ku_x(bu_x + 2cu_y)\,dy$$
$$- \int\limits_{C_{B\delta}} b\,(2u_x u_y\,dx + u_y^2)\,dy = \int\limits_{C_{B\delta}} c(Ku_x^2 - u_y^2)\,dx$$
$$+ Ku_x(bu_x + 2cu_y)\,dy + x(B)\int\limits_{C_{B\delta}} u_y^2\,dy - x(B)\int\limits_{C_{B\delta}} u_y\,du \tag{2.32}$$
$$- \int\limits_{C_{B\delta}} (x - x(B))\,(u_y^2\,dy + 2u_x u_y\,dx).$$

The first and last terms on the right tend to zero as $\delta \longrightarrow 0$, while the second is nonnegative. Since $u(x, y)$ vanishes at the endpoints of $C_{B\delta}$, we see, integrating the third term in (2.32) by parts, that

$$x(B)\int\limits_{C_{B\delta}} u\,du_y = x(B)\left[\int\limits_{y_1}^{\delta^{1/2}} uu_{yy}\,dy + \int\limits_{x(B)-\delta}^{x_1} uu_{xy}\,dx\right]$$

or, by (2.1),

$$x(B)\int\limits_{C_{B\delta}} u\,du_y = x(B)\left[-\int\limits_{y_1}^{\delta^{1/2}} K(y)\,uu_{xx}\,dy + \int\limits_{x(B)-\delta}^{x_1} uu_{xy}\,dx\right],$$

where the points $(x(B) - \delta, y_1)$ and $(x_1, \delta^{1/2})$ lie on the curves C_2 and Γ, respectively.

By (2.16) we have $u = O(|BP|^{1-\beta})$ and, using (2.17),

$$x(B)\left|\int\limits_{C_{B\delta}} u\,du_y\right|$$
$$\leqslant M\left\{\int\limits_{y_1}^{\delta^{1/2}} y^{1-2\beta}\,dy + \int\limits_{x(B)-\delta}^{x_1} [\delta + (x - x(B))^2]^{-\beta}\,dx\right\}$$
$$< M_1\left\{\frac{y^{2-2\beta}}{2-2\beta}\Bigg|_{y_1}^{\delta^{1/2}} + \delta^{\frac{1}{2}-\beta}\operatorname{arctg}\frac{x - x(B)}{\delta^{1/2}}\Bigg|_{x(B)-\delta}^{x_1}\right\}$$

or, by (2.18),

$$x(B)\left|\int\limits_{C_{B\delta}} u\,du_y\right| < M_1\left[\frac{1}{2-2\beta}(\delta^{1-\beta} + y_1^{2-2\beta}) + \delta^{1-\beta} + B_1\delta^{\beta_1-\beta}\right].$$

Hence, for $\delta \longrightarrow 0$, $y_1 \longrightarrow 0$ and $\beta_1 - \beta > 0$, we have

$$\lim_{\delta\to 0} x(B) \int_{C_{B\delta}} u\,du_y = 0.$$

Thus the first three terms in (2.23) are nonnegative thanks to the choice of the functions $b(x, y)$ and $c(x, y)$, while the integrals along $\gamma_{O\delta}$, $C_{A\delta}$ and $C_{B\delta}$ are either nonnegative or tend to zero as $\delta \to 0$. Since the sum of these integrals is zero, each of them is also zero. But then it follows from (2.31) that $u_x = u_y = 0$ in D_δ; and since $u(x, y) = 0$ on $\Gamma + C_1 + C_2$, it follows that $u(x, y) \equiv 0$ in D, Q.E.D.

§3. Existence theorem for the generalized Tricomi problem([3])

In this section we consider the existence problem for solutions of the generalized Tricomi problem for the Čaplygin equation

$$K(u) \equiv K(\sigma)\, u_{\theta\theta} + u_{\sigma\sigma} = 0. \tag{3.1}$$

1. *Auxiliary inequality.* Let D_γ be the domain bounded by 1) a curve Γ in the upper half-plane with endpoints A (0, 0) and B (1, 0), 2) a smooth curve γ, emanating from A and cutting the characteristic $\eta = 1$ at a point C_1, and 3) the segment C_1B of the characteristic $\eta = 1$.

We shall assume that the curve Γ satisfies the conditions of §4 of Chapter II, and also condition (5.70) of Chapter II. Moreover, we stipulate that along γ

$$0 \geqslant \frac{d\sigma}{d\theta} \geqslant -\frac{1}{\sqrt{-K(\sigma)}}, \tag{3.2}$$

i.e., the slope of γ at any point does not exceed that of the characteristics passing through that point.

Let $u(\theta, \sigma)$ be a solution of (3.1), continuous in $\bar{D}_\gamma$, in class $\mathfrak{R}$, vanishing on Γ and satisfying

$$u\,|_\gamma = F(s), \tag{3.3}$$

where s is arc-length along γ, reckoned from the origin.

Set $p = 1 - \theta$ and $q = \sigma$ in (5.67) of Chapter II; then, by (3.3),

$$\begin{aligned}\int_\gamma (1-\eta)\, u_\theta^2 \left(\frac{d\theta}{d\sigma} + K\frac{d\sigma}{d\theta}\right) d\theta \geqslant \int_\gamma (1-\eta)\, F'^2(s) \frac{ds}{d\sigma}\, ds \\ + \iint_{D_\gamma^+} \sigma K_\sigma u_\theta^2\, d\theta\, d\sigma + \iint_{D_\gamma^-} (-\sqrt{K}\, u_\theta + u_\sigma)^2\, d\theta\, d\sigma.\end{aligned} \tag{3.4}$$

Condition (3.2) can be rewritten as

([3]) §3 presents results of Babenko (see [4a]).

$$K(\sigma)\frac{d\sigma}{d\theta}+\frac{d\theta}{d\sigma}\leqslant 0. \tag{3.2'}$$

Let us assume that (3.2′) is a strict inequality along γ. Then, from (3.4),

$$-\int_\gamma (1-\eta)\,u_\theta^2\left(\frac{d\theta}{d\sigma}+K\,\frac{d\sigma}{d\theta}\right)d\theta\leqslant-\int_\gamma (1-\eta)\,F'^2(s)\,\frac{ds}{d\sigma}\,ds.$$

We transform from (θ, σ) to new variables (x, y) by

$$\theta = x, \qquad y^{3/2}=\frac{3}{2}\int_0^\sigma \sqrt{K(t)}\,dt.$$

Then the last inequality becomes

$$\begin{aligned}-\int_\gamma (1-\eta)\sqrt{\frac{K}{y}}\left(\frac{dx}{dy}+y\,\frac{dy}{dx}\right)u_x^2\,dx \\ \leqslant-\int_\gamma (1-\eta)\,F'^2(s)\,\frac{ds}{dy}\sqrt{\frac{K}{y}}\,ds.\end{aligned} \tag{3.5}$$

By (3.2), we can use the characteristic coordinate η as a parameter on γ, and then (3.5) becomes

$$\begin{aligned}-\int_\gamma (1-\eta)\sqrt{\frac{K}{y}}\left(1+y\left(\frac{dy}{dx}\right)^2\right)u_x^2\,\frac{dx}{dy}\,dx \\ \leqslant-\int_\gamma (1-\eta)\,F'^2(\eta)\,\frac{d\eta}{dy}\sqrt{\frac{K}{y}}\,d\eta.\end{aligned} \tag{3.6}$$

By virtue of the conditions imposed on $K(\sigma)$, there exist two constants $\alpha > 0$ and $\beta > 0$ such that $\alpha < \sqrt{K/y} < \beta$ everywhere in D. Next,

$$\frac{d\eta}{dx}=1-\sqrt{-y}\,\frac{dy}{dx},$$

and so $1 < d\eta/dx \leqslant 2$; similarly,

$$\frac{d\eta}{dy}=\frac{dx}{dy}-\sqrt{-y},$$

so that $|d\eta/dy| > 2\sqrt{-y}$.

We assume that

$$-\frac{dx}{dy}\leqslant C\sqrt{-y}, \qquad 1<C<\infty. \tag{3.7}$$

Then

$$2\sqrt{-y} < \left|\frac{d\eta}{dy}\right| < (C+1)\sqrt{-y}.$$

Using these inequalities, we obtain from (3.6)

$$\frac{\alpha}{2}\int_{\gamma}(1-\eta)\left(1+y\left(\frac{dy}{dx}\right)^2\right)u_x^2\sqrt{-y}\,d\eta \leqslant \beta(C+1)\int_0^1(1-\eta)F'^2(\eta)\sqrt{-y}\,d\eta. \tag{3.8}$$

Set

$$\varepsilon = 1 + y\left(\frac{dy}{dx}\right)^2. \tag{3.9}$$

By (3.7), we have

$$\frac{d\xi}{d\eta} \leqslant \frac{C-1}{C+1} < 1.$$

Therefore, the equality

$$\sqrt{-y} = \left(\frac{3}{4}\right)^{1/3}(\eta-\xi)^{1/3}$$

implies that

$$C_1\eta^{1/3} < \sqrt{-y} < C_2\eta^{1/3}. \tag{3.10}$$

By (3.9) and (3.10), it follows from (3.8) that

$$\int_{\gamma}\varepsilon(1-\eta)\eta^{1/3}u_x^2\,d\eta \leqslant C_3^2\int_0^1(1-\eta)\eta^{1/3}F'^2(\eta)\,d\eta. \tag{3.11}$$

Transforming (3.1) to the form (see §5 of Chapter II)

$$F(z) \equiv yz_{xx} + z_{yy} + c(y)z = 0, \tag{3.12}$$

we set

$$u = ze^{-\frac{1}{2}\int_0^y b(y)\,dy}. \tag{3.13}$$

Let $z|_{\gamma} = \Phi(\eta)$. Then

$$F'(\eta) = \Phi'(\eta)e^{-\frac{1}{2}\int_0^y b(y)\,dy} - \frac{1}{2}\Phi(\eta)b(y)e^{-\frac{1}{2}\int_0^y b(y)\,dy}\frac{dy}{d\eta}. \tag{3.14}$$

Substituting (3.13) and (3.14) into (3.11), and using Minkowski's inequality, we get

$$\left(\int_\gamma \varepsilon(1-\eta)\,\eta^{1/3} z_x^2\,d\eta\right)^{1/2}$$

$$\leqslant C_4\left[\left(\int_0^1 (1-\eta)\,\eta^{1/3}\Phi'^2(\eta)\,d\eta\right)^{1/2} + \left(\int_0^1 (1-\eta)\,\eta^{-1/3}\Phi^2(\eta)\,d\eta\right)^{1/2}\right].$$

Since $\Phi(0)=0$, it follows that

$$\int_0^1 (1-\eta)\,\eta^{-1/3}\Phi^2(\eta)\,d\eta \leqslant \frac{9}{4}\int_0^1 (1-\eta)\,\eta^{1/3}\Phi'^2(\eta)\,d\eta$$

and so

$$\int_\gamma \varepsilon(1-\eta)\,\eta^{1/3} z_x^2\,d\eta \leqslant C_5^2\int_0^1 (1-\eta)\,\eta^{1/3}\Phi'^2(\eta)\,d\eta. \tag{3.15}$$

Now, since

$$z_y^2\,dy = -z_x z_y\,dx + z_y\Phi'(\eta)\,d\eta,$$

everywhere on γ, it follows that

$$\int_\gamma \varepsilon(1-\eta)\,z_y^2\left|\frac{dy}{d\eta}\right|d\eta$$

$$\leqslant \int_\gamma \varepsilon(1-\eta)\,|z_x z_y|\frac{dx}{d\eta}\,d\eta + \int_\gamma \varepsilon(1-\eta)\,|z_y\Phi'(\eta)|\,d\eta. \tag{3.16}$$

Applying Cauchy's inequality to the right-hand side and using (3.15), we see that

$$\int_\gamma \varepsilon(1-\eta)\,\eta^{-1/3} z_y^2\,d\eta \leqslant C_5^2\int_0 (1-\eta)\,\eta^{1/3}\Phi'^2(\eta)\,d\eta. \tag{3.17}$$

Adding (3.15) and (3.17), we get

$$\int_0^1 \varepsilon(1-\eta)\,\eta^{1/3}(z_x^2+\eta^{-2/3}z_y^2)\,d\eta \leqslant C_6^2\int_0^1 (1-\eta)\,\eta^{1/3}\Phi'^2(\eta)\,d\eta. \tag{3.18}$$

Replacing the derivatives z_x and z_y in (3.18) by their values

$$z_x = z_\xi + z_\eta, \qquad z_y = \sqrt{-y}\,(z_\xi - z_\eta)$$

and taking (3.10) into account, we finally obtain

$$\int_{\gamma} \varepsilon (1-\eta)\,\eta^{1/3}\,(z_\xi^2 + z_\eta^2)\,d\eta < A^2 \int_0^1 (1-\eta)\,\eta^{1/3}\,\Phi'^2(\eta)\,d\eta. \tag{3.19}$$

2. *Derivation of a fundamental inequality.* Let $z(x, y) \in C^2(D)$ be a solution of (3.12) which vanishes on Γ. Let G be the domain bounded by the curve γ and the characteristics $\xi = 0$ and $\eta = 1$.

Let $\xi = \gamma(\eta)$ be the equation of γ in characteristic coordinates. Set $\xi_0 = \gamma(\widetilde{\eta})$ and $\bar{\xi} = \gamma(\eta)$.

Let $v(\xi, \eta; \xi_0, \eta_0)$ be the Riemann function of equation (3.12). It satisfies the equation

$$\begin{aligned} v(\xi, \eta; \xi_0, \eta_0) &= v_0(\xi, \eta; \xi_0, \eta_0) \\ &+ \frac{1}{4}\left(\frac{4}{3}\right)^{2/3} \int_{\xi_0}^{\xi}\int_{\eta_0}^{\eta} \frac{C(\xi', \eta')}{(\eta' - \xi')^{2/3}}\, v_0(\xi, \eta; \xi', \eta')\, v(\xi', \eta'; \xi_0, \eta_0)\, d\xi'\, d\eta', \end{aligned} \tag{3.20}$$

where

$$\begin{aligned} v_0(\xi, \eta; \xi_0, \eta_0) &= \frac{(\eta - \xi)^{1/3}}{(\eta_0 - \xi)^{1/6}(\eta - \xi_0)^{1/6}}\, F\left(\frac{1}{6}, \frac{1}{6}, 1; \sigma\right), \\ \sigma &= \frac{(\xi - \xi_0)(\eta_0 - \eta)}{(\eta_0 - \xi)(\eta - \xi_0)}. \end{aligned} \tag{3.21}$$

Applying the Riemann formula to the curvilinear triangle bounded by the $\xi = \xi_0$, $\eta = \eta_0$ and the curve γ, we get

$$z(\xi_0, \eta_0) = \frac{z(\xi_0, \tilde{\eta})}{2}\left(\frac{\tilde{\eta} - \xi_0}{\eta_0 - \xi_0}\right)^{1/6} + \frac{z(\xi, \eta_0)}{2}\left(\frac{\eta - \xi}{\eta_0 - \xi_0}\right)^{1/6}$$

$$+ \int_{\tilde{\eta}}^{\eta_0} \left[\frac{zv}{6(\eta - \xi)}\left(1 + \frac{d\xi}{d\eta}\right) + \frac{1}{2} v\left(z_\eta - z_\xi \frac{d\xi}{d\eta}\right) + \frac{z}{2}\left(v_\xi \frac{d\xi}{d\eta} - v_\eta\right)\right] d\eta.$$

Setting $\xi_0 = 0$ and noting that $z|_\gamma = \Phi(\eta)$ and $\Phi(0) = 0$, we find that

$$\begin{aligned} z(0, \eta_0) &= \frac{\Phi(\eta_0)}{2}\left(\frac{\eta_0 - \xi}{\eta_0}\right)^{1/6} + \int_0^{\eta_0} \left[\frac{\Phi(\eta)}{6(\eta - \xi)}\, v(\xi, \eta; 0, \eta_0)\left(1 + \frac{d\xi}{d\eta}\right)\right. \\ &\left. + \frac{1}{2} v\left(z_\eta - z_\xi \frac{d\xi}{d\eta}\right) + \frac{\Phi(\eta)}{2}\left(v_\xi \frac{d\xi}{d\eta} - v_\eta\right)\right] d\eta. \end{aligned}$$

Along γ, we have

$$z_\eta = \Phi'(\eta) - z_\xi \frac{d\xi}{d\eta}.$$

Consequently,

$$z(0, \eta_0) = \frac{\Phi(\eta_0)}{2}\left(\frac{\eta_0 - \xi}{\eta_0}\right)^{1/6}$$
$$+ \int_0^{\eta_0} \Big\{ v(\xi, \eta; 0, \eta_0) \Big[\frac{\Phi(\eta)}{6(\eta - \xi)}\Big(1 + \frac{d\xi}{d\eta}\Big) + \frac{\Phi'(\eta)}{2} - z_\xi \frac{d\xi}{d\eta}\Big] \tag{3.22}$$
$$+ \frac{\Phi(\eta)}{2}\Big(v_\xi \frac{d\xi}{d\eta} - v_\eta\Big)\Big\} d\eta.$$

Set

$$I_1 = \int_0^{\eta_0} v(\xi, \eta; 0, \eta_0)\Big[\frac{\Phi(\eta)}{6(\eta - \xi)}\Big(1 + \frac{d\xi}{d\eta}\Big) + \frac{\Phi'(\eta)}{2} - z_\xi \frac{d\xi}{d\eta}\Big] d\eta$$
$$= \int_0^{\eta_0} v(\xi, \eta; 0, \eta_0) f(\eta)\, d\eta,$$

$$I_2 = \frac{1}{2}\int_0^{\eta_0} \Phi(\eta)\Big(v_\xi \frac{d\xi}{d\eta} - v_\eta\Big) d\eta.$$

Using (5.45) of Chapter II, we get

$$\left|\frac{\partial v(\xi, \eta; 0, \eta_0)}{\partial \eta_0}\right| \leqslant C_1 \frac{(\eta - \xi)^{1/3}}{(\eta_0 - \xi)^{7/6}\eta^{1/6}}.$$

Therefore

$$\left|\frac{dI_1}{d\eta_0}\right| \leqslant \left(\frac{\eta_0 - \xi}{\eta_0}\right)^{1/6} |f(\eta_0)|$$
$$+ C_1 \int_0^{\eta_0} \frac{(\eta - \xi)^{1/3} |f(\eta)|}{(\eta_0 - \xi)^{7/6}\eta^{1/6}} d\eta \leqslant |f(\eta_0)| + \frac{C_2}{\eta}\int_0^{\eta_0} |f(\eta)|\, d\eta. \tag{3.23}$$

Applying Minkowski's inequality, we get

$$\left(\int_0^1 (1 - \eta)\eta^{1/3}\Big(\frac{dI_1}{d\eta}\Big)^2 d\eta\right)^{1/2} \leqslant \left(\int_0^1 (1 - \eta)\eta^{1/3} f^2(\eta)\, d\eta\right)^{1/2}$$
$$+ \left(\int_0^1 (1 - \eta)\eta^{1/3}\Big(\frac{C_2}{\eta}\int_0^\eta |f(t)|\, dt\Big)^2 d\eta\right)^{1/2}.$$

However, by Hardy's inequality (see [81]),

$$\int_0^1 (1 - \eta)\eta^{1/3}\left(\frac{1}{\eta}\int_0^\eta |f(t)|\, dt\right)^2 d\eta \leqslant C_3^2 \int_0^1 (1 - \eta)\eta^{1/3} f^2(\eta)\, d\eta.$$

Therefore

$$\left(\int_0^1 (1-\eta)\,\eta^{1/3}\left(\frac{dI_1}{d\eta}\right)^2 d\eta\right)^{1/2} \leqslant C_4\left(\int_0^1 (1-\eta)\,\eta^{1/3} f^2(\eta)\,d\eta\right)^{1/2}.$$

Inserting the value of $f(\eta)$ and using Minkowski's inequality, we get

$$\left(\int_0^1 (1-\eta)\,\eta^{1/3}\left(\frac{dI_1}{d\eta}\right)^2 d\eta\right)^{1/2} \leqslant C_5\left\{\left(\int_0^1 (1-\eta)\,\eta^{1/3}\frac{\Phi^2(\eta)}{(\eta-\xi)^2}\,d\eta\right)^{1/2}\right.$$

$$\left.+\left(\int_0^1 (1-\eta)\,\eta^{1/3}\Phi'^2(\eta)\,d\eta\right)^{1/2} + \left(\int_0^1 (1-\eta)\,\eta^{1/3}\left(\frac{d\xi}{d\eta}\right)^2 z_\xi^2\,d\eta\right)^{1/2}\right\}.$$

Again by Hardy's inequality,

$$\int_0^1 (1-\eta)\,\eta^{1/3}\frac{\Phi^2(\eta)}{(\eta-\xi)^2}\,d\eta \leqslant C_6\int_0^1 (1-\eta)\,\eta^{1/3}\left(\frac{1}{\eta}\int_0^\eta |\Phi'(t)|\,dt\right)^2 d\eta$$

$$\leqslant C_7\int_0^1 (1-\eta)\,\eta^{1/3}\Phi'^2(\eta)\,d\eta.$$

Now,

$$\int_0^1 (1-\eta)\,\eta^{1/3}\left(\frac{d\xi}{d\eta}\right)^2 z_\xi^2\,d\eta \leqslant \int_0^1 (1-\eta)\,\eta^{1/3}\frac{d\xi}{d\eta}\,z_\xi^2\,d\eta$$

$$= \int_0^1 \varepsilon\,\frac{(1-\eta)\,\eta^{1/3}}{\left(1-\sqrt{-y}\,\dfrac{dy}{dx}\right)^2}\,z_\xi^2\,d\eta \leqslant \int_0^1 \varepsilon\,(1-\eta)\,\eta^{1/3} z_\xi^2\,d\eta,$$

since $dy/dx < 0$.

Thus, using (3.19), we obtain

$$\int_0^1 (1-\eta)\,\eta^{1/3}\left(\frac{dI_1}{d\eta}\right)^2 d\eta \leqslant A_1^2\int_0^1 (1-\eta)\,\eta^{1/3}\Phi'^2(\eta)\,d\eta. \tag{3.24}$$

Consider $dI_2/d\eta_0$. The derivatives of the Riemann function satisfy the following estimates (see [66g]):

$$\left|\frac{\partial v(\xi,\eta;\xi_0,\eta_0)}{\partial \xi_0}\right| \leqslant \frac{(\eta-\xi)^{1/3}}{(\eta-\xi_0)^{7/6}(\eta_0-\xi)^{1/6}}\,C_1(\xi,\eta;\xi_0,\eta_0),$$

$$\left|\frac{\partial v(\xi,\eta;\xi_0,\eta_0)}{\partial \eta_0}\right| \leqslant \frac{(\eta-\xi)^{1/3}}{(\eta-\xi_0)^{1/6}(\eta_0-\xi)^{7/6}}\,C_2(\xi,\eta;\xi_0,\eta_0),$$

where C_1 and C_2 are bounded functions. By these estimates

$$\left|\frac{\partial}{\partial\eta_0}\left(\frac{\partial v}{\partial\xi}\frac{d\xi}{d\eta}-\frac{\partial v}{\partial\eta}\right)\right|<C_8(\eta-\xi)^{-2/3}(\eta_0-\xi)^{-7/6}\eta^{-1/6},$$

$$\left|\left(\frac{\partial v}{\partial\xi}\frac{d\xi}{d\eta}-\frac{\partial v}{\partial\eta}\right)_{\eta=\eta_0}\right|<C_9\eta_0^{1/6}(\eta_0-\xi)^{-7/6}.$$

Consequently

$$\left|\frac{dI_2}{d\eta_0}\right|\leqslant\frac{C_9}{2}\frac{\eta_0^{1/6}}{(\eta_0-\xi)^{7/6}}|\Phi(\eta_0)|+\frac{C_8}{2}\int_0^{\eta_0}(\eta-\xi)^{-2/3}(\eta_0-\xi)^{-7/6}\eta^{-1/6}|\Phi(\eta)|\,d\eta$$

$$\leqslant C_{10}\left\{\frac{|\Phi(\eta_0)|}{\eta_0}+\frac{1}{\eta_0}\int_0^{\eta_0}\frac{|\Phi(\eta)|}{\eta}\,d\eta\right\}.$$

As before, we find that

$$\int_0^1(1-\eta)\eta^{1/3}\left(\frac{dI_2}{d\eta}\right)^2d\eta\leqslant A_2^2\int_0^1(1-\eta)\eta^{1/3}\Phi'^2(\eta)\,d\eta. \tag{3.25}$$

It is readily seen that

$$\int_0^1(1-\eta)\eta^{1/3}\left\{\frac{d}{d\eta}\left[\left(\frac{\eta-\xi}{\eta}\right)^{1/6}\Phi(\eta)\right]\right\}^2d\eta$$
$$\leqslant A_3^2\int_0^1(1-\eta)\eta^{1/3}\Phi'^2(\eta)\,d\eta. \tag{3.26}$$

By (3.24)–(3.26), we infer from (3.22) that

$$\int_0^1(1-\eta)\eta^{1/3}\left[\frac{\partial z(0,\eta)}{\partial\eta}\right]^2d\eta\leqslant A_4^2\int_0^1(1-\eta)\eta^{1/3}\Phi'^2(\eta)\,d\eta. \tag{3.27}$$

We now use (5.9) of Chapter II, where we must set $B=0$ since $z|_\Gamma=0$. Comparing it with (3.27), we get

$$\int_0^1(1-x)\mu^2(x)\,dx\leqslant A_5^2\int_0^1(1-x)x^{1/3}\Phi'^2(x)\,dx. \tag{3.28}$$

We now prove the inequality which is the converse of (3.28).

A solution $z(x, y)$ of (3.12) in the domain D^- can be expressed (see (2.5) of Chapter II) as

$$z(\xi,\eta)=z_0(\xi,\eta)+\int_\xi^\eta d\xi'\int_{\xi'}^\eta C(\xi',\eta')v(\xi',\eta';\xi,\eta)z_0(\xi',\eta')\,d\eta'$$
$$=z_0(\xi,\eta)+I(\xi,\eta), \tag{3.29}$$

where

$$z_0(\xi, \eta) = \gamma_1 \int_\xi^\eta \frac{(\eta - \xi)^{2/3} \tau(t)\, dt}{(\eta - t)^{5/6} (t - \xi)^{5/6}} - \gamma_2 \int_\xi^\eta \frac{\nu(t)\, dt}{(t - \xi)^{1/6} (\eta - t)^{1/6}}. \tag{3.30}$$

The derivatives of the integral I are continuous and satisfy the inequalities (see (5.51) of Chapter II)

$$\left| \frac{\partial I}{\partial \xi} \right| < C(\eta - \xi)^{1/3}, \quad \left| \frac{\partial I}{\partial \eta} \right| < C(\eta - \xi)^{1/3}.$$

Let us investigate the behavior of $z_0(\xi, \eta)$ on γ. Let

$$\tau(x) = \int_0^x (x - t)^{-1/6} \varphi(t)\, dt, \quad \nu(x) = \int_0^x (x - t)^{-5/6} \psi(t)\, dt, \tag{3.31}$$

We know (see Chapter II, §5) that

$$\tau(x) = \int_0^x (x - t)^{-1/3} \varphi_1(t)\, dt,$$

$$\varphi_1(x) = x^{1/6} \int_0^x (x - t)^{-5/6} t^{-1/6} \lambda_1(t)\, dt, \tag{3.32}$$

$$\nu(x) = x^{1/6} \int_0^x (x - t)^{-5/6} t^{-1/6} \mu(t)\, dt,$$

whence

$$\varphi(x) = \frac{\Gamma\left(\frac{1}{6}\right)\Gamma\left(\frac{2}{3}\right)}{\Gamma\left(\frac{5}{6}\right)} \left\{ \lambda_1(x) + \frac{x^{-5/6}}{36} \int_0^x \frac{\lambda_1(t)}{t^{1/6}} F\left(\frac{5}{6}, \frac{5}{6}, 2; 1 - \frac{t}{x}\right) dt \right\}, \tag{3.33}$$

$$\psi(x) = \mu(x) + \frac{x^{-5/6}}{36} \int_0^x \frac{\mu(t)}{t^{1/6}} F\left(\frac{5}{6}, \frac{5}{6}, 2; 1 - \frac{t}{x}\right) dt. \tag{3.34}$$

For convenience, we shall work for the moment with the functions $\varphi(x)$ and $\psi(x)$, and later, using (3.33) and (3.34), go over to $\lambda_1(x)$ and $\mu(x)$.

Set $\Phi_0(\eta) = z_0|_\gamma = z_0(\gamma(\eta), \eta)$. Substituting (3.31) into (3.30), changing the order of integration and using the integral representation of the hypergeometric function, we obtain

$$z_0(\xi,\eta)=\int_0^{\xi}\psi_1(x)(\eta-x)^{-1/6}F\left(\frac{1}{6},\frac{1}{6},\frac{1}{3};\frac{\eta-\xi}{\eta-x}\right)dx$$
$$+\int_{\xi}^{\eta}\psi_2(x)(\eta-\xi)^{-1/6}F\left(\frac{1}{6},\frac{5}{6},1;\frac{\eta-x}{\eta-\xi}\right)dx \tag{3.35}$$
$$+\int_0^{\xi}\psi_3(x)(\eta-x)^{-1/6}F\left(\frac{1}{6},\frac{1}{6},1;\frac{\xi-x}{\eta-x}\right)dx,$$

where

$$\psi_1(x)=\varphi(x)-\sqrt{3}\,\frac{\Gamma\left(\frac{1}{6}\right)\Gamma\left(\frac{2}{3}\right)}{\Gamma\left(\frac{5}{6}\right)}\gamma_2\psi(x),$$
$$\psi_2(x)=2\pi\left[\gamma_1\varphi(x)-\gamma_2\psi(x)\right], \tag{3.36}$$
$$\psi_3(x)=2\pi\sqrt{3}\,\gamma_2\psi(x).$$

Denote the sum of the first two integrals in (3.35) by $z_1(\xi,\eta)$ and the last integral by $z_2(\xi,\eta)$. Then $z_0(\xi,\eta)=z_1(\xi,\eta)+z_2(\xi,\eta)$. Using the formula

$$F(a,b,a+b,t)=-\frac{\Gamma(a+b)}{\Gamma(a)\Gamma(b)}F(a,b,1;1-t)\ln(1-t)$$
$$+\frac{\Gamma(a+b)}{\Gamma^2(a)\Gamma^2(b)}\sum_{k=0}^{\infty}\frac{\Gamma(a+k)\Gamma(b+k)}{(k!)^2} \tag{3.37}$$
$$\times\left[2\frac{\Gamma'(1+k)}{\Gamma(1+k)}-\frac{\Gamma'(a+k)}{\Gamma(a+k)}-\frac{\Gamma'(b+k)}{\Gamma(b+k)}\right](1-t)^k,$$

we find that

$$\frac{\Gamma^2\left(\frac{1}{6}\right)}{\Gamma\left(\frac{1}{3}\right)}F\left(\frac{1}{6},\frac{1}{6},\frac{1}{3};t\right)=\ln\frac{1}{1-t}+\theta_1(t),$$
$$2\pi F\left(\frac{1}{6},\frac{5}{6},1;t\right)=\ln\frac{1}{1-t}+\theta_2(t), \tag{3.38}$$

where $\theta_1(t)$ and $\theta_2(t)$ are differentiable functions and

$$|\theta_1'(t)|<C\left(\ln\frac{1}{1-t}+1\right),\qquad |\theta_2'(t)|<C\left(\ln\frac{1}{1-t}+1\right).$$

Moreover, from (3.37) we have

$$\theta_1(1) = 2\Gamma'(1) - 2\frac{\Gamma'\left(\frac{1}{6}\right)}{\Gamma\left(\frac{1}{6}\right)},$$

$$\theta_2(1) = 2\Gamma'(1) - \frac{\Gamma'\left(\frac{5}{6}\right)}{\Gamma\left(\frac{5}{6}\right)} - \frac{\Gamma'\left(\frac{1}{6}\right)}{\Gamma\left(\frac{1}{6}\right)}, \qquad \theta_2(0) = 2\pi.$$

Setting $(\eta - \xi)/(\eta - x) = \zeta$ and $\gamma_1\varphi(x) - \gamma_2\psi(x) = f(x)$, we get

$$\begin{aligned}(\eta-\xi)^{1/6} z_1(\xi, \eta) &= \frac{\Gamma^2\left(\frac{1}{6}\right)}{\Gamma\left(\frac{1}{3}\right)} \int_0^\xi (\zeta^{1/6} - 1) F\left(\frac{1}{6}, \frac{1}{6}, \frac{1}{3}; \zeta\right) f(x)\,dx \\ &+ \int_0^\xi [\ln(\eta - x) + \theta_1(\zeta)] f(x)\,dx - \int_0^\eta f(x) \ln|x - \xi|\,dx \\ &+ \int_\xi^\eta \left[\ln(\eta - \xi) + \theta_2\left(\frac{1}{\zeta}\right)\right] f(x)\,dx.\end{aligned} \tag{3.39}$$

Differentiation of (3.39) with respect to η gives

$$\begin{aligned}&(\eta-\xi)^{1/6}\frac{\partial z_1}{\partial \eta} \\ &= 2\pi f(\eta) + \int_0^\xi \theta_3(\zeta) f(x) \frac{dx}{\eta - x} + \frac{1}{\eta - \xi}\int_0^\eta \theta_4\left(\frac{1}{\zeta}\right) f(x)\,dx,\end{aligned} \tag{3.40}$$

where

$$\begin{aligned}\theta_3(\zeta) &= \theta_1'(\zeta)(1-\zeta) + 1 + \frac{\Gamma^2\left(\frac{7}{6}\right)}{\Gamma\left(\frac{4}{3}\right)} F\left(\frac{1}{6}, \frac{1}{6}, \frac{4}{3}; \zeta\right)(\zeta^{1/6} - 1) - \frac{1}{6}\zeta^{1/6} F_1(\zeta), \\ \theta_4\left(\frac{1}{\zeta}\right) &= \theta_2'\left(\frac{1}{\zeta}\right)\left(1 - \frac{1}{\zeta}\right) + 1 - \frac{1}{6} F_2\left(\frac{1}{\zeta}\right), \\ F_1(\zeta) &= \frac{\Gamma^2\left(\frac{1}{6}\right)}{\Gamma\left(\frac{1}{3}\right)} F\left(\frac{1}{6}, \frac{1}{6}, \frac{1}{3}; \zeta\right), \\ F_2\left(\frac{1}{\zeta}\right) &= 2\pi F\left(\frac{1}{6}, \frac{5}{6}, 1; \frac{1}{\zeta}\right).\end{aligned} \tag{3.41}$$

Similarly, differentiation of (3.39) with respect to ξ gives

$$(\eta-\xi)^{1/6}\frac{\partial z_1}{\partial \xi}=\pi\sqrt{3}f(\xi)+\int_0^\eta \frac{f(x)\,dx}{x-\xi} \\ +\int_0^\xi \theta_5(\zeta)f(x)\frac{dx}{\eta-x}+\frac{1}{\eta-\xi}\int_\xi^\eta \theta_6\left(\frac{1}{\zeta}\right)f(x)\,dx, \tag{3.42}$$

where

$$\theta_5(\zeta)=\theta_1'(\zeta)+\frac{\Gamma^2\left(\frac{7}{6}\right)}{\Gamma\left(\frac{4}{3}\right)}\frac{\zeta^{1/6}-1}{1-\zeta}F\left(\frac{1}{6},\ \frac{1}{6},\ \frac{4}{3};\ \zeta\right), \\ \theta_6\left(\frac{1}{\zeta}\right)=\frac{1}{\zeta}\theta_2'\left(\frac{1}{\zeta}\right)+\frac{1}{6}F_2\left(\frac{1}{\zeta}\right)-1. \tag{3.43}$$

Multiplying (3.42) by $d\xi/d\eta$ and adding the result to (3.40), we get

$$(\eta-\xi)^{1/6}\Phi_1'(\eta)=2\pi f(\eta)+\pi\sqrt{3}f(\xi)\frac{d\xi}{d\eta} \\ +\frac{d\xi}{d\eta}\int_0^\eta\frac{f(x)\,dx}{x-\xi}+\int_0^\xi\left[\theta_3(\zeta)+\theta_5(\zeta)\frac{d\xi}{d\eta}\right]\frac{f(x)\,dx}{\eta-x} \\ +\frac{1}{\eta-\xi}\int_\xi^\eta\left[\theta_4\left(\frac{1}{\xi}\right)+\theta_6\left(\frac{1}{\zeta}\right)\frac{d\xi}{d\eta}\right]f(x)\,dx, \tag{3.44}$$

where $\Phi_1'(\eta)=z_1|_\gamma=z_1(\gamma(\eta),\eta)$.

By (3.33) and (3.34) we have

$$f(x)=\gamma_2\left(\frac{\gamma_1}{\gamma_2}\varphi(x)-\psi(x)\right)=\gamma_2\left\{\frac{\Gamma\left(\frac{1}{6}\right)\Gamma\left(\frac{2}{3}\right)}{\Gamma\left(\frac{5}{6}\right)}\frac{\gamma_1}{\gamma_2}\lambda_1(x)-\mu(x)\right. \\ \left.+\frac{x^{-5/6}}{36}\int_0^x t^{-1/6}\left[\frac{\Gamma\left(\frac{1}{6}\right)\Gamma\left(\frac{2}{3}\right)}{\Gamma\left(\frac{5}{6}\right)}\frac{\gamma_1}{\gamma_2}\lambda_1(t)-\mu(t)\right]F\left(\frac{5}{6},\ \frac{5}{6},\ 2;\ 1-\frac{t}{x}\right)dt\right\},$$

but

$$k=\frac{\Gamma\left(\frac{5}{6}\right)}{\Gamma\left(\frac{1}{6}\right)\Gamma\left(\frac{2}{3}\right)}\frac{\gamma_2}{\gamma_1},\qquad \frac{1}{k}\lambda_1(x)-\mu(x)=\lambda(x). \tag{3.45}$$

Thus,

$$\frac{1}{\gamma_2}f(x)=\lambda(x)+\frac{x^{-5/6}}{36}\int_0^x\frac{\lambda(t)}{t^{1/6}}F\left(\frac{5}{6},\ \frac{5}{6},\ 2;\ 1-\frac{t}{x}\right)dt. \tag{3.46}$$

We now estimate the integral

$$\int_0^1 \eta^{1/3}(1-\eta)\,\Phi_1'^2(\eta)\,d\eta.$$

By (3.38) it follows from (3.41) and (3.43) that

$$\begin{aligned} |\theta_i(\zeta)| &< C_1\left(\ln\frac{1}{1-\zeta}+1\right) \qquad (i=3,\ 5),\\ \left|\theta_i\left(\frac{1}{\zeta}\right)\right| &< C_1\left(\ln\frac{1}{1-\frac{1}{\zeta}}+1\right) \qquad (i=4,\ 6). \end{aligned} \tag{3.47}$$

Here $\zeta=(\eta-\xi)/(\eta-x)$. Denote the sum of the last two integrals in (3.44) by J_1. Noting that

$$\frac{d\xi}{d\eta}\leqslant\frac{c-1}{c+1}=1-\delta, \qquad \xi\leqslant(1-\delta)\,\eta,$$

and using (3.47), we get

$$|J_1|<\frac{2C_1}{\delta}\left\{\frac{1}{\eta}\int_0^\xi |f(x)|\left(\ln\frac{1}{1-\zeta}+1\right)dx+\frac{1}{\eta}\int_\xi^\eta |f(x)|\left(\ln\frac{1}{1-\frac{1}{\zeta}}+1\right)dx\right\}$$

$$=\frac{2C_1}{\eta\delta}\left\{\int_0^\xi |f(x)|\ln\frac{\eta-x}{\xi-x}\,dx+\int_\xi^\eta |f(x)|\ln\frac{\eta-\xi}{x-\xi}\,dx+\int_0^\eta |f(x)|\,dx\right\}.$$

Set $f_1(x)=(1-x)^{1/2}|f(x)|$. Then

$$(1-\eta)^{1/2}|J_1|\leqslant\frac{2C_1}{\eta\delta}\left\{\int_0^\xi f_1(x)\ln\frac{\eta-x}{\xi-x}\,dx\right.$$

$$\left.+\int_\xi^\eta f_1(x)\ln\frac{\eta-\xi}{x-\xi}\,dx+\int_0^\eta f_1(x)\,dx\right\}.$$

Consider the integral

$$\begin{aligned}\int_\xi^\eta (t-\xi)^{-1/3}\,dt \int_0^t (t-x)^{-2/3} f_1(x)\,dx &= \int_0^\xi f_1(x)\,dx \int_\xi^\eta (t-\xi)^{-1/3}(t-x)^{-2/3}\,dt \\ &\quad + \int_\xi^\eta f_1(x)\,dx \int_x^\eta (t-\xi)^{-1/3}(t-x)^{-2/3}\,dt \end{aligned} \tag{3.48}$$

$$= 3\int_0^\xi f_1(x)\,dx \int_0^{\left(\frac{\eta-\xi}{\eta-x}\right)^{1/3}} \frac{z\,dz}{1-z^3} + 3\int_\xi^\eta f_1(x)\,dx \int_0^{\left(\frac{\eta-x}{\eta-\xi}\right)^{1/3}} \frac{dz}{1-z^3}.$$

Elementary manipulations give

$$3\int_0^{\left(\frac{\eta-\xi}{\eta-x}\right)^{1/3}} \frac{z\,dz}{1-z^3} = -\ln\frac{\xi-x}{\eta-x} + O(1), \qquad 0 < x < \xi,$$

$$3\int_0^{\left(\frac{\eta-x}{\eta-\xi}\right)^{1/3}} \frac{dz}{1-z^3} = -\ln\frac{x-\xi}{\eta-\xi} + O(1), \qquad \xi < x < \eta.$$

Substituting this into (3.48), we find

$$\begin{aligned}\int_\xi^\eta (t-\xi)^{-1/3}\,dt \int_0^t (t-x)^{-2/3} f_1(x)\,dx &= \int_0^\xi f_1(x)\ln\frac{\eta-x}{\xi-x}\,dx \\ &\quad + \int_\xi^\eta f_1(x)\ln\frac{\eta-\xi}{x-\xi}\,dx + \int_0^\eta f_1(x)\,O(1)\,dx.\end{aligned}$$

Thus,

$$\begin{aligned}(1-\eta)^{1/2}|J_1| \leqslant \frac{2C_1}{\eta\delta}\int_\xi^\eta (t-\xi)^{-1/3}\,dt \int_0^t (t-x)^{-2/3} f_1(x)\,dx \\ + \frac{C_2}{\delta\eta}\int_0^\eta f_1(x)\,dx,\end{aligned} \tag{3.49}$$

where C_2 does not depend on $\xi = \gamma(\eta)$.

LEMMA 1. *The expression*

$$\frac{1}{\eta}\int_{\xi}^{\eta}(t-\xi)^{-1/3}\,dt\int_{0}^{t}(t-x)^{-2/3}f_1(x)\,dx$$

is majorized by a square-integrable function of η *on* $(0, 1)$.

In fact,

$$\int_{\xi}^{\eta}(t-\xi)^{-1/3}\,dt\int_{0}^{t}(t-x)^{-2/3}f_1(x)\,dx=\int_{0}^{\eta-\xi}y^{-1/3}\,dy\int_{0}^{\xi+y}z^{-2/3}f_1(\xi+y-z)\,dz$$

$$=\int_{0}^{\eta-\xi}y^{-1/3}\,dy\int_{0}^{\xi}z^{-2/3}f_1(\xi+y-z)\,dz+\int_{0}^{\eta-\xi}y^{-1/3}\,dy\int_{\xi}^{\xi+y}z^{-2/3}f_1(\xi+y-z)\,dz$$

$$=\int_{0}^{\xi}z^{-2/3}\,dz\int_{0}^{\eta-\xi}y^{-1/3}f_1(\xi-z+y)\,dy+$$

$$+\int_{\xi}^{\eta}z^{-2/3}\,dz\int_{z-\xi}^{\eta-\xi}y^{-1/3}f_1(\xi+y-z)\,dy=I_1+I_2.$$

Next,

$$I_2=\int_{\xi}^{\eta}z^{-2/3}\,dz\int_{0}^{\eta-z}(u+z-\xi)^{-1/3}f_1(u)\,du$$

$$\leqslant\int_{\xi}^{\eta}z^{-2/3}\,dz\int_{0}^{\eta-z}u^{-1/3}f_1(u)\,du\leqslant 3\eta^{1/3}\int_{0}^{\eta}x^{-1/3}f_1(x)\,dx.$$

Now consider I_1. If $g_{\xi-z}(y)$ is a nonincreasing function, equimeasurable(4) with $f_1(\xi-z+y)$, $0\leqslant y\leqslant 1-\xi+z$, then

$$\int_{0}^{\eta-\xi}y^{-1/3}f_1(\xi-z+y)\,dy\leqslant\int_{0}^{\eta-\xi}y^{-1/3}g_{\xi-z}(y)\,dy$$

(see [81], §9.42). Let $g(x)$ be a nonincreasing function, equimeasurable with $f_1(x)$ $(0\leqslant x\leqslant 1)$. Obviously, $g_{\xi-z}(x)\leqslant g(x)$, so that

$$\int_{0}^{\eta-\xi}y^{-1/3}g_{\xi-z}(y)\,dy\leqslant\int_{0}^{\eta-\xi}y^{-1/3}g(y)\,dy,$$

and

(4) Two functions $f(x)$ and $\varphi(x)$, $a\leqslant x\leqslant b$, are said to be *equimeasurable* if $mE(f>y)=mE(\varphi>y)$ for all y.

$$I_1 \leqslant \int_0^{\xi} z^{-2/3}\,dz \int_0^{\eta-\xi} y^{-1/3} g(y)\,dy < 3\eta^{1/3} \int_0^{\eta} x^{-1/3} g(x)\,dx.$$

Thus,

$$\begin{aligned}\int_{\xi}^{\eta} (t-\xi)^{-1/3}\,dt \int_0^t (t-x)^{-2/3} f_1(x)\,dx \\ \leqslant 3\eta^{1/3} \int_0^{\eta} x^{-1/3} [f_1(x)+g(x)]\,dx.\end{aligned} \tag{3.50}$$

Q.E.D.

We now return to (3.49). By (3.50), it follows from (3.49) that

$$(1-\eta)^{1/2} |J_1| \leqslant \frac{6C_1}{\delta}\left(\eta^{-2/3}\int_0^{\eta} x^{-1/3} f_1(x)\,dx + \eta^{-2/3}\int_0^{\eta} x^{-1/3} g(x)\,dx\right) + \frac{C_2}{\delta\eta}\int_0^{\eta} f_1(x)\,dx.$$

Using the inequalities of Minkowski and Hardy (see [81]), we find that

$$\begin{aligned}\left(\int_0^1 (1-\eta) J_1^2\,d\eta\right)^{1/2} &\leqslant C_3\left(\int_0^1 f_1^2(x)\,dx\right)^{1/2} + C_3\left(\int_0^1 g^2(x)\,dx\right)^{1/2} \\ &= 2C_3\left(\int_0^1 f_1^2(x)\,dx\right)^{1/2} = 2C_3\left(\int_0^1 (1-\eta) f^2(\eta)\,d\eta\right)^{1/2}.\end{aligned} \tag{3.51}$$

Consider the integral

$$J_2 = \frac{d\xi}{d\eta}\int_0^{\eta} \frac{f(x)\,dx}{x-\xi}.$$

Let $\xi_1 = \gamma(1)$ and $\eta \leqslant \xi_1$. Then

$$J_2 = \frac{d\xi}{d\eta}\int_0^{\xi_1} \frac{f(x)\,dx}{x-\xi} - \frac{d\xi}{d\eta}\int_{\eta}^{\xi_1} \frac{f(x)\,dx}{x-\xi}.$$

Hence, setting

$$\varphi(t) = \int_0^{\xi_1} \frac{f(x)\,dx}{x-t},$$

we get

$$|J_2| \leqslant \frac{d\xi}{d\eta}\, |\varphi(\xi(\eta))| + \frac{d\xi}{d\eta} \int_\eta^\xi \frac{|f(x)|}{x-\xi}\, dx$$

$$\leqslant \frac{d\xi}{d\eta}\, |\varphi(\xi(\eta))| + \frac{1}{\delta} \int_\eta^{\xi_1} \frac{|f(x)|}{x}\, dx,$$

since $x - \xi(\eta) > \delta x$. Applying Minkowski's inequality, we get

$$\left(\int_0^{\xi_1} (1-\eta) J_2^2\, d\eta \right)^{1/2} \leqslant \left(\int_0^{\xi_1} \left(\frac{d\xi}{d\eta} \right)^2 \varphi^2(\xi(\eta))\, d\eta \right)^{1/2} + \frac{1}{\delta} \left(\int_0^{\xi_1} \left(\int_\eta^{\xi_1} \frac{|f(x)|\, dx}{x} \right)^2 d\eta \right)^{1/2}. \tag{3.52}$$

But

$$\int_0^{\xi_1} \left(\frac{d\xi}{d\eta} \right)^2 \varphi^2(\xi(\eta))\, d\eta \leqslant \int_0^{\xi_1} \varphi^2(t) \frac{d\xi}{d\eta}\, dt \leqslant \int_0^{\xi_1} \varphi^2(t)\, dt$$

or, by the boundedness theorem for operators with Cauchy kernels (see, for example, [55b], §2, Theorem 1), we have

$$\int_0^{\xi_1} \left(\frac{d\xi}{d\eta} \right)^2 \varphi^2(\xi(\eta))\, d\eta \leqslant \pi^2 \int_0^{\xi_1} f^2(x)\, dx$$
$$\leqslant \frac{\pi^2}{1-\xi_1} \int_0^{\xi_1} (1-x) f^2(x)\, dx \leqslant \frac{\pi^2}{1-\xi_1} \int_0^1 (1-x) f^2(x)\, dx. \tag{3.53}$$

We claim that

$$\int_0^{\xi_1} \left(\int_\eta^{\xi_1} \frac{|f(x)|}{x}\, dx \right)^2 d\eta \leqslant 4 \int_0^{\xi_1} f^2(x)\, dx$$
$$\leqslant \frac{4}{1-\xi_1} \int_0^1 (1-x) f^2(x)\, dx. \tag{3.54}$$

To prove this, it suffices to note that, for any function $g(x) \in L_2(0, \xi_1)$,

$$\left| \int_0^{\xi_1} g(\eta) \left(\int_\eta^{\xi_1} \frac{|f(x)|}{x}\, dx \right) d\eta \right| = \left| \int_0^{\xi_1} \frac{|f(x)|}{x}\, dx \int_0^x g(\eta)\, d\eta \right|$$
$$\leqslant \left(\int_0^{\xi_1} f^2(x)\, dx \right)^{1/2} \left(\int_0^{\xi_1} \left(\frac{1}{x} \int_0^x g(\eta)\, d\eta \right)^2 dx \right)^{1/2}.$$

Applying Hardy's inequality, we get

$$\left|\int_0^{\xi_1} g(\eta)\left(\int_\eta^{\xi_1}\frac{|f(x)|}{x}\,dx\right)d\eta\right| \leqslant 2\left(\int_0^{\xi_1} f^2(x)\,dx\right)^{1/2}\left(\int_0^{\xi_1} g^2(x)\,dx\right)^{1/2}.$$

Thus it follows from (3.52), via (3.53) and (3.54), that

$$\left(\int_0^{\xi_1}(1-\eta)J_2^2\,d\eta\right)^{1/2} \leqslant \frac{\pi+2}{\delta(1-\xi_1)^{1/2}}\left(\int_0^1(1-x)f^2(x)\,dx\right)^{1/2}. \tag{3.55}$$

If $\eta > \xi_1$, then

$$J_2=\frac{d\xi}{d\eta}\varphi(\xi(\eta))+\frac{d\xi}{d\eta}\int_{\xi_1}^{\eta}\frac{f(x)}{x-\xi(\eta)}\,dx.$$

Hence

$$(1-\eta)^{1/2}|J_2|<\frac{d\xi}{d\eta}(1-\eta)^{1/2}|\varphi(\xi(\eta))|$$

$$+\frac{d\xi}{d\eta}\int_{\xi_1}^{\eta}\frac{(1-x)^{1/2}|f(x)|\,dx}{x-\xi(\eta)}<\frac{d\xi}{d\eta}(1-\eta)^{1/2}|\varphi(\xi(\eta))|$$

$$+\frac{d\xi}{d\eta}\int_{\xi_1}^{1}\frac{(1-x)^{1/2}|f(x)|\,dx}{x-\xi(\eta)}=\frac{d\xi}{d\eta}(1-\eta)^{1/2}|\varphi(\xi,\eta)|+\frac{d\xi}{d\eta}|\varphi_1(\xi(\eta))|,$$

where

$$\varphi_1(t)=\int_{\xi_1}^{1}\frac{(1-x)^{1/2}|f(x)|}{x-t}\,dx.$$

Thus,

$$\left(\int_{\xi_1}^{1}(1-\eta)J_2^2\,d\eta\right)^{1/2}\leqslant\left(\int_{\xi_1}^{1}(1-\eta)\varphi^2(\xi(\eta))\frac{d\xi}{d\eta}\,d\eta\right)^{1/2}$$

$$+\left(\int_{\xi_1}^{1}\varphi_1^2(\xi(\eta))\,d\eta\right)^{1/2}\leqslant\left(\int_0^{\xi_1}\varphi^2(t)\,dt\right)^{1/2}+\left(\int_0^{\xi_1}\varphi_1^2(t)\,dt\right)^{1/2} \tag{3.56}$$

$$\leqslant\pi\left(\int_0^{\xi_1}f^2(x)\,dx\right)^{1/2}+\pi\left(\int_0^{\xi_1}(1-x)f^2(x)\,dx\right)^{1/2}$$

$$\leqslant\pi\left[(1-\xi_1)^{-1/2}+1\right]\left(\int_0^1(1-x)f^2(x)\,dx\right)^{1/2}.$$

Adding (3.55) and (3.56), we finally obtain

$$\int_0^1 (1-\eta)\, J_2^2\, d\eta \leqslant \frac{8\pi^2}{\delta^2(1-\xi_1)} \int_0^1 (1-x) f^2(x)\, dx. \tag{3.57}$$

Now, everywhere on γ

$$\left(\frac{\eta-\xi}{\eta}\right)^{1/6} > \delta^{1/6} \quad \text{and} \quad 1-\xi_1 > \delta,$$

and so we conclude from (3.44), using (3.51) and (3.57) and rearranging, that

$$\int_0^1 \eta^{1/3}(1-\eta)\, \Phi_1'^2(\eta)\, d\eta \leqslant \frac{C_4^2}{\delta^{10/3}} \int_0^1 (1-x) f^2(x)\, dx. \tag{3.58}$$

Previously, we set

$$z_2(\xi, \eta) = \int_0^\xi \psi_3(x)(\eta - x)^{-1/6} F\left(\frac{1}{6}, \frac{1}{6}, 1; \frac{\xi - x}{\eta - x}\right) dx. \tag{3.59}$$

Hence

$$(\eta-\xi)^{1/6} \frac{\partial z_2}{\partial \xi}$$

$$= \psi_3(\xi) + \frac{1}{36} \int_0^\xi \psi_3(x) \frac{(\eta-\xi)^{1/6}}{(\eta-x)^{7/6}} F\left(\frac{7}{6}, \frac{7}{6}, 2; \frac{\xi-x}{\eta-x}\right) dx$$

$$= \psi_3(\xi) + \frac{1}{36} \int_0^\xi \psi_3(x) \frac{(\eta-\xi)^{-1/6}}{(\eta-x)^{5/6}} F\left(\frac{5}{6}, \frac{5}{6}, 2; \frac{\xi-x}{\eta-x}\right) dx,$$

$$(\eta-\xi)^{1/6} \frac{\partial z_2}{\partial \eta} = -\frac{1}{6} \int_0^\xi \psi_3(x) \frac{(\eta-\xi)^{1/6}}{(\eta-\xi)^{7/6}} F\left(\frac{7}{6}, \frac{1}{6}, 1; \frac{\xi-x}{\eta-x}\right) dx$$

$$= -\frac{(\eta-\xi)^{-1/6}}{6} \int_0^\xi \frac{\psi_3(x)}{(\eta-x)^{5/6}} F\left(-\frac{1}{6}, \frac{5}{6}, 1; \frac{\xi-x}{\eta-x}\right) dx.$$

Multiplying the first inequality by $d\xi/d\eta$ and adding the result to the second, we get

$$(\eta-\xi)^{1/6} \Phi_2'(\eta)$$

$$= \psi_3(\xi) \frac{d\xi}{d\eta} - \frac{1}{6} \int_0^\xi \frac{\psi_3(x)}{(\eta-\xi)^{1/6}(\eta-x)^{5/6}} \left[F\left(-\frac{1}{6}, \frac{5}{6}, 1; \frac{\xi-x}{\eta-x}\right) \right. \tag{3.60}$$

$$\left. - \frac{1}{6} \frac{d\xi}{d\eta} F\left(\frac{5}{6}, \frac{5}{6}, 2; \frac{\xi-x}{\eta-x}\right) \right] dx,$$

where $\Phi_2(\eta) = z_2(\gamma(\eta), \eta)$.

Proceeding from (3.60), we obtain the estimate

$$(\eta-\xi)^{1/6}\,|\Phi_2'(\eta)|\leqslant|\psi_3(\xi)|\,\frac{d\xi}{d\eta}+\frac{C_5}{\eta}\int_0^\eta|\psi_3(x)|\,dx.$$

Hence, as before,

$$\begin{aligned}\int_0^1(1-\eta)\,\eta^{1/3}\Phi_2'^2(\eta)\,d\eta&\leqslant\frac{C_6^2}{\delta^{1/3}}\int_0^1(1-x)\,\psi_3^2(x)\,dx\\&=\frac{C_7^2}{\delta^{1/3}}\int_0^1(1-x)\,\psi^2(x)\,dx.\end{aligned}\tag{3.61}$$

Thus, (3.58) and (3.61) give

$$\int_0^1(1-\eta)\,\eta^{1/3}\Phi_0'^2(\eta)\,d\eta\leqslant\frac{C_8^2}{\delta^{10/3}}\left[\int_0^1(1-x)\,f^2(x)\,dx+\int_0^1(1-x)\,\psi^2(x)\,dx\right],$$

but

$$\begin{aligned}\int_0^1(1-x)\,f^2(x)\,dx&\leqslant C_9^2\int_0^1(1-x)\,\lambda^2(x)\,dx,\\\int_0^1(1-x)\,\psi^2(x)\,dx&\leqslant C_{10}^2\int_0^1(1-x)\,\mu^2(x)\,dx,\end{aligned}\tag{3.62}$$

as is readily deduced from (3.46) and (3.34). Thus,

$$\int_0^1(1-\eta)\,\eta^{1/3}\Phi_0'^2(\eta)\,d\eta\leqslant\frac{C_{10}^2}{\delta^{10/3}}\int_0^1(1-x)\,[\lambda^2(x)+\mu^2(x)]\,dx.\tag{3.63}$$

We now return to (3.29). By (5.46) of Chapter II, we have

$$\begin{aligned}&\left|\frac{d}{d\eta}I(\gamma(\eta),\,\eta)\right|\\&\qquad\leqslant C_{11}^2\max_{\xi\leqslant\xi'\leqslant\eta'\leqslant\eta}\left|z_0(\xi',\,\eta')\,(\eta'-\xi')^{1/6}\right|(\eta-\xi)^{1/6}.\end{aligned}\tag{3.64}$$

But in view of (3.39) it follows from (3.35) that

$$\begin{aligned}(1-\eta')^{1/2}\left|(\eta'-\xi')^{1/6}z_0(\xi',\,\eta')\right|&\leqslant C_{12}^2\left(\int_0^{\eta'}(1-x)\,f^2(x)\,dx\right)^{1/2}\\&\quad+C_{13}^2\left(\int_0^{\eta'}(1-x)\,\psi^2(x)\,dx\right)^{1/2}.\end{aligned}\tag{3.65}$$

Hence

$$\int_0^1 (1-\eta)\,\eta^{1/3} \left| \frac{dI}{d\eta} \right|^2 d\eta \leqslant C_{14}^2 \int_0^1 (1-x)\,[\lambda^2(x)+\mu^2(x)]\,dx. \tag{3.66}$$

It now follows from (3.63) and (3.66) that

$$\int_0^1 (1-\eta)\,\eta^{1/3}\Phi'^2(\eta)\,d\eta \leqslant \frac{C_{15}^2}{\delta^{10/3}} \int_0^1 (1-x)\,[\lambda^2(x)+\mu^2(x)]\,dx.$$

Applying Lemma 2 of Chapter II, §5, to the right-hand side of this inequality, we finally obtain

$$\int_0^1 (1-\eta)\,\eta^{1/3}\Phi'^2(\eta)\,d\eta \leqslant M^2 \int_0^1 (1-x)\,\mu^2(x)\,dx.$$

Together with (3.28), this gives

$$\begin{aligned} m^2 &\int_0^1 (1-x)\,\mu^2(x)\,dx \\ &\leqslant \int_0^1 (1-\eta)\,\eta^{1/3}\Phi'^2(\eta)\,d\eta \leqslant M^2 \int_0^1 (1-x)\,\mu^2(x)\,dx. \end{aligned} \tag{3.67}$$

The left inequality (3.67) has been proved for twice continuously differentiable solutions of equation (3.12) with $z|_\Gamma = 0$. If $z \in \mathfrak{R}$ and $z|_\Gamma = 0$, then by Lemmas 4 and 5 of Chapter II, §5, there exists a sequence of solutions $\{z_n(\xi, \eta)\}$, $z_n|_\Gamma = 0$, for which (3.67) holds and

$$\lim_{n\to\infty} \int_0^1 (1-x)\,|\mu(x)-\mu_n(x)|^2\,dx = 0.$$

Hence it follows from the right inequality of (3.67) that

$$\begin{aligned} &\int_0^1 (1-\eta)\,\eta^{1/3}\,|\Phi'(\eta)-\Phi_n'(\eta)|^2\,d\eta \\ &\qquad\qquad \leqslant M^2 \int_0^1 (1-x)\,|\mu(x)-\mu_n(x)|^2\,dx \leqslant M^2\varepsilon_n \\ &\qquad\qquad (\varepsilon_n \to 0 \quad \text{as} \quad n\to\infty), \end{aligned}$$

and, consequently, we can let $n \longrightarrow \infty$ in the inequality

$$m^2 \int_0^1 (1-x)\,\mu_n^2(x)\,dx \leqslant \int_0^1 (1-\eta)\,\eta^{1/3}\Phi_n'^2(\eta)\,d\eta$$

obtaining the left inequality (3.67).

We have thus proved

LEMMA 2. *Let the curve* Γ *satisfy conditions* (5.70) *of Chapter* II, *and let inequalities* (3.2) *hold along the curve* γ. *If* $z(x, y) \in \mathfrak{R}$ *is a solution of equation* (3.12), *vanishing on* Γ, *then inequalities* (3.67) *hold.*

3. *Existence theorem for the generalized Tricomi problem when* γ *coincides with the characteristic* $\xi = 0$ *in a small neighborhood of the point* A. We first need an expression for $\mu(x)$ in terms of $\lambda(x)$, which we shall derive from (5.33) of Chapter II. This formula can be rewritten as

$$\mu(x)+\frac{1}{\pi\sqrt{3}}\int_0^1\left[\frac{x(1-t)}{t(1-x)}\right]^{1/6}\left(\frac{1}{t-x}+\frac{1-2t}{t+x-2xt}\right)\mu(t)\,dt$$
$$= -\frac{2}{3}\lambda(x)-\int_0^1\left[\frac{x(1-t)}{t(1-x)}\right]^{1/6}K(t,x)\mu(t)\,dt \equiv -f(x). \tag{3.68}$$

By Lemma 7 of Chapter II, §5,

$$\mu(x) = -\frac{3}{4}\left\{f(x)-\frac{1}{\pi\sqrt{3}}\int_0^1\left[\frac{t(1-t)}{x(1-x)}\right]^{1/2}\left(\frac{1}{t-x}-\frac{1}{t+x-2xt}\right)f(t)\,dt\right\}$$

or, replacing $f(x)$ by its expression,

$$\mu(x)+\int_0^1 K_1(x,t)\mu(t)\,dt$$
$$= -\frac{1}{2}\lambda(x)+\frac{1}{2\pi\sqrt{3}}\int_0^1\left[\frac{t(1-t)}{x(1-x)}\right]^{1/2}\left(\frac{1}{t-x}-\frac{1}{t+x-2xt}\right)\lambda(t)\,dt, \tag{3.69}$$

where

$$K_1(x,t)=\frac{3}{4}\left\{\left[\frac{x(1-t)}{t(1-x)}\right]^{1/6}K(t,x)\right.$$
$$-\frac{1}{\pi\sqrt{3}}\left(\frac{1-t}{t}\right)^{1/6}\int_0^1\left[\frac{\zeta(1-\zeta)}{x(1-x)}\right]^{1/2}\left(\frac{\zeta}{1-\zeta}\right)^{1/6} \tag{3.70}$$
$$\left.\times\left(\frac{1}{\zeta-x}-\frac{1}{\zeta+x-2x\zeta}\right)K(t,\zeta)\,d\zeta\right\}.$$

The estimate

$$|K(t,\zeta)|\leqslant K\left(\left|\ln\frac{1}{t+\zeta-2t\zeta}\right|+\zeta^{-1/6}\right)$$

(see (5.34) of Chapter II) and Lemma 3 of Chapter II, §5, gives us

$$\left(\int_0^1 (1-x)\,K_1^2(x,\,t)\,dx\right)^{1/2}$$
$$\leqslant C\left(\frac{1-t}{t}\right)^{1/6}\left(\int_0^1 x^{1/3}(1-x)^{2/3}K^2(t,\,x)\,dx\right)^{1/2}\leqslant C_1\left(\frac{1-t}{t}\right)^{1/6}.$$

Hence

$$\int_0^1 dt\int_0^1 \frac{1-x}{1-t}\,K_1^2(x,\,t)\,dx<C_2.$$

We can thus apply Fredholm theory to the equation

$$(1-x)^{1/2}\mu(x)+\int_0^1\left(\frac{1-x}{1-t}\right)^{1/2}K_1(x,\,t)(1-t)^{1/2}\mu(t)\,dt=\chi(x). \quad (3.71)$$

We shall show that the homogeneous equation

$$h(x)+\int_0^1\left(\frac{1-x}{1-t}\right)^{1/2}K_1(x,\,t)\,h(t)\,dt=0 \quad (3.72)$$

has no nontrivial solutions in $L_2(0,\,1)$. Suppose the contrary: let $h(x)$ be a nontrivial solution of (3.72). Then the function

$$\mu(x)=(1-x)^{-1/2}h(x) \quad (3.73)$$

satisfies the equation

$$\mu(x)+\frac{1}{\pi\sqrt{3}}\int_0^1\left[\frac{x(1-t)}{t(1-x)}\right]^{1/6}\left(\frac{1}{t-x}+\frac{1-2t}{t+x-2xt}\right)\mu(t)\,dt$$
$$+\int_0^1\left[\frac{x(1-t)}{t(1-x)}\right]^{1/6}K(t,\,x)\,\mu(t)\,dt=0, \quad (3.74)$$

and $\lambda(x)\equiv 0$. Recalling the expression for $\lambda(x)$ (see (5.32) of Chapter II) and formulas (5.15) and (5.16) of Chapter II, we get

$$\frac{\sqrt{3}}{2\pi}\,\frac{d}{dx}\int_0^x (x-t)^{-2/3}\tau(t)\,dt=k\nu(x).$$

Hence

$$\tau(x)=k\int_0^x (x-t)^{-1/3}\nu(t)\,dt. \quad (3.75)$$

Thus there exists a solution $z(x, y) \in \mathfrak{R}$ of equation (3.12), vanishing on Γ, for which (3.75) holds.

Set

$$g(x) = -\int_0^1 \left[\frac{x(1-t)}{t(1-x)}\right]^{1/6} K(t, x)\mu(t)\,dt.$$

From the estimate for $K(t, x)$ we obtain

$$|g(x)| \leqslant K(1-x)^{-1/6}\int_0^1 \left(\frac{1-t}{t}\right)^{1/6}\left(\left|\ln\frac{1}{x+t-2xt}\right| + 1\right)|\mu(t)|\,dt$$

$$\leqslant K(1-x)^{-1/6}\left(\int_0^1 (1-t)\mu^2(t)\,dt\right)^{1/2}$$

$$\times\left(\int_0^1 t^{-1/3}(1-t)^{-2/3}\left(\left|\ln\frac{1}{x+t-2xt}\right| + 1\right)^2 dt\right)^{1/2} \leqslant K(1-x)^{-1/6}.$$

Therefore, $g(x) \in L_p(0, 1)$ for $p < 6$.

Applying Lemma 7 of Chapter II, §5, to equation (3.74), we get

$$\mu(x) = \frac{3}{4}\left\{g(x) - \frac{1}{\pi\sqrt{3}}\int_0^1 \left[\frac{t(1-t)}{x(1-x)}\right]^{1/2}\left(\frac{1}{t-x} - \frac{1}{t+x-2xt}\right) g(t)\,dt\right\}.$$

By Theorem 3 of Chapter II, §5, the function $x^{\epsilon/2}(1-x)^{\epsilon/2}\mu(x)$ is square-integrable on (0, 1). But on the other hand $(1-x)^{1/2}\mu(x) \in L_2(0, 1)$, and so

$$(1-x)^{1/2}\mu(x) \in L_2(0, 1), \quad \text{то} \quad (1-x)^{\varepsilon/2}\mu(x) \in L_2(0, 1),$$

where ϵ is an arbitrarily small number. But then

$$\nu(x) = x^{1/6}\int_0^x (x-t)^{-5/6} t^{-1/6}\mu(t)\,dt$$

is square-integrable on (0, 1).

Let $\{\mu_n(x)\}$ be a sequence of twice continuously differentiable functions, vanishing in some neighborhood of the points $x = 0$ and $x = 1$, such that

$$\lim_{n\to\infty}\int_0^1 (1-x)^{\varepsilon}|\mu - \mu_n|^2\,dx = 0.$$

Then

$$\lim_{n\to\infty}\int_0^1 |\nu-\nu_n|^2\,dx=0 \quad\text{and}\quad \lim_{n\to\infty}\tau_n(x)=\tau(x)$$

uniformly for $0\leqslant x\leqslant 1$, where $\tau_n(x)$ and $\tau(x)$ are defined by (3.75).

Now construct the sequence $\{z_n(x, y)\}$ of solutions of (3.12), such that $z_n|_\Gamma = 0$, corresponding to the functions $\nu(x)$. Define

$$u_n(\theta,\ \sigma)=z_n e^{-\frac{1}{2}\int_0^y b(y)\,dy}.$$

Since

$$K(\sigma)\frac{\partial^2 u_n}{\partial\theta^2}+\frac{\partial^2 u_n}{\partial\sigma^2}=0,$$

it follows that in D^+

$$\left(Ku_n\frac{\partial u_n}{\partial\theta}\right)_\theta+\left(u_n\frac{\partial u_n}{\partial\sigma}\right)_\sigma-K\left(\frac{\partial u_n}{\partial\theta}\right)^2-\left(\frac{\partial u_n}{\partial\sigma}\right)^2=0.$$

Integrating this over D^+, applying Green's formula, and using the fact that $u_n|_\Gamma = 0$, we obtain

$$\int_0^1 u_n(\theta,\ 0)\frac{\partial u_n(\theta,\ 0)}{\partial\sigma}\,d\theta+\iint_{D^+}\left[K(\sigma)\left(\frac{\partial u_n}{\partial\theta}\right)^2+\left(\frac{\partial u_n}{\partial\sigma}\right)^2\right]d\sigma\,d\theta=0.$$

Hence

$$\int_0^1 u_n(\theta,\ 0)\frac{\partial u_n(\theta,\ 0)}{\partial\sigma}\,d\theta<0.$$

But

$$u_n(\theta,\ 0)=z_n(x,\ 0)=\tau_n(x),$$

$$\frac{\partial u_n(\theta,\ 0)}{\partial\sigma}=\left[\nu_n(x)-\frac{1}{2}b(0)\,\tau_n(x)\right]\frac{dy}{d\sigma}\bigg|_{\sigma=0}.$$

Since $dy/d\sigma|_{\sigma=0}>0$, it follows that

$$\int_0^1 \tau_n(x)\left[\nu_n(x)-\frac{b(0)}{2}\tau_n(x)\right]dx<0$$

or

$$\int_0^1 \tau_n(x)\,\nu_n(x)\,dx<\frac{b(0)}{2}\int_0^1 \tau_n^2(x)\,dx\leqslant 0\quad(b(0)\leqslant 0).$$

Letting $n \longrightarrow \infty$, we get

$$\int_0^1 \tau(x)\,\nu(x)\,dx < 0. \tag{3.76}$$

On the other hand, as was shown by Tricomi (see [69a]), it follows from (3.75) that

$$\int_0^1 \tau(x)\,\nu(x)\,dx \geqslant 0,$$

contradicting (3.76). Thus $\mu(x) \equiv 0$. Thus, $\lambda = -1$ is not an eigenvalue of the kernel

$$\left(\frac{1-x}{1-t}\right)^{1/2} K_1(x, t).$$

Let $R(x, t; \lambda)$ be its resolvent. Set $R(x, t; -1) = R(x, t)$. Then the solution of (3.69) can be written as

$$\mu(x) = -\frac{1}{2}\lambda(x) + \frac{1}{2\pi\sqrt{3}}\int_0^1 \left[\frac{t(1-t)}{x(1-x)}\right]^{1/2}\left(\frac{1}{t-x} - \frac{1}{t+x-2xt}\right)\lambda(t)\,dt + \int_0^1 \left(\frac{1-t}{1-x}\right)^{1/2} K_2(x, t)\,\lambda(t)\,dt, \tag{3.77}$$

where

$$K_2(x, t) = \frac{R(x, t)}{2} - \frac{1}{2\pi\sqrt{3}}\int_0^1 \left(\frac{t}{\zeta}\right)^{1/2}\left(\frac{1}{t-\zeta} - \frac{1}{t+\zeta-2\zeta t}\right) R(x, \zeta)\,d\zeta.$$

Note that the function

$$g_1(x) = \int_0^1 \left(\frac{x}{\zeta}\right)^{1/2}\left(\frac{1}{x-\zeta} - \frac{1}{x+\zeta-2x\zeta}\right) f(\zeta)\,d\zeta$$

is square-integrable if $f(\zeta) \in L_2(0, 1)$. Indeed, let $h(x)$ be an arbitrary smooth function on $(0, 1)$. Then

$$\int_0^1 g_1(x)\,h(x)\,dx = \int_0^1 f(\zeta)\,d\zeta \int_0^1 \left(\frac{\zeta}{x}\right)^{1/2}\left(\frac{1}{\zeta-x} - \frac{1}{x+\zeta-2x\zeta}\right) h(x)\,dx.$$

Consequently, by Lemma 3 of Chapter II, §5,

$$\left|\int_0^1 g_1(x)\,h(x)\,dx\right| \leqslant \left(\int_0^1 f^2(x)\,dx\right)^{1/2}\left(\int_0^1 h^2(x)\,dx\right)^{1/2}.$$

Hence $\int_0^1 g_1^2(x)\,dx \leqslant \int_0^1 f^2(x)\,dx$. Thus

$$\left(\int_0^1 K_2^2(x,t)\,dt\right)^{1/2} \leqslant \frac{1}{2}\left(\int_0^1 R^2(x,t)\,dt\right)^{1/2} + \frac{1}{2\pi\sqrt{3}}\left(\int_0^1 R^2(x,t)\,dt\right)^{1/2}$$

$$\leqslant \left(\int_0^1 R^2(x,t)\,dt\right)^{1/2}$$

and so

$$\int_0^1 dx \int_0^1 K_2^2(x,t)\,dt \leqslant \int_0^1 dx \int_0^1 R^2(x,t)\,dt < \infty. \tag{3.78}$$

We now consider the situation on the curve γ.

Let γ be a smooth curve, satisfying conditions (3.2). In addition, we assume that for $0 \leqslant \eta \leqslant \epsilon$ the curve γ coincides with the characteristic $\xi = 0$.

For $0 \leqslant \eta \leqslant \epsilon$ we have (5.46) of Chapter II:

$$\eta^{1/6}\Phi'(\eta) = 2\pi\gamma_2\left[\frac{1}{k}\lambda_1(\eta) - \mu(\eta)\right]$$

$$+ \int_0^\eta \left(\frac{\eta}{x}\right)^{1/6}\frac{\partial g_2(\eta,x)}{\partial\eta}\lambda_1(x)\,dx - \int_0^\eta \left(\frac{\eta}{x}\right)^{1/6}\frac{\partial h_1(\eta,x)}{\partial\eta}\mu(x)\,dx$$

or, by (3.45),

$$\frac{1}{2\pi\gamma_2}\eta^{1/6}\Phi'(\eta) = \lambda(\eta) + \int_0^\eta g_3(\eta,x)\,\lambda(x)\,dx + \int_0^\eta h_2(\eta,x)\,\mu(x)\,dx,$$

where

$$g_3(\eta,x) = \frac{k}{2\pi\gamma_2}\left(\frac{\eta}{x}\right)^{1/6}\frac{\partial g_2(\eta,x)}{\partial\eta}, \quad h_2(\eta,x) = \frac{1}{2\pi\gamma_2}\left(\frac{\eta}{x}\right)^{1/6}\left(k\frac{\partial g_2}{\partial\eta} - \frac{\partial h_1}{\partial\eta}\right).$$

By the estimates (see Chapter II, §5.3)

$$\left|\frac{\partial g_2}{\partial\eta}\right| < C_2\eta^{1/3}, \quad \left|\frac{\partial h_1}{\partial\eta}\right| < C_2\eta^{1/3}$$

we have $|g_3|, |h_2| < C_3\eta^{1/2}x^{-1/6}$. Set

$$(1-\eta)^{1/2}\lambda(\eta) = \theta(\eta). \tag{3.79}$$

Then

$$\frac{(1-\eta)^{1/2}\eta^{1/6}}{2\pi\gamma_2}\Phi'(\eta) = \theta(\eta) + \int_0^\eta \left(\frac{1-\eta}{1-x}\right)^{1/2} g_3(\eta, x)\,\theta(x)\,dx$$
$$+ \int_0^\eta (1-\eta)^{1/2} h_2(\eta, x)\,\mu(x)\,dx.$$

Inserting the value of $(1-x)^{1/2}\mu(x)$ according to (3.77), we get

$$\frac{(1-\eta)^{1/2}\eta^{1/6}}{2\pi\gamma_2}\Phi'(\eta) = \theta(\eta) + \int_0^1 K_4(\eta, x)\,\theta(x)\,dx. \tag{3.80}$$

Here

$$K_4(\eta, x) = \begin{cases} \left(\frac{1-\eta}{1-x}\right)^{1/2}\left[g_3(\eta, x) - \frac{1}{2}h_2(\eta, x)\right] + K_3(\eta, x) & \text{for } x < \eta, \\ K_3(\eta, x) & \text{for } x > \eta, \end{cases}$$

where

$$K_3(\eta, x) = \frac{1}{2\pi\sqrt{3}}\int_0^\eta \left(\frac{1-\eta}{1-t}\right)^{1/2}\left(\frac{x}{t}\right)^{1/2}\left(\frac{1}{x-t} - \frac{1}{x+t-2xt}\right) h_2(\eta, t)\,dt$$
$$+ \int_0^\eta \left(\frac{1-\eta}{1-t}\right)^{1/2} h_2(\eta, t)\,K_2(t, x)\,dt.$$

It can be shown that

$$\int_0^\varepsilon d\eta \int_0^1 K_4^2(\eta, x)\,dx < \infty.$$

Now let $\epsilon < \eta < 1$. Then, by (3.44) and (3.60),

$$
\begin{aligned}
&(\eta-\xi)^{1/6}\Phi_0'(\eta)\\
&= 2\pi f(\eta)+\pi\sqrt{3}f(\xi)\frac{d\xi}{d\eta}+\psi_3(\xi)\frac{d\xi}{d\eta}+\frac{d\xi}{d\eta}\int_0^\eta\frac{f(x)\,dx}{x-\xi}\\
&+\int_0^\xi\left[\theta_3(\zeta)+\theta_5(\zeta)\frac{d\xi}{d\eta}\right]\frac{f(x)\,dx}{\eta-x}\\
&+\frac{1}{\eta-\xi}\int_\xi^\eta\left[\theta_4\left(\frac{1}{\zeta}\right)+\theta_6\left(\frac{1}{\zeta}\right)\frac{d\xi}{d\eta}\right]f(x)\,dx \qquad (3.81)\\
&-\frac{1}{6}\int_0^\xi\left[F\left(-\frac{1}{6},\frac{1}{6},1;\frac{\xi-x}{\eta-x}\right)-\frac{1}{6}F\left(\frac{5}{6},\frac{5}{6},2;\frac{\xi-x}{\eta-x}\right)\frac{d\xi}{d\eta}\right]\\
&\qquad\times\frac{\psi_3(x)\,dx}{(\eta-\xi)^{1/6}(\eta-x)^{5/6}};
\end{aligned}
$$

or, replacing $f(x)$ and $\psi_3(x)$ by their expressions according to (3.46), (3.36) and (3.34),

$$
\begin{aligned}
\frac{(\eta-\xi)^{1/6}}{2\pi\gamma_2}\Phi_0'(\eta)&=\lambda(\eta)+\frac{\sqrt{3}}{2}\frac{d\xi}{d\eta}\lambda(\xi)+\sqrt{3}\frac{d\xi}{d\eta}\mu(\xi)\\
&+\frac{1}{2\pi}\frac{d\xi}{d\eta}\int_0^\eta\frac{\lambda(x)\,dx}{x-\xi}+I_1+I_2,
\end{aligned}
$$

where I_1 and I_2 denote the sums of the integrals depending on λ and μ, respectively. Hence, by (3.79),

$$
\begin{aligned}
\frac{(\eta-\xi)^{1/6}}{2\pi\gamma_2}(1-\eta)^{1/2}\Phi_0'(\eta)&=\theta(\eta)+\frac{\sqrt{3}}{\pi}\frac{d\xi}{d\eta}\left(\frac{1-\eta}{1-\xi}\right)^{1/2}\theta(\xi)\\
&+\sqrt{3}\frac{d\xi}{d\eta}\left(\frac{1-\eta}{1-\xi}\right)^{1/2}(1-\xi)^{1/2}\mu(\xi) \qquad (3.82)\\
+\frac{1}{2\pi}\frac{d\xi}{d\eta}\int_0^\eta\left(\frac{1-\eta}{1-x}\right)^{1/2}&\frac{\theta(x)\,dx}{x-\xi}+(1-\eta)^{1/2}I_1+(1-\eta)^{1/2}I_2.
\end{aligned}
$$

We have

$$
\begin{aligned}
(1-\eta)^{1/2} I_1 = \Bigg\{ & \frac{\eta^{-5/6}}{36} \int_0^{\eta} \frac{\lambda(t)}{t^{1/6}} F\left(\frac{5}{6}, \frac{5}{6}, 2;\ 1-\frac{t}{\eta}\right) dt \\
& + \frac{\sqrt{3}}{2} \frac{\xi^{-5/6}}{36} \frac{d\xi}{d\eta} \int_0^{\xi} \frac{\lambda(t)}{t^{1/6}} F\left(\frac{5}{6}, \frac{5}{6}, 2;\ 1-\frac{t}{\xi}\right) dt \\
& + \frac{1}{72\pi} \frac{d\xi}{d\eta} \int_0^{\eta} \frac{x^{-5/6}\, dx}{x-\xi} \int_0^{x} \frac{\lambda(t)}{t^{1/6}} F\left(\frac{5}{6}, \frac{5}{6}, 2;\ 1-\frac{t}{x}\right) dt \\
& + \int_0^{\xi} \left[\theta_3(\zeta) + \theta_5(\zeta) \frac{d\xi}{d\eta}\right] \frac{f(x)\, dx}{\eta-\xi} \\
& + \frac{1}{\eta-\xi} \int_{\xi}^{\eta} \left[\theta_4\left(\frac{1}{\zeta}\right) + \theta_6\left(\frac{1}{\zeta}\right) \frac{d\xi}{d\eta}\right] f(x)\, dx \Bigg\} (1-\eta)^{1/2} \\
& = \int_0^{\eta} \Psi_1(\eta, t)\, \theta(t)\, dt.
\end{aligned}
\tag{3.83}
$$

We claim that under suitable assumptions

$$
\int_{\varepsilon}^{1} d\eta \int_0^{1} \Psi_1^2(\eta, t)\, dt < \infty. \tag{3.84}
$$

We shall examine all the integrals in (3.83) one by one. For the first integral, the last inequality is obvious. For the second, we have

$$
\begin{aligned}
& \int_{\varepsilon}^{1} d\eta \int_0^{\xi} \xi^{-5/3} \left(\frac{d\xi}{d\eta}\right)^2 F^2\left(\frac{5}{6}, \frac{5}{6}, 2;\ 1-\frac{t}{\xi}\right) \frac{1-\eta}{1-t}\, t^{-1/3}\, dt \\
& \qquad < \frac{3}{2} F^2\left(\frac{5}{6}, \frac{5}{6}, 2;\ 1\right) \int_{\varepsilon}^{1} \frac{\left(\frac{d\xi}{d\eta}\right)^2}{\xi}\, d\eta.
\end{aligned}
$$

Let us assume that for sufficiently small $\delta > 0$

$$
\int_{\varepsilon}^{1} \frac{\left(\frac{d\xi}{d\eta}\right)^2}{\xi^{1+\delta}}\, d\eta < \infty. \tag{3.85}
$$

Furthermore,

$$(1-\eta)^{1/2}\frac{d\xi}{d\eta}\int_0^{\eta} x^{-5/6}\frac{dx}{x-\xi}\int_0^{x}\frac{\lambda(t)}{t^{1/6}}F\left(\frac{5}{6},\frac{5}{6},2;\,1-\frac{t}{x}\right)dt$$

$$=(1-\eta)^{1/2}\frac{d\xi}{d\eta}\left[\int_0^{\xi} t^{-1/6}\lambda(t)\,dt\int_t^{\eta}\frac{F\left(\frac{5}{6},\frac{5}{6},2;\,1-\frac{t}{x}\right)dx}{x^{5/6}(x-\xi)}\right.$$

$$\left.+\int_{\xi}^{\eta} t^{-1/6}\lambda(t)\,dt\int_t^{\eta}\frac{F\left(\frac{5}{6},\frac{5}{6},2;\,1-\frac{t}{x}\right)dx}{x^{5/6}(x-\xi)}\right].$$

But for $t<\xi$

$$\int_t^{\eta}\frac{F\left(\frac{5}{6},\frac{5}{6},2;\,1-\frac{t}{x}\right)}{x^{5/6}(x-\xi)}\,dx< C\xi^{-5/6}|\ln(\xi-t)|,$$

and for $t>\xi$

$$\int_t^{\eta}\frac{F\left(\frac{5}{6},\frac{5}{6},2;\,1-\frac{t}{x}\right)}{x^{5/6}(x-\xi)}\,dx< Ct^{-5/6}|\ln(t-\xi)|.$$

Thus, by (3.85),

$$\int_{\varepsilon}^{1}\frac{(1-\eta)\left(\frac{d\xi}{d\eta}\right)^2}{\xi^{5/3}}\,d\eta\int_0^{\xi}\frac{|\ln(\xi-t)|^2}{t^{1/3}(1-t)}\,dt$$

$$+\int_{\varepsilon}^{1}(1-\eta)\left(\frac{d\xi}{d\eta}\right)^2 d\eta\int_{\xi}^{\eta}\frac{|\ln(t-\xi)|^2}{t^2(1-t)}\,dt<\infty.$$

The last two integrals in (3.83) are treated in similar fashion. Now consider $(1-\eta)^{1/2}I_2$:

$$(1-\eta)^{1/2}I_2$$

$$=(1-\eta)^{1/2}\left[\frac{\sqrt{3}}{36}\xi^{-5/6}\frac{d\xi}{d\eta}\int_0^{\xi}\frac{\mu(t)}{t^{1/6}}F\left(\frac{5}{6},\frac{5}{6},2;\,1-\frac{t}{\xi}\right)dt\right.$$

$$+\sqrt{3}\int_0^{\xi}\Psi(\xi,\eta,t)\mu(t)\,dt \tag{3.86}$$

$$\left.+\frac{\sqrt{3}}{36}\int_0^{\xi}\frac{\mu(t)}{t^{1/6}}\,dt\int_t^{\xi}\frac{F\left(\frac{5}{6},\frac{5}{6},2;\,1-\frac{t}{x}\right)}{x^{5/6}}\Psi(\xi,\eta,x)\,dx\right]$$

$$=\int_0^{\xi}(1-t)^{1/2}\Psi_2(\eta,t)\mu(t)\,dt,$$

where

$$\Psi(\xi, \eta, x) = -\frac{1}{6}(\eta-\xi)^{-1/6}(\eta-x)^{-5/6}\Big[F\Big(-\frac{1}{6}, \frac{1}{6}, 1; \frac{\xi-x}{\eta-x}\Big) - \frac{1}{6}F\Big(\frac{5}{6}, \frac{5}{6}, 2; \frac{\xi-x}{\eta-x}\Big)\frac{d\xi}{d\eta}\Big].$$

It is readily seen that

$$\int_\varepsilon^1 d\eta \int_0^1 \Psi_2^2(\eta, t)\, dt < \infty.$$

In view of (3.83) and (3.86), we can write (3.82) as

$$\begin{aligned}\frac{(\eta-\xi)^{1/6}}{2\pi\gamma_2}&(1-\eta)^{1/2}\Phi_0'(\eta)\\ &= \theta(\eta) + \sqrt{3}\Big(\frac{1-\eta}{1-\xi}\Big)^{1/2}\frac{d\xi}{d\eta}\Big[\frac{1}{2}\theta(\xi) + (1-\xi)^{1/2}\mu(\xi) + \frac{1}{2\pi}\int_0^\eta\Big(\frac{1-\xi}{1-x}\Big)^{1/2}\frac{\theta(x)\,dx}{x-\xi}\Big]\\ &+ \int_0^\eta \Psi_1(\eta, t)\theta(t)\,dt + \int_0^\xi (1-t)^{1/2}\Psi_2(\eta, t)\mu(t)\,dt\end{aligned}$$

or, replacing $\mu(x)$ by (3.77),

$$\begin{aligned}\frac{(\eta-\xi)^{1/6}}{2\pi\gamma_2}&(1-\eta)^{1/2}\Phi_0'(\eta) = \theta(\eta)\\ &+ \Big(\frac{1-\eta}{1-\xi}\Big)^{1/2}\frac{d\xi}{d\eta}\Big[\frac{1}{2\pi}\int_0^1\Big(\frac{t}{\xi}\Big)^{1/2}\Big(\frac{1}{t-\xi} - \frac{1}{t+\xi-2\xi t}\Big)\theta(t)\,dt\\ &+ \frac{1}{2\pi}\int_0^\eta\Big(\frac{1-\xi}{1-t}\Big)^{1/2}\frac{\theta(t)\,dt}{t-\xi}\Big]\\ &+ \int_0^\eta \Psi_1(\eta, t)\theta(t)\,dt + \int_0^1 \Psi_3(\eta, t)\theta(t)\,dt.\end{aligned} \tag{3.87}$$

As before, we can prove that

$$\int_\varepsilon^1 d\eta \int_0^1 \Psi_3^2(\eta, t)\, dt < \infty. \tag{3.88}$$

Next,

$$\frac{d\xi}{d\eta}\int_0^{\eta}\left(\frac{1-\xi}{1-t}\right)^{1/2}\frac{\theta(t)\,dt}{t-\xi}$$

$$= \frac{d\xi}{d\eta}\int_0^{\eta}\left(\frac{t}{\xi}\right)^{1/2}\frac{\theta(t)\,dt}{t-\xi}+\frac{d\xi}{d\eta}\int_0^{\eta}\left[\left(\frac{1-\xi}{1-t}\right)^{1/2}-\left(\frac{t}{\xi}\right)^{1/2}\right]\frac{\theta(t)\,dt}{t-\xi} \tag{3.89}$$

$$= \frac{d\xi}{d\eta}\int_0^{1}\left(\frac{t}{\xi}\right)^{1/2}\left(\frac{1}{t-\xi}-\frac{1}{t+\xi-2t\xi}\right)\theta(t)\,dt+\int_0^{1}\Psi_4(\eta,t)\,\theta(t)\,dt.$$

It is not hard to see that if (3.85) holds, then

$$\int_{\varepsilon}^{1}d\eta\int_0^{1}\Psi_4^2(\eta,x)\,dx<\infty.$$

Inserting (3.89) into (3.87), we get

$$\frac{(\eta-\xi)^{1/6}}{2\pi\gamma_2}(1-\eta)^{1/2}\Phi_0'(\eta)$$

$$= \theta(\eta)+\left(\frac{1-\eta}{1-\xi}\right)^{1/2}\frac{d\xi}{d\eta}\frac{1}{\pi}\int_0^{1}\left(\frac{t}{\xi}\right)^{1/2}\left(\frac{1}{t-\xi}-\frac{1}{t+\xi-2\xi t}\right)\theta(t)\,dt \tag{3.90}$$

$$+\int_0^{1}K_1(\eta,t)\,\theta(t)\,dt,$$

where

$$\int_{\varepsilon}^{1}d\eta\int_0^{1}K_1^2(\eta,t)\,dt<\infty. \tag{3.91}$$

Finally, to get an expression for

$$\frac{(\eta-\xi)^{1/6}}{2\pi\gamma_2}(1-\eta)^{1/2}\Phi'(\eta)\quad(\varepsilon<\eta<1),$$

we must still consider the double integral in (3.29): $z=z_0+I$, where I has the form (see (5.38) of Chapter II)

$$I(\xi,\eta)=\int_{\xi}^{\eta}g_0(\xi,\eta,t)\,\tau(t)\,dt-\int_{\xi}^{\eta}h_0(\xi,\eta,t)\,\nu(t)\,dt.$$

Differentiating with respect to η, we get

$$\frac{dI(\gamma(\eta),\eta)}{d\eta}=\int_{\xi}^{\eta}\left(\frac{\partial g_0}{\partial\xi}\frac{d\xi}{d\eta}+\frac{\partial g_0}{\partial\eta}\right)\tau(t)\,dt-\int_{\xi}^{\eta}\left(\frac{\partial h_0}{\partial\xi}\frac{d\xi}{d\eta}+\frac{\partial h_0}{\partial\eta}\right)\nu(t)\,dt.$$

Using (5.39), (5.40), and (5.45) of Chapter II, we easily see that

$$\left|\frac{\partial g_0}{\partial \eta}\right| < C(\eta - t)^{-5/6}, \quad \left|\frac{\partial g_0}{\partial \xi}\right| < C(t - \xi)^{-5/6},$$
$$\left|\frac{\partial h_0}{\partial \eta}\right| < C(\eta - t)^{-1/6}, \quad \left|\frac{\partial h_0}{\partial \xi}\right| < C(t - \xi)^{-1/6}. \tag{3.92}$$

Therefore

$$\frac{(\eta-\xi)^{1/6}}{2\pi\gamma_2}(1-\eta)^{1/2}\frac{dI(\gamma(\eta),\eta)}{d\eta} = \int_0^1 \Psi_5(\eta, t)\,\theta(t)\,dt$$

and

$$\int_0^1 d\eta \int_0^1 \Psi_5^2(\eta, t)\,dt < \infty.$$

Thus, finally, we see that for $0 < \eta < 1$

$$\frac{(\eta-\xi)^{1/6}}{2\pi\gamma_2}(1-\eta)^{1/2}\Phi'(\eta) = \theta(\eta) + \int_0^1 K(\eta, t)\,\theta(t)\,dt$$
$$+\left(\frac{1-\eta}{1-\xi}\right)^{1/2}\frac{d\xi}{d\eta}\frac{1}{\pi}\int_0^1\left(\frac{t}{\xi}\right)^{1/2}\left(\frac{1}{t-\xi} - \frac{1}{t+\xi-2t\xi}\right)\theta(t)\,dt, \tag{3.93}$$

where

$$\int_0^1\int_0^1 K^2(\eta, t)\,d\eta\,dt < \infty.$$

Assume that

$$\int_0^1 (1-\eta)\,\eta^{1/3}\Phi'^2(\eta)\,d\eta < \infty.$$

Write (3.93) in the form

$$A\theta = \frac{(\eta-\xi)^{1/6}}{2\pi\gamma_2}(1-\eta)^{1/2}\Phi'(\eta), \tag{3.93'}$$

where A is a linear operator in $L_2(0, 1)$. This operator is bounded above and below, by virtue of (3.67). We claim that it can be regularized.

Define an operator D in $L_2(0, 1)$ by

$$D\theta = \theta(\eta) + \left(\frac{1-\eta}{1-\xi}\right)^{1/2} \frac{d\xi}{d\eta} \frac{1}{\pi} \int_0^1 \left(\frac{t}{\xi}\right)^{1/2} \left(\frac{1}{t-\xi} - \frac{1}{t+\xi-2t\xi}\right) \theta(t)\, dt \qquad (3.94)$$
$$= \theta(\eta) + \vartheta(\eta).$$

This operator has an inverse. Indeed, let

$$\hat{\theta}(x) = \frac{1}{\pi} \int_0^1 \left(\frac{t}{x}\right)^{1/2} \left(\frac{1}{t-x} - \frac{1}{t+x-2xt}\right) \theta(t)\, dt.$$

Then

$$\int_0^1 \vartheta^2(\eta)\, d\eta < \int_0^1 \hat{\theta}^2(\xi(\eta)) \left(\frac{d\xi}{d\eta}\right)^2 d\eta$$
$$\leqslant \max\left|\frac{d\xi}{d\eta}\right| \int_0^{\xi(1)} \hat{\theta}^2(x)\, dx < \max\left|\frac{d\xi}{d\eta}\right| \int_0^1 \hat{\theta}^2(x)\, dx$$

or, by Lemma 3 of Chapter II, §5,

$$\int_0^1 \vartheta^2(\eta)\, d\eta < \max\left|\frac{d\xi}{d\eta}\right| \int_0^1 \theta^2(x)\, dx.$$

Now, since $\max |d\xi/d\eta| \leqslant 1 - \delta$, it follows from (3.94) that

$$\| (D-E)\theta \| < (1-\delta)^{1/2} \| \theta \|,$$

i.e., the norm of the operator $D - E$ is less than unity. Consequently, there exists a bounded inverse D^{-1}.

Applying D^{-1} to both sides of (3.93′) and using (3.93) and (3.94), we get

$$\theta(\eta) + D^{-1} \int_0^1 K(\eta, t)\, \theta(t)\, dt = D^{-1} \frac{(\eta-\xi)^{1/6}}{2\pi\gamma_2} (1-\eta)^{1/2} \Phi'(\eta),$$

or

$$\theta(\eta) + B\theta = \varphi(\eta). \qquad (3.95)$$

The operator B is compact in $L_2(0, 1)$, as the product of a bounded operator and a compact one. Hence D^{-1} regularizes the operator A. Since D^{-1} has no zeros other than $x = 0$, equations (3.93′) and (3.95) are equivalent for any free term

$$f(\eta) = \frac{(\eta-\xi)^{1/6}}{2\pi\gamma_2} (1-\eta)^{1/2} \Phi'(\eta).$$

By the uniqueness theorem for the generalized Tricomi problem, equation (3.95) is solvable for any $\varphi \in L_2(0, 1)$. But then (3.93) has a solution for any $f \in L_2(0, 1)$ and, by (3.67), there exists a bounded inverse A^{-1}. Thus we arrive at

LEMMA 3. *Let the curve γ coincide with the characteristic $\xi = 0$ for $0 \leqslant \eta \leqslant \epsilon$ and satisfy conditions* (3.2) *and* (3.85). *Then equation* (3.12) *has a unique solution $z \in \mathfrak{R}$ in D, satisfying the boundary conditions $z|_\Gamma = 0$ and $z|_\gamma = \Phi(\eta)$, if*

$$\int_0^1 (1-\eta)\, \eta^{1/3} \Phi'^2(\eta)\, d\eta < \infty.$$

4. *General case.* We now consider the generalized Tricomi problem without any additional conditions. Let γ be an arbitrary curve satisfying (3.2). We construct a sequence of curves $\{\gamma_n\}$ according to the following rule. Let $\{\epsilon_n\}$ be a monotone decreasing sequence of numbers tending to zero. Define δ_n by setting $2\epsilon_n = \gamma(\delta_n)$. Obviously, $\lim_{n\to\infty}\delta_n = 0$.

Set $\gamma_n(\eta) = \gamma(\eta) - \epsilon_n$ for $\delta_n \leqslant \eta \leqslant 1$. Define $\gamma_n(\eta)$ on the interval $[0, \delta_n]$ so that: 1) $\gamma_n(\eta)$ coincides with the characteristic $\xi = 0$ in a sufficiently small neighborhood of the point (0, 0), 2) condition (3.85) holds, and 3) $0 \leqslant \gamma_n'(\eta) \leqslant 1 - \delta$.

The curve γ_n determines a corresponding operator A_n in $L_2(0, 1)$. Define a sequence of operators $B_n, \mathfrak{L}_2(0, 1) \to \mathfrak{L}_2(0, 1)$, by $B_n\lambda = (1-\eta)^{-1/2}A_n\theta$, or, by (3.79), $B_n\lambda = (1-\eta)^{-1/2}A_n(1-\eta)^{1/2}\lambda$. We claim that this sequence is strongly convergent in $\mathfrak{L}_2(0, 1)$. The operator B_n can be represented as the sum of two operators B_{0n} and I_n:

$$B_{0n}\lambda = \frac{[\eta - \gamma_n(\eta)]^{1/6}}{2\pi\gamma_2}\Phi'_{0n}(\eta), \qquad I_n\lambda = \frac{[\eta-\gamma_n(\eta)]^{1/6}}{2\pi\gamma_2}\frac{dI_n}{d\eta}.$$

Let $\lambda(\eta)$ be a continuously differentiable function vanishing in a sufficiently small neighborhood of $\eta = 0$. Then the function $f(\eta)$ defined by (3.46) also has this property. We already have (see (3.81))

$$(\eta - \xi_n)^{1/6}\Phi'_{0n}(\eta) = 2\pi f(\eta) + \pi\sqrt{3}\, f(\xi_n)\frac{d\xi_n}{d\eta} + \frac{d\xi_n}{d\eta}\int_0^\eta \frac{f(x)\,dx}{x-\xi_n}$$

$$+\int_0^{\xi_n}\left[\theta_3(\zeta) + \theta_5(\zeta)\frac{d\xi_n}{d\eta}\right]\frac{f(x)\,dx}{\eta - x} + \frac{1}{\eta-\xi_n}\int_{\xi_n}^{\eta}\left[\theta_4\left(\frac{1}{\zeta}\right) + \theta_6\left(\frac{1}{\xi}\right)\frac{d\xi_n}{d\eta}\right]f(x)\,dx \tag{3.96}$$

$$+\psi_3(\xi_n)\frac{d\xi_n}{d\eta} + \int_0^{\xi_n}\Psi(\xi_n, \eta, x)\,\psi_3(x)\,dx,,$$

where $\xi_n = \gamma_n(\eta)$ and $\Psi(\xi_n, \eta, x)$ is a linear combination of hypergeometric functions. Hence it is clear that all the terms that depend on $f(x)$ are uniformly convergent $(0 \leqslant \eta \leqslant 1)$ to

$$2\pi f(\eta) + \pi\sqrt{3}\, f(\xi)\frac{d\xi}{d\eta} + \frac{d\xi}{d\eta}\int_0^\eta \frac{f(x)\,dx}{x-\xi} + \int_0^\xi \left[\theta_3(\zeta) + \theta_5(\xi)\frac{d\xi}{d\eta}\right]\frac{f(x)\,dx}{\eta - x}$$

$$+ \frac{1}{\eta-\xi}\int_\xi^\eta \left[\theta_4\left(\frac{1}{\zeta}\right) + \theta_6\left(\frac{1}{\zeta}\right)\frac{d\xi}{d\eta}\right] f(x)\,dx.$$

Consider the last two integrals in (3.96). Note that, by (3.36) and (3.62), $\psi_3(x) \in \mathfrak{L}_2(0, 1)$. Let

$$S = \int_0^1 (1-\eta)\left|\psi_3(\xi_n)\frac{d\xi_n}{d\eta} - \psi_3(\xi_m)\frac{d\xi_m}{d\eta}\right|^2 d\eta$$

$$= \left(\int_0^h + \int_h^1\right)(1-\eta)\left|\psi_3(\xi_n)\frac{d\xi_n}{d\eta} - \psi_3(\xi_m)\frac{d\xi_m}{d\eta}\right|^2 d\eta.$$

For sufficiently large n and m, we have $\delta_n < h$ and $\delta_m < h$; hence

$$S < 2\int_0^h (1-\eta)\psi_3^2(\xi_n)\left(\frac{d\xi_n}{d\eta}\right)^2 d\eta + 2\int_0^h (1-\eta)\psi_3^2(\xi_m)\left(\frac{d\xi_m}{d\eta}\right)^2 d\eta$$

$$+ \int_h^1 (1-\eta)|\psi_3(\xi - \varepsilon_n) - \psi_3(\xi - \varepsilon_m)|^2\left(\frac{d\xi}{d\eta}\right)^2 d\eta$$

$$< 2\int_0^{\xi_n(h)} (1-\eta)\psi_3^2(\eta)\,d\eta + 2\int_0^{\xi_m(h)} (1-\eta)\psi_3^2(\eta)\,d\eta$$

$$+ \int_{\xi(h)}^1 (1-\eta)|\psi_3(\eta - \varepsilon_n) - \psi_3(\eta - \varepsilon_m)|^2\,d\eta \leqslant \delta_1,$$

where δ_1 may be made arbitrarily small by taking h sufficiently small and n and m sufficiently large.

The kernel $\Psi(\xi, \eta, x)$ of the last integral in (3.96) is a continuous function of x, ξ and η, $\eta > 0$. Consequently, if $\eta > 0$, then

$$\lim_{n\to\infty}\int_0^{\xi_n} \Psi(\xi_n, \eta, x)\psi_3(x)\,dx = \int_0^\xi \Psi(\xi, \eta, x)\psi_3(x)\,dx.$$

Since

$$\left|\int_0^{\xi_n} \Psi(\xi_n, \eta, x)\,\psi_3(x)\,dx\right| \leqslant \frac{c}{\eta}\int_0^{\eta} |\psi_3(x)|\,dx,$$

and

$$\frac{1}{\eta}\int_0^{\eta} |\psi_3(x)|\,dx \in \mathscr{L}_2(0, 1),$$

it follows that the sequence

$$\left\{\int_0^{\xi} \Psi(\xi_n, \eta, x)\,\psi_3(x)\,dx\right\}$$

is convergent in $\mathfrak{L}_2(0, 1)$.

Thus the sequence $B_{0n}\lambda$ is convergent on the set of continuously differentiable functions $\lambda(\eta)$ on $(0, 1)$ that vanish in a neighborhood of $\eta = 0$. But this set is dense in $\mathfrak{L}_2(0, 1)$, and the norms of the operators B_{0n} are uniformly bounded. Consequently, the sequence B_{0n} is strongly convergent in $\mathfrak{L}_2(0, 1)$.

Now consider $I_n\lambda$ (see (3.91)):

$$2\pi\gamma_2 I_n\lambda = [\eta - \gamma_n(\eta)]^{1/6}\frac{dI_n}{d\eta}$$
$$= (\eta - \xi_n)^{1/6}\left[\int_{\xi_n}^{\eta} \tilde{g}(\xi_n, \eta, t)\,\tau(t)\,dt - \int_{\xi_n}^{\eta} \tilde{h}(\xi_n, \eta, t)\,\nu(t)\,dt\right],$$

where

$$\tilde{g}(\xi_n, \eta, t) = \frac{\partial g_0}{\partial \xi_n}\frac{d\xi_n}{d\eta} + \frac{\partial g_0}{\partial \eta}, \qquad \tilde{h}(\xi_n, \eta, t) = \frac{\partial h_0}{\partial \xi_n}\frac{d\xi_n}{d\eta} + \frac{\partial h_0}{\partial \eta}.$$

For $m > n$,

$$2\pi\gamma_2 |I_n\lambda - I_m\lambda| \leqslant \int_{\xi_n}^{\xi_m} |\tilde{g}(\xi_n, \eta, t)|\,|\tau(t)|\,dt$$
$$+ \int_{\xi_m}^{\eta} |\tilde{g}(\xi_n, \eta, t) - \tilde{g}(\xi_m, \eta, t)|\,|\tau(t)|\,dt + \int_{\xi_n}^{\xi_m} |\tilde{h}(\xi_n, \eta, t)|\,|\nu(t)|\,dt$$
$$+ \int_{\xi_m}^{\eta} |\tilde{h}(\xi_n, \eta, t) - \tilde{h}(\xi_m, \eta, t)|\,|\nu(t)|\,dt.$$

Hence, after simple maniuplations, we get

$$2\pi\gamma_2 \, | I_n\lambda - I_m\lambda |$$

$$\leqslant \max_{0\leqslant t\leqslant \eta} |\tau(t)| \left[\int_{\xi_n}^{\xi_m} |\tilde{g}(\xi_n, \eta, t)| \, dt + \int_{\xi_m}^{\eta} |\tilde{g}(\xi_n, \eta, t) - \tilde{g}(\xi_m, \eta, t)| \, dt \right]$$

$$+ \left(\int_0^{\eta} \nu^2(t) \, dt \right)^{1/2} \left[\left(\int_{\xi_n}^{\xi_m} \tilde{h}^2(\xi_n, \eta, t) \, dt \right)^{1/2} \right.$$

$$\left. + \left(\int_{\xi_m}^{\eta} |\tilde{h}(\xi_n, \eta, t) - \tilde{h}(\xi_m, \eta, t)|^2 \, dt \right)^{1/2} \right].$$

The integrals depending on $\tilde{g}$ and $\tilde{h}$ can be made arbitrarily small by taking n sufficiently large, as the integrands are continuous functions of ξ, η and t ($\xi < t < \eta$) and estimates (3.92) are true. Thus

$$2\pi\gamma_2 | I_n\lambda - I_m\lambda | \leqslant \omega_n \left[\max_{0\leqslant t\leqslant \eta} |\tau(t)| + \left(\int_0^{\eta} \nu^2(t) \, dt \right)^{1/2} \right].$$

Hence

$$2\pi\gamma_2 (1-\eta)^{1/2} | I_n\lambda - I_m\lambda |$$

$$\leqslant \omega_n \left[\max_{0\leqslant t\leqslant \eta} |(1-t)^{1/2} \tau(t)| + \left(\int_0^{\eta} (1-t) \nu^2(t) \, dt \right)^{1/2} \right]$$

$$\leqslant C\omega_n \left[\left(\int_0^1 (1-t) \lambda^2(t) \, dt \right)^{1/2} + \left(\int_0^1 (1-t) \mu^2(t) \, dt \right)^{1/2} \right]$$

$$\leqslant C_1 \omega_n \left(\int_0^1 (1-t) \lambda^2(t) \, dt \right)^{1/2}.$$

Consequently, the sequence I_n is convergent in norm in $\mathfrak{L}_2(0, 1)$.

We proved above that the sequence B_{0n} is strongly convergent in $\mathfrak{L}_2(0, 1)$. Thus, the sequence $B_n = B_{0n} + I_n$ converges strongly in $\mathfrak{L}_2(0, 1)$ to the operator B.

To complete the proof of the existence theorem, we show that the sequence of adjoints B_n^* is strongly convergent in $\mathfrak{L}_2(0, 1)$. The adjoint is defined, as usual, by

$$\int_0^1 (1-\eta) A\lambda \cdot g(\eta) \, d\eta = \int_0^1 (1-\eta) \lambda(\eta) A^* g \, d\eta.$$

We have $B_n = B_{0n} + I_n$ and $B_n^* = B_{0n}^* + I_n^*$. Since the sequence I_n converges in norm, so does I_n^*. Thus, we need only show that the sequence B_{0n}^* is strongly convergent in $\mathfrak{L}_2(0, 1)$.

This will be done by determining the adjoint of each term in (3.96) and then establishing the convergence of each. The procedure is elementary, and we shall give the details only for a few terms.

Consider the term $f(\xi_n)d\xi_n/d\eta$. Define operators C_n and H by

$$C_n\varphi = \varphi(\xi_n)\frac{d\xi_n}{d\eta},$$

$$H\lambda = \gamma_2\left[\lambda(x) + \frac{x^{-5/6}}{36}\int_0^x \frac{\lambda(t)}{t^{1/6}} F\left(\frac{5}{6}, \frac{5}{6}, 2; 1 - \frac{t}{x}\right)dt\right] = f(x).$$

Then

$$f(\xi_n)\frac{d\xi_n}{d\eta} = C_nH\lambda, \qquad (C_nH)^* = H^*C_n^*.$$

Since H^* is a bounded operator, it suffices to establish convergence for the operators C_n^*. We have

$$\int_0^1 (1-\eta)\varphi(\xi_n)\frac{d\xi_n}{d\eta}g(\eta)\,d\eta = \int_0^{\xi_n(1)} (1-\xi_n)\varphi(\xi_n)\frac{1-\zeta_n(\xi_n)}{1-\xi_n}g(\zeta_n(\xi_n))\,d\xi_n,$$

where $\eta = \zeta_n(\xi_n)$ is the inverse function of $\gamma_n(\eta) = \xi_n$. It is evident from the above expression that the sequence C_n^*g will converge if g is continuous. But the norms of the operators C_n are uniformly bounded and the set of continuous functions is dense in $\mathfrak{L}_2(0, 1)$; hence the operators C_n^* are strongly convergent in $\mathfrak{L}_2(0, 1)$. Therefore the sequence $(C_nH)^*$ is strongly convergent in $\mathfrak{L}_2(0, 1)$.

The treatment of $\psi_3(\xi_n)d\xi_n/d\eta$ is similar.

Consider the operator

$$\frac{d\xi_n}{d\eta}\int_0^\eta \frac{f(x)\,dx}{x-\xi_n}.$$

Let

$$F_n\varphi = \frac{d\xi_n}{d\eta}\int_0^\eta \frac{\varphi(x)\,dx}{x-\xi_n}.$$

Then

$$\frac{d\xi_n}{d\eta}\int_0^\eta \frac{f(x)\,dx}{x-\xi_n} = F_nH\lambda.$$

It will thus suffice to prove convergence for the operator F_n^*. But

$$F_n^* g = \frac{1}{1-x} \int_x^1 \frac{(1-\eta)\, g(\eta)}{x - \xi_n(\eta)} \frac{d\xi_n}{d\eta}\, d\eta = g(x) \ln | x - \xi_n(x) |$$

$$+ \frac{1}{1-x} \int_x^1 \ln | x - \xi_n(\eta) | \frac{d}{d\eta} [(1-\eta)\, g]\, d\eta$$

and the sequence $F_n^* g$ is uniformly convergent on the interval $0 \leqslant x \leqslant 1$, if $g(\eta)$ is continuously differentiable and vanishes together with its derivative in a small neighborhood of $\eta = 0$. The set of all such functions is dense in $\mathfrak{L}_2(0, 1)$, and the norms of the operators F_n^* are uniformly bounded; hence the sequence F_n^* is strongly convergent in $\mathfrak{L}_2(0, 1)$.

The remaining terms in (3.96) are examined in the same way.

Thus, the sequence of adjoints B_n^* converges strongly in $\mathfrak{L}_2(0, 1)$ to the operator B^*.

We now prove that the equation

$$B_n \lambda = f \tag{3.97}$$

is solvable for any $f \in \mathfrak{L}_2(0, 1)$.

Let B_n be the sequence of operators constructed above. By virtue of the assertion proved in the preceding subsection, the equation $B_n \lambda = f$ is solvable for any $f \in \mathfrak{L}_2(0, 1)$, and there exists a bounded inverse B_n^{-1}. Its norm satisfies the inequality $\|B_n^{-1}\| \leqslant m_n^{-1}$, where m_n is defined by the left inequality of (3.67). The sequence $\{\gamma_n(\eta)\}$ is such that

$$\frac{d\gamma_n}{d\eta} \leqslant 1 - \delta \qquad (n = 1, 2, 3, \ldots).$$

Since the constant m_n in (3.67) depends only on δ, it follows that all the constants m_n are bounded away from zero; i.e. $m_n \geqslant m' > 0$. But

$$(B_n^*)^{-1} = (B_n^{-1})^*, \quad \|(B_n^{-1})^*\| = \|B_n^{-1}\| \leqslant m_n^{-1}$$

and so $\|B_n^* \lambda\| \geqslant m_n \|\lambda\| \geqslant m' \|\lambda\|$. Letting $n \longrightarrow \infty$, we get

$$\| B^* \lambda \| \geqslant m' \| \lambda \|.$$

Thus the equation $B^* \lambda = 0$ has no nontrivial solutions in $\mathfrak{L}_2(0, 1)$. But then, by a standard theorem of functional analysis, equation (3.97) is solvable for any $f \in \mathfrak{L}_2(0, 1)$.

We have thus proved

LEMMA 4 [4a]. *Let Γ satisfy the condition*

$$(1 - \theta) \frac{d\sigma}{ds} + \sigma \frac{d\theta}{ds} \leqslant 0, \tag{3.98}$$

and let conditions (3.2) *hold along* γ. *Then equation* (3.12), *i.e.*

$$yz_{xx} + z_{xy} + c(y)z = 0,$$

has a solution $z \in \mathfrak{R}$ *in D, satisfying the boundary conditions* $z|_\Gamma = 0$ *and* $z|_\gamma = \Phi(\eta)$, *if*

$$\int_0^1 (1-\eta)\eta^{1/3}\Phi'^2(\eta)\,d\eta < \infty. \tag{3.99}$$

We shall now eliminate the boundary condition $z|_\Gamma = 0$. Let $z|_\Gamma = \varphi(s)$, $0 \leqslant s \leqslant l$, where $\varphi(s)$ satisfies a Hölder condition with exponent $\alpha > 0$. Without loss of generality, we may assume that $\varphi(0) = 0$.

The function (see (4.16) of Chapter II)

$$z_1(x, y) = -\int_0^l \varphi(s)\tilde{\rho}(s, x, y)\,ds \tag{3.100}$$

is a solution of (3.12), continuous in $\overline{D}^+$, satisfying the boundary conditions

$$z_1|_\Gamma = \varphi(s), \qquad \lim_{y\to 0}\frac{\partial z_1}{\partial y} = 0 \qquad (0 < x < 1).$$

On the parabolic curve, we have

$$z_1(x, 0) = -\int_0^l \varphi(s)\tilde{\rho}(s, x, 0)\,ds.$$

Solving the Cauchy problem for (3.12), with initial data

$$z_1(x, 0) = \tau_1(x), \qquad \lim_{y\to 0}\frac{\partial z_1}{\partial y} = 0,$$

we obtain via (5.38) of Chapter II a solution $z_1(x, y)$ of (3.12) in the domain D^-. It remains to show that the values of $z_1(x, y)$ on the curve γ satisfy (3.99). Proceeding in the same way as in the proof of the right inequality (3.67), and noting that $\nu_1(x) \equiv 0$, we obtain

$$\int_0^1 \eta^{1/3}(1-\eta)\Phi_1'^2(\eta)\,d\eta < C\int_0^1 (1-x)\lambda_1^2(x)\,dx,$$

where $\Phi_1(\eta) = z_1|_\gamma$. By (3.12) we have

$$x^{-1/6}\lambda_1(x) = \frac{1}{2\pi}\frac{d}{dx}\int_0^x (x-t)^{-1/6}t^{-1/6}\varphi_1(t)\,dt. \tag{3.101}$$

As shown above (see (5.15) of Chapter II),

$$\varphi_1(x)=\frac{\sqrt{3}}{\pi}\frac{d}{dx}\int_0^x (x-t)^{-2/3}\tau_1(t)\,dt=\frac{\sqrt{3}}{2\pi}\int_0^x (x-t)^{-2/3}\tau_1'(t)\,dt.$$

Inserting this into (3.101), we get

$$\lambda_1(x)=x^{1/6}\frac{\sqrt{3}}{4\pi^2}\frac{d}{dx}\int_0^x (x-t)^{-1/6}t^{-1/6}\,dt\int_0^t (t-y)^{-2/3}\tau_1'(y)\,dy$$

$$=\frac{\sqrt{3}}{4\pi^2}\frac{\Gamma\left(\frac{5}{6}\right)\Gamma\left(\frac{1}{3}\right)}{\Gamma\left(\frac{1}{6}\right)}\int_0^x\left(\frac{y}{x}\right)^{1/6}(x-y)^{-5/6}\tau_1'(y)\,dy,$$

but, by Lemma 7 of Chapter II, §4, it follows from (3.100) that

$$|\tau_1'(x)|<C_1\left(x^{-1/9}+(1-x)^{-1/3}\right).$$

Thus, $\lambda_1(x)\in\mathfrak{L}_2(0,1)$, and so

$$\int_0^1 \eta^{1/3}(1-\eta)\,\Phi_1'^2(\eta)\,d\eta<\infty.$$

Our arguments imply that the generalized Tricomi problem for equation (3.12) has a solution $z=z_1+z_2$, where $z_2(x,y)$ is a solution of the generalized Tricomi problem with boundary conditions $z_2|_\Gamma=0$ and $z_2|_\gamma=\Phi(\eta)-\Phi_1(\eta)$, and $z_1(x,y)$ is the solution of (3.12) constructed above.

THEOREM 3.3 [4a]. *Assume that the following conditions hold*:

1) *Condition* (3.98) *holds along the curve* Γ, *and conditions* (3.2) *along* γ.

2) *The function* $\varphi(s)$ *satisfies the conditions of Lemma* 7 *of Chapter* II, §4.

3) $\int_0^1(1-\eta)\eta^{1/3}\psi'^2(\eta)d\eta<\infty$.

Then equation (3.12) *has a solution* $z\in\mathfrak{R}$ *in the domain* D_γ, *satisfying the boundary conditions* $z|_\Gamma=\varphi(s)$ *and* $z|_\gamma=\psi(\eta)$.

CHAPTER IV

THE FRANKL′ PROBLEM

§1. Statement of the problem

Consider the Čaplygin equation

$$K(y)\,u_{xx}+u_{yy}=0, \tag{1.1}$$

where $K(y)$ is a continuously differentiable function, $K(y)>0$ if $y>0$, $K(y)<0$ if $y<0$, and $K(-y)=-K(y)$.

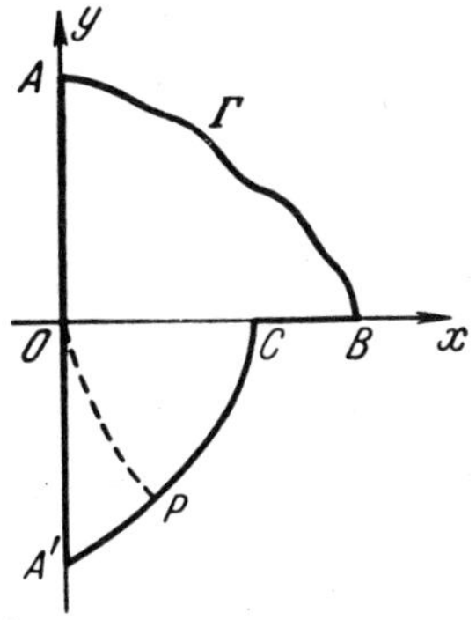

FIGURE 9.

Let D be the domain bounded by an interval $A'A$ of the y-axis, $-1\leqslant y\leqslant 1$, a smooth curve Γ in the upper half-plane with endpoints A (0, 1) and $B(a, 0)$, a characteristic arc $A'C$ of equation (1.1), A' (0, 1), $C(a_1, 0)$, where $a_1\leqslant a$, and the interval CB of the x-axis, $a_1\leqslant x\leqslant a$ (Figure 9).

In 1956, Frankl′ [22e] suggested the following problem:

Find a solution $u(x, y)$ of equation (1.1) *in the domain D, continuous in $\bar{D}$ and satisfying the boundary conditions*

$$u\,|_{\Gamma}=\psi_1(s),\quad 0\leqslant s\leqslant l, \tag{1.2}$$

$$u\,|_{CB}=\psi_2(x),\qquad a_1\leqslant x\leqslant a, \tag{1.3}$$

$$\left.\frac{\partial u}{\partial x}\right|_{A'A} = 0, \tag{1.4}$$

$$u(0, y) - u(0, -y) = f(y), \quad -1 \leqslant y \leqslant 1, \tag{1.5}$$

where ψ_1 *and* ψ_2 *are prescribed functions satisfying a Hölder condition,* $\psi_1(0) = \psi_2(a)$, *and* $f(y)$ *has a first derivative satisfying a Hölder condition on* $(-1, +1)$.

§2. Uniqueness of the solution of the Frankl′ problem

1. *Method of auxiliary functions.*(1) Let us assume that, in addition to the usual smoothness condition (Ljapunov condition), the curve Γ satisfies the following conditions:

a) In a neighborhood of the point $A(0, 1)$,

$$\frac{dy}{dx} \geqslant 0, \tag{2.1}$$

where $x = x(s)$, $y = y(s)$ are parametric equations of Γ and s is arc-length reckoned from the point $B\ (a, 0)$.

b) If θ is the angle between the positive x-axis and the tangent to Γ in the direction of increasing arc-length s, then

$$0 \leqslant \theta \leqslant 2\pi. \tag{$*$}$$

We shall prove that the homogeneous Frankl′ problem ($\psi_1 = \psi_2 = f \equiv 0$) has no nontrivial solutions.

Let $v(x, y)$ denote a function which, together with a solution $u(x, y)$ of the Frankl′ problem, satisfies the system of equations

$$K(y)u_x - v_y = 0, \qquad u_y + v_x = 0. \tag{2.2}$$

If $v(0, 0) = 0$, the function $v(x, y)$ is uniquely determined by $u(x, y)$ and equations (2.2). By (1.4),

$$v(0, y) = 0 \quad \text{for} \quad -1 \leqslant y \leqslant 1. \tag{2.3}$$

We now stipulate that $u(x, y)$ also satisfies conditions guaranteeing the continuity of $v(x, y)$ in the closed domain $\bar{D}$.

Following Morawetz [56b], we define a function in the domain D by

$$F(x, y) = \int_{(0,0)}^{(x,y)} 2uv\, dx + (Ku^2 - v^2)\, dy. \tag{2.4}$$

By (2.2), the integral (2.4) is obviously independent of the integration path, provided that the path lies entirely in D.

---

(1) See [16b].

Let D^+ be the region of D in which $y > 0$, and D^- the region of D in which $y < 0$. In D^+, the function $F(x, y)$ satisfies

$$K(y) F_{xx} + F_{yy} = K'(y) u^2.$$

Using (2.4) to express u^2 in terms of the derivatives of $F(x, y)$, we get

$$K(y) F_{xx} + F_{yy} = \alpha F_x + \beta F_y, \tag{2.5}$$

where

$$\alpha = \frac{K'(y) F_x}{2\sqrt{KF_x^2 + F_y^2}}, \qquad \beta = \frac{K'(y)}{2K}\left(1 + \frac{F_y}{\sqrt{KF_x^2 + F_y^2}}\right).$$

The coefficients α and β are bounded in any closed subdomain of D^+, and equation (2.5) is elliptic in D^+. Consequently, $F(x, y)$ satisfies the maximum principle in the domain D^+, i.e. it achieves its maximum on the boundary of D^+, and at the point where this maximum is achieved

$$\frac{\partial F}{\partial n} < 0,\text{(2)} \tag{2.6}$$

where n is the interior normal to the boundary of D^+.

$F(x, y)$ does not increase as a function of y in D^-, since

$$F_y = Ku^2 - v^2 \leqslant 0, \tag{2.7}$$

and so the maximum of this function is attained on the boundary of D.

The function $F(x, y)$ is constant on the segment CB of the boundary of D, since condition (1.3) ($\psi_2 \equiv 0$) gives $F_x = 2uv = 0$. On the characteristic arc $A'C$ we have $dx/ds = \sqrt{-K}dy/ds$, and therefore

$$\left.\frac{\partial F}{\partial s}\right|_{A'C} = -\left(\sqrt{-K}\, u - v\right)^2 \frac{dy}{ds}, \tag{2.8}$$

while $F_y = Ku^2 \leqslant 0$ on OA'.

Consequently, the only point in the hyperbolic region where $F(x, y)$ may attain a maximum is A' $(0, -1)$. But $F(0, -1) = F(0, 1)$. Indeed, by the properties of $K(y)$ and by conditions (1.4) and (1.5) ($f \equiv 0$), we have

$$F(0, -1) = \int_0^{-1} K(y) u^2(0, y)\, dy$$

$$= -\int_0^{-1} K(-y) u^2(0, -y)\, dy = \int_0^1 K(y) u^2(0, y)\, dy = F(0, 1).$$

---

(2) This assertion remains valid if the equation is degenerate on the boundary of the domain, provided the direction of the boundary is not characteristic.

On the curve Γ,

$$\frac{\partial F}{\partial s} = -v^2 \frac{dy}{ds},$$

since $u|_\Gamma = 0$. By (2.1), $F(x, y)$ is not an increasing function of s on Γ in the neighborhood of A, and so the maximum of $F(x, y)$ must be achieved at an interior point of Γ. At this point, $v^2 dy/ds = 0$. If $v = 0$, then

$$\frac{dF}{dn} = -v^2 \cos(ny) = 0,$$

at this point, contradicting (2.6). If $dy/ds = 0$, then, by $(*)$, we have $dF/dn = v^2 \geqslant 0$, again contradicting (2.6). Thus $F(x, y)$ must be a constant in D^+, and since $F(0, 0) = 0$, it follows that $F(x, y) \equiv 0$ in D^+. By (2.7) and (2.8), we have in D^-

$$F(x, 0) \leqslant F(x, y) \leqslant F(0, -1), \qquad 0 \leqslant x \leqslant a_1,$$

but

$$F(x, 0) = 0, \qquad F(0, -1) = F(0, 1) = 0,$$

so that $F(x, y) \equiv 0$ in D^-. Therefore, $F(x, y) \equiv 0$ in D, and then it follows from (2.4) that $u = v \equiv 0$ in D, proving uniqueness for the Frankl′ problem (1.1)–(1.5).

2. *The a, b, c-method.*(3) We now use the a, b, c-method to prove the uniqueness theorem for quasi-regular solutions of the Frankl′ problem with $K(y) + K(-y) \leqslant 0$. Note that the derivatives u_x and u_y of a solution of the Frankl′ problem may be discontinuous on the characteristic OP (see Figure 9). It is sufficient to assume that u_x and u_y are left and right continuous on OP. Then, if $u(x, y)$ is continuous on OP, it follows that the derivative $(-u_x\sqrt{-K} + u_y)$ is continuous on OP.

Define the function $v(x, y)$ adjoint to $u(x, y)$ by

$$v(x, y) = \int_{(0,\,0)}^{(x,\,y)} u_y\,dx - Ku_x\,dy. \tag{2.9}$$

We first define this integral separately in the domains $A'OPA'$ and $OPCBAO$. It is evident from (1.1) that the integrals in both domains are independent of the integration path. On the characteristic OP we have $dx = \sqrt{-K}dy$, and so

$$dv = u_y\,dx - Ku_x\,dy = \sqrt{-K}(u_y\,dy + u_x\,dx) = \sqrt{-K}\,du.$$

Since $u(x, y)$ is continuous on OP, it follows that

(3) See [51].

$$d(v^+ - v^-) = \sqrt{-K}\, d(u^+ - u^-) = 0.$$

Hence $v^+ - v^- = \text{const} = 0$, since $v^+ = v^-$ at $(0, 0)$. Thus $v(x, y)$ is also continuous in $\bar{D}$.

The functions u and v must satisfy the system of equations

$$K(y)\, u_x + v_y = 0, \qquad v_x - u_y = 0. \tag{2.10}$$

Let $b(x, y)$ and $c(x, y)$ be arbitrary piecewise-smooth functions in $\bar{D}$. Obviously,

$$\iint_D [(cKu - bv)(v_x - u_y) + (bu + cv)(Ku_x + v_y)]\, dx\, dy = 0.$$

Using Green's formula, we get

$$\begin{gathered}\iint_D [((cK)_y - Kb_x)\, u^2 - 2(Kc_x + b_y)\, uv + (b_x - c_y)\, v^2]\, dx\, dy \\ + \int_S (bKu^2 + 2cKuv - bv^2)\, dy + (cKu^2 - 2buv - cv^2)\, dx = 0,\end{gathered} \tag{2.11}$$

where S is the boundary of D.

It follows from (1.4), (2.10) and (2.9) that $v_y|_{AA'} = 0$, and thus $v(0, y) = 0$. The boundary conditions (1.2), (1.3) and (1.5) (where $\psi_1 = \psi_2 = f \equiv 0$) give us

$$u|_\Gamma = 0, \quad u|_{CB} = 0, \quad u(0, y) = u(0, -y).$$

Then we can write (2.11) as

$$\begin{gathered}\iint_D [((cK)_y - Kb_x)\, u^2 - 2(Kc_x + b_y)\, uv + (b_x - c_y)\, v^2]\, dx\, dy \\ + \int_\Gamma - v^2(c\, dx + b\, dy) + \int_{AA'} bKu^2\, dy \\ - \int_{A'C} (\sqrt{-K}\, u + v)^2 (c\, dx + b\, dy) + \int_{CB} - cv^2\, dx = 0.\end{gathered} \tag{2.12}$$

We now stipulate certain conditions on the functions b and c, guaranteeing that all the integrals in (2.12) are nonnegative. It is sufficient to assume that

$$\left.\begin{aligned} (cK)_y - Kb_x &\geqslant 0, \\ b_x - c_y &\geqslant 0, \quad \text{in } D, \\ (Kc_x + b_y)^2 &\leqslant (b_x - c_y)(-Kb_x + (cK)_y), \end{aligned}\right\} \tag{2.13}$$

$$c\frac{dx}{ds} + b\frac{dy}{ds} \leqslant 0 \quad \text{on} \quad \Gamma, \tag{2.14}$$

$$\int_0^1 [b(0, y)K(y) + b(0, -y)K(-y)]\, u^2(0, y)\, dy \leqslant 0 \tag{2.15}$$

or

$$b(0, y)K(y) + b(0, -y)K(-y) \leqslant 0, \tag{2.15'}$$

$$b + c\sqrt{-K} \leqslant 0 \quad \text{on} \quad A'C, \tag{2.16}$$

$$c(x, 0) \leqslant 0 \quad \text{on} \quad CB. \tag{2.17}$$

For example, for $y \geqslant 0$, put $b = x - m$ and $c = y$, where m is a constant and $m > \max_D x$, and for $y \leqslant 0$ put $b = x - m$ and $c = 0$. Then (2.13), (2.16) and (2.17) are clearly satisfied, and (2.14) and (2.15) become

$$(x - m)\frac{dy}{ds} + y\frac{dx}{ds} \leqslant 0 \quad \text{on} \quad \Gamma \tag{2.14'}$$

and

$$K(y) + K(-y) \leqslant 0, \qquad 0 \leqslant y \leqslant 1. \tag{2.15*}$$

Since the integrals in (2.12) are nonnegative and their sum is zero, it follows that each of them vanishes. Hence, since the first integral in (2.12) vanishes, it follows that $u(x, y) \equiv 0$ in D.

We have thus proved the uniqueness of a quasi-regular solution to the Frankl′ problem, provided that conditions (2.14) and (2.15) are satisfied.

§3. Existence of the solution of the Frankl′ problem

We shall prove the existence of a solution of the Frankl′ problem in the case $K(y) = \operatorname{sgn} y|y|^m$ (see [16b]).

In this case equation (1.1) is

$$\operatorname{sgn} y\,|y|^m u_{xx} + u_{yy} = 0. \tag{3.1}$$

Here $A'C$ is a segment of the characteristic

$$x + \frac{2}{m+2}(-y)^{\frac{m+2}{2}} = \frac{2}{m+2}.$$

We claim that there exists a function $u(x, y)$ with the following properties:

1) It is a regular solution of equation (3.1) in D.

2) It is continuous in the closed domain $\bar{D}$.

3) Its partial derivatives u_x and u_y are continuous in $\bar{D}$, with the possible exception of the points $O(0, 0)$, $A(0, 1)$, $A'(0, -1)$, $C(2/(m+2), 0)$ and $B(a, 0)$, where they may have poles of order less than unity.

4) It satisfies the boundary conditions (1.2)–(1.5).

Let D^* denote the domain symmetric to D with respect to the y-axis, and

$D_1 = D + D^* + A'A$. The domain D_1^* is bounded by $\Gamma + \Gamma^*$ and the segment $[-a, a]$ of the x-axis, where Γ^* is the curve symmetric to Γ with respect to the y-axis.

Define the function $\psi_1^*(s)$ on Γ^* to be equal to $\psi_1(l - s)$, where l is the length of $\Gamma + \Gamma^*$.

We wish to find a solution of (3.1) in D_1, satisfying the boundary conditions

$$u|_\Gamma = \psi_1(s), \qquad u|_{\Gamma^*} = \psi_1^*(s), \tag{3.2}$$

$$u(x, 0) = \psi_2(x), \qquad \frac{2}{m+2} < x < a,$$
$$u(x, 0) = \psi_2(-x), \qquad -a < x < -\frac{2}{m+2} \tag{3.3}$$

and condition (1.5); we shall assume that the functions $\tau(x) = u(x, 0)$ and $\nu(x) = \partial u(x, 0)/\partial y$ are even for $-2/(m + 2) < x < 2/(m + 2)$. Then condition (1.4) is satisfied. Obviously, any solution of this auxiliary problem for $x \geqslant 0$ is a solution of the Frankl′ problem.

We shall assume that the curve $\Gamma + \Gamma^*$ satisfies the conditions of the uniqueness theorem, and that the coordinates of its points, as functions of s, have continuous second derivatives which satisfy a Hölder condition on $[0, 1]$. Moreover, in the neighborhood of the point B on Γ we demand that

$$\left|\frac{dx}{ds}\right| \leqslant C y^{m+1}(s).$$

Without loss of generality, we may assume that $\psi_1 = \psi_2 \equiv 0$. In fact, let $w(x, y)$ be a solution of (3.1) for $y > 0$, satisfying the boundary conditions

$$w|_\Gamma = \psi_1(s), \qquad w|_{\Gamma^*} = \psi_1^*(s),$$
$$w(x, 0) = 0, \qquad -\frac{2}{m+2} < x < \frac{2}{m+2},$$
$$w(x, 0) = \psi_2(x), \qquad \frac{2}{m+2} < x < a, \qquad w(x, 0) = \psi_2(-x),$$
$$-a < x < -\frac{2}{m+2},$$
$$\psi_2\left(\frac{2}{m+2}\right) = 0, \qquad \psi_1(0) = \psi_2(a).$$

Let $w(x, y)$ be a solution of the Cauchy problem for equation (3.1) in the hyperbolic region of D_1 with data $w(x, 0)$ and $\partial w(x, y)/dy$. Then the condition $w_x(0, y) = 0$ will be satisfied on $A'A$. The difference $w(0, y) - w(0, -y) = f_1(y)$ is obviously odd. Thus the function $u(x, y) - w(x, y)$ is a solution of the Frankl′ problem satisfying conditions (1.2)–(1.5) with $\psi_2 = \psi_1 \equiv 0$ and $f(y)$ replaced by $f(y) - f_1(y)$.

The solution of equation (3.1) in the hyperbolic region of D_1, satisfying the conditions $u(x, 0) = \tau(x)$ and $\partial u(x, 0)/\partial y = \nu(x)$, is given by (see [66g])

$$u(x, y) = \gamma_1 \int_0^1 \tau\left[x + \frac{2}{m+2}(-y)^{\frac{m+2}{2}}(2t-1)\right] t^{\beta-1}(1-t)^{\beta-1}\,dt$$
$$+\gamma_2 y \int_0^1 \nu\left[x + \frac{2}{m+2}(-y)^{\frac{m+2}{2}}(2t-1)\right] t^{-\beta}(1-t)^{-\beta}\,dt, \tag{3.4}$$

where

$$\gamma_1 = \frac{\Gamma(2\beta)}{\Gamma^2(\beta)}, \quad \gamma_2 = \frac{\Gamma(2-2\beta)}{\Gamma^2(1-\beta)}, \quad \beta = \frac{m}{2(m+2)}.$$

We now put $x = 0$ in (3.4) and, using (1.5), solve the resulting equation for $\tau(x)$. To this end, we replace $(-y)$ by $(y/(1-2\beta))^{1-2\beta}$, multiply both sides by $y^{2\beta}/(x^2 - y^2)^\beta$, integrate with respect to y from 0 to x, and finally differentiate with respect to x. A few manipulations then give

$$2^{1-2\beta}\gamma_1 \frac{\pi}{\sin\beta\pi}\tau(x) = (1-2\beta)^{2\beta}\int_0^x \frac{\nu(t)\,dt}{(x-t)^{2\beta}}$$
$$+\frac{d}{dx}\int_0^x \frac{y^{2\beta}\varphi\left[\left(\frac{y}{1-2\beta}\right)^{1-2\beta}\right]dy}{(x^2-y^2)^\beta} - \frac{d}{dx}\int_0^x \frac{y^{2\beta} f\left[\left(\frac{y}{1-2\beta}\right)^{1-2\beta}\right]}{(x^2-y^2)^\beta}\,dy, \tag{3.5}$$

where $\varphi(y) = u(0, y)$, $0 \leqslant y \leqslant 1$.

The solution of the Dirichlet problem for equation (3.1) in the domain D_1^+ with boundary conditions (3.2), (3.3) ($\psi_1 = \psi_2 = 0$) and

$$u(x, 0) = \tau(x), \qquad -\frac{2}{m+2} < x < \frac{2}{m+2}, \tag{3.6}$$

is given by (see [66g])

$$u(x, y) = \int_{-(1-2\beta)}^{1-2\beta} \tau(\xi)\left[\frac{\partial G_0(\xi, 0; x, y)}{\partial\eta} + \frac{\partial H(\xi, 0; x, y)}{\partial\eta}\right]d\xi. \tag{3.7}$$

Here $G_0(\xi, \eta; x, y)$ is the Green's function of the Dirichlet problem in the normal domain D_0:

$$G_0(\xi, \eta; x, y) = q_2(\xi, \eta; x, y) - \left(\frac{a}{R}\right)^{2\beta} q_2(\xi, \eta; \bar{x}, \bar{y}), \tag{3.8}$$

where

$$q_2(\xi, \eta; x, y)$$
$$= k_2\left(\frac{4}{m+2}\right)^{4\beta-2} r_1^{-2\beta}(1-\sigma)^{1-2\beta} F(1-\beta, 1-\beta, 2-2\beta; 1-\sigma) \tag{3.9}$$

is a fundamental solution of (3.1). We have used the notation

$$\left.\begin{matrix} r^2 \\ r_1^2 \end{matrix}\right\} = (x-\xi)^2 + \frac{4}{(m+2)^2}\left(y^{\frac{m+2}{2}} \mp \eta^{\frac{m+2}{2}}\right)^2, \quad \sigma = \frac{r^2}{r_1^2},$$

$$k_2 = \frac{1}{4\pi}\left(\frac{4}{m+2}\right)^{2-2\beta} \frac{\Gamma^2(1-\beta)}{\Gamma(2-2\beta)}, \qquad R^2 = x^2 + \frac{4}{(m+2)^2} y^{m+2},$$

$$\bar{x} = \frac{a^2}{R^2} x, \qquad \bar{y} = \left(\frac{a^2}{R^2}\right)^{\frac{2}{m+2}} y.$$

The function $H(\xi, \eta; x, y)$ is defined by

$$H(\xi, \eta; x, y) = \int_0^l \rho(s; \xi, \eta)\, G_0(x(s), y(s); x, y)\, ds, \tag{3.10}$$

where $\rho(s; \xi, \eta)$ is a solution of the integral equation

$$\rho(s; \xi, \eta) - 2\int_0^l K_2(t, s)\rho(t; \xi, \eta)\, dt = 2A_s[q_2(x(s), y(s); \xi, \eta)],$$

$$A_s[q_2] \equiv y^m \frac{dy}{ds}\frac{\partial q_2}{\partial x} - \frac{dx}{ds}\frac{\partial q_2}{\partial y},$$

$$K_2(t, s) = A_s[q_2(\xi(t), \eta(t); x(s), y(s))].$$

Since $\tau(x)$ is an even function, we can determine $\nu(x)$ from (3.7):

$$\nu(x) = \nu_0(x) + \int_0^{1-2\beta} \tau(\xi)\left[\frac{\partial^2 H(-\xi, 0; x, 0)}{\partial\eta\, \partial y} + \frac{\partial^2 H(\xi, 0; x, 0)}{\partial\eta\, \partial y}\right] d\xi, \tag{3.11}$$

where

$$\nu_0(x) = \left[\frac{\partial}{\partial y}\int_{-(1-2\beta)}^{1-2\beta} \tau(\xi)\frac{\partial G_0(\xi, 0; x, y)}{\partial\eta}\, d\xi\right]_{y=0}$$

$$= \left[\frac{\partial}{\partial y}\int_{-(1-2\beta)}^{1-2\beta} k_2 y \left\{\left[(x-\xi)^2 + \frac{4}{(m+2)^2} y^{m+2}\right]^{\beta-1}\right.\right.$$

$$\left.\left. - \left[\left(a - \frac{\xi x}{a}\right)^2 + \frac{4\xi^2 y^{m+2}}{(m+2)^2 a^2}\right]^{\beta-1}\right\}\tau(\xi)\, d\xi\right]_{y=0}$$

$$= \frac{k_2}{1-2\beta}\int_0^{1-2\beta} \frac{\tau'(\xi)\, d\xi}{(x+\xi)^{1-2\beta}} - \frac{k_2}{1-2\beta}\int_0^{1-2\beta} \frac{(x-\xi)\,\tau'(\xi)\, d\xi}{|x-\xi|^{2-2\beta}}$$

$$- k_2\int_0^1 \tau(\xi)\left[\frac{1}{\left(a - \frac{x\xi}{a}\right)^{2-2\beta}} + \frac{1}{\left(a + \frac{x\xi}{a}\right)^{2-2\beta}}\right] d\xi.$$

Now insert the expressions for $\nu(x)$ and $\varphi(y) = u(0, y)$ into (3.5). Then, noting that

$$\int_0^x \frac{\nu_0(t)\,dt}{(x-t)^{2\beta}} = \frac{k_2}{1-2\beta}\left[-\pi \tan \beta\pi\, \tau(x) + \int_0^{1-2\beta} \left(\frac{x}{\xi}\right)^{1-2\beta} \frac{2\xi\tau(\xi)\,d\xi}{\xi^2 - x^2} - \int_0^{1-2\beta} \tau(\xi)\left(\frac{x}{a}\right)^{1-2\beta} \frac{2a\,d\xi}{a^2 - \frac{x^2\xi^2}{a^2}}\right],$$

and rearranging, we obtain

$$\tau(x) - 4\lambda \int_0^{1-2\beta} x^{1-2\beta}\left(\frac{\xi^{2+2\beta}}{\xi^4 - x^4} - \frac{a^{2+2\beta}}{a^4 - \frac{x^4\xi^4}{a^4}}\right)\tau(\xi)\,d\xi - 2\lambda \int_0^{1-2\beta} H_1(\xi, x)\,\tau(\xi)\,d\xi = F(x), \tag{3.12}$$

where

$$\lambda = \frac{\cos \beta\pi}{\pi(1 + \sin \beta\pi)}, \tag{3.13}$$

$$H_1(\xi, x) = \frac{(1-2\beta)^{1-2\beta}}{2k_2} \int_0^l \left[\frac{\partial\rho(s;\,\xi,0)}{\partial\eta} + \frac{\partial\rho(s;\,-\xi,0)}{\partial\eta}\right] \times \left[\frac{d}{dx}\int_0^x \frac{y^{2\beta} G_0\left(x(s),\, y(s);\, 0,\, \left(\frac{y}{1-2\beta}\right)^{1-2\beta}\right)}{(x^2-y^2)^\beta}\,dy + (1-2\beta)^{2\beta}\int_0^x \frac{\partial G_0(x(s),\, y(s);\, t,\, 0)}{\partial y}\frac{dt}{(x-t)^{2\beta}}\right] ds, \tag{3.14}$$

$$F(x) = -\frac{\lambda(1-2\beta)}{2k_2}\frac{d}{dx}\int_0^x \frac{y^{2\beta} f\left[\left(\frac{y}{1-2\beta}\right)^{1-2\beta}\right] dy}{(x^2-y^2)^\beta}. \tag{3.15}$$

It is easy to see that the function $H_1(\xi, x)$ is bounded and continuous, and $F(x)$ is continuous and satisfies a Hölder condition on $(-1, 1)$.

Replacing x by $4\alpha\sqrt[4]{x}$ and ξ by $4\alpha\sqrt[4]{\xi}$ in (3.12) and setting $\tau_1(x) = \tau(4\alpha\sqrt[4]{x})$, we finally obtain

$$x^{-\alpha}\tau_1(x) - \lambda \int_0^1 \left(\frac{1}{\xi - x} - \frac{b}{b^2 - \xi x}\right) \xi^{-\alpha}\tau_1(\xi)\, d\xi$$

$$= x^{-\alpha} F\left(4\alpha \sqrt[4]{x}\right) + \lambda \int_0^1 b\, \frac{1 - \left(\frac{b}{\xi}\right)^{\frac{1+\beta}{2}}}{b^2 - x\xi}\, \xi^{-\alpha}\tau_1(\xi)\, d\xi \tag{3.16}$$

$$+ \lambda \int_0^1 2\alpha \xi^{-3/4} \left(\frac{\xi}{x}\right)^{\alpha} H_1\left(4\alpha \sqrt[4]{\xi},\, 4\alpha \sqrt[4]{x}\right) \xi^{-\alpha}\tau_1(\xi)\, d\xi,$$

where $b = (a/(1-2\beta))^4$ and $\alpha = (1-2\beta)/4$.

Thus, the existence problem for the Frankl′ problem is equivalent to the solvability of the singular equation (3.16).

Consider the equation

$$\varphi(x) - \lambda \int_0^1 \left(\frac{1}{\xi - x} - \frac{b}{b^2 - x\xi}\right) \varphi(\xi)\, d\xi = \psi(x). \tag{3.17}$$

We shall assume that $\psi(x)$ satisfies a Hölder condition on $[0, 1]$. A solution $\varphi(x)$ of equation (3.17) will be sought in the class of functions satisfying the following conditions:

1) $\varphi(x)$ satisfies a Hölder condition on any interval $[a, b]$, $0 < a < b < 1$.

2) In the neighborhood of either endpoint of the interval $[0, 1]$, $\varphi(x)$ has the representation

$$\varphi(x) = \frac{\varphi^*(x)}{|x - c|^{\gamma}}, \qquad 0 \leqslant \gamma < 1,$$

where $\varphi^*(x)$ satisfies a Hölder condition on $[0, 1]$.

Let z be any point of the complex plane. Following Carleman [11], we set

$$\Phi(z) = \frac{1}{2\pi i} \int_0^1 \left(\frac{1}{\xi - z} - \frac{b}{b^2 - z\xi}\right) \varphi(\xi)\, d\xi. \tag{3.18}$$

$\Phi(z)$ is obviously holomorphic throughout the complex plane slit along the interval $[0, 1]$ and the ray (b^2, ∞) of the real axis.

For $0 < x < 1$ we have

$$\begin{gathered}\Phi^+(x) - \Phi^-(x) = \varphi(x),\\ \Phi^+(x) + \Phi^-(x) = \frac{1}{\pi i} \int_0^1 \left(\frac{1}{\xi - x} - \frac{b}{b^2 - x\xi}\right) \varphi(\xi)\, d\xi.\end{gathered} \tag{3.19}$$

We may thus reduce (3.17) to the following equation:

$$(1 - i\pi\lambda)\Phi^+(x) - (1 + i\pi\lambda)\Phi^-(x) = \psi(x), \quad 0 < x < 1. \tag{3.20}$$

It is easy to see from (3.18) that

$$\Phi\left(\frac{b^2}{z}\right)=\frac{z}{b}\,\Phi(z). \tag{3.21}$$

The mapping $t = b^2/z$ maps the upper half-plane onto the lower and conversely; the interval $(0, 1)$ is thereby mapped onto the ray (b^2, ∞). Replacing x by b^2/x in (3.20) and using (3.21), we obtain

$$(1+i\pi\lambda)\,\Phi^+(x)-(1-i\pi\lambda)\,\Phi^-(x)=-\frac{b}{x}\,\psi\left(\frac{b^2}{x}\right),\qquad b^2<x<\infty. \tag{3.22}$$

We now define

$$G(x)=\begin{cases}\dfrac{1+i\pi\lambda}{1-i\pi\lambda}=e^{2i\alpha\pi}, & 0<x<1,\\[2mm] \dfrac{1-i\pi\lambda}{1+i\pi\lambda}=e^{-2i\alpha\pi}, & b^2<x<\infty;\end{cases}$$

$$h(x)=\begin{cases}e^{i\pi\alpha}\cos\alpha\pi\,\psi(x) & 0<x<1,\\[2mm] -e^{-i\pi\alpha}\cos\alpha\pi\,\dfrac{b}{x}\,\psi\left(\dfrac{b^2}{x}\right), & b^2<x<\infty.\end{cases}$$

Then (3.20) and (3.22) can be combined:

$$\Phi^+(x)-G(x)\,\Phi^-(x)=h(x),\quad 0<x<1,\quad b^2<x<\infty. \tag{3.23}$$

It is readily seen that the function

$$X(z)=\frac{(1-z)^{\alpha}}{z^{\alpha}}\,(b^2-z)^{\alpha} \tag{3.24}$$

is a particular solution of the homogeneous problem

$$X^+(x)=G(x)\,X^-(x),\quad 0<x<1,\quad b^2<x<\infty. \tag{3.25}$$

By (3.25), the boundary condition (3.23) can be rewritten as

$$\frac{\Phi^+(x)}{X^+(x)}-\frac{\Phi^-(x)}{X^-(x)}=\frac{h(x)}{X^+(x)}. \tag{3.26}$$

This immediately yields a particular solution:

$$\begin{aligned}\Phi(z)&=\frac{X(z)}{2\pi i}\left[\int_0^1\frac{h(t)\,dt}{X^+(t)\,(t-z)}+\int_{b^2}^{\infty}\frac{h(t)\,dt}{X^+(t)\,(t-z)}\right]\\ &=\frac{e^{i\pi\alpha}\cos\alpha\pi\,X(z)}{2\pi i}\int_0^1\frac{\psi(t)}{X^+(t)}\left(\frac{1}{t-z}-\frac{b}{b^2-zt}\right)dt.\end{aligned} \tag{3.27}$$

To find the general solution, let us consider the homogeneous equation

$$\frac{\Phi^+(x)}{X^+(x)}-\frac{\Phi^-(x)}{X^-(x)}=0.$$

This equation shows that the function $\chi(z) = \Phi(z)/X(z)$ is holomorphic throughout the plane, with the possible exception of the points $z = 0$, $z = 1$ and $z = b^2$, which can only be poles. Assuming that the holomorphic function $\Phi(z)$ has singularities of order less than one as $z \longrightarrow 0$, $z \longrightarrow 1$ and $z \longrightarrow b^2$, we get

$$\chi(z) = \frac{C}{(1-z)(b^2-z)}.$$

Thus the general solution of the nonhomogeneous problem (3.23) is

$$\Phi(z) = \frac{e^{i\pi\alpha}\cos\alpha\pi\, X(z)}{2\pi i}\int_0^1 \frac{\psi(t)}{X^+(t)}\left(\frac{1}{t-z} - \frac{b}{b^2-zt}\right)dt + \frac{C}{z^\alpha(1-z)^{1-\alpha}(b^2-z)^{1-\alpha}}. \tag{3.28}$$

The solution of (3.17) is now obtained from the formula $\varphi(x) = \Phi^+(x) - \Phi^-(x)$. The required manipulations give

$$\varphi(x) = \frac{\psi(x)}{1+\lambda^2\pi^2} + \frac{\lambda}{1+\lambda^2\pi^2}\frac{(1-x)^\alpha(b^2-x)^\alpha}{x^\alpha} \times \int_0^1 \frac{t^\alpha\psi(t)}{(1-t)^\alpha(b^2-t)^\alpha}\left(\frac{1}{t-x} - \frac{b}{b^2-xt}\right)dt + \frac{C_1}{x^\alpha(1-x)^{1-\alpha}(b^2-x)^{1-\alpha}}. \tag{3.29}$$

Returning to (3.16), we shall solve it treating the right-hand side as a known function. Setting $x^{-\alpha}\tau_1(x) = \varphi(x)$ and using (3.29), we obtain an integral equation

$$\tau_1(x) - \lambda_1\int_0^1 K_1(\xi, x)\tau_1(\xi)\,d\xi = F_1(x)$$

or, transforming to $\tau(x)$,

$$\tau(x) - \lambda_1\int_0^{1-2\beta} K(x, \xi)\tau(\xi)\,d\xi = F_2(x), \tag{3.30}$$

where

$$K(x, \xi) = \frac{16a^4(\xi^{2+2\beta} - a^{2+2\beta})}{a^8 - x^4\xi^4} \int_0^{1-2\beta} \left[\frac{((1-2\beta)^4 - x^4)(a^4b - x^4)}{((1-2\beta)^4 - t^4)(a^4b - t^4)}\right]^\alpha$$

$$\times \left[\frac{a^8 - x^4\xi^4 - \left(\frac{x}{t}\right)^4 (a^8 - x^4\xi^4)\left(\frac{a^4b - t^4}{a^4b - x^4}\right)^\alpha}{t^4 - x^4} - \frac{a^4(a^8 - x^4\xi^4)}{a^8 - x^4t^4}\right]$$

$$\times \frac{t^{4-2\beta}\,dt}{a^8 - t^4\xi^4} + 8\int_0^{1-2\beta} \left[\frac{((1-2\beta)^4 - x^4)(a^4b - x^4)}{((1-2\beta)^4 - t^4)(a^4b - t^4)}\right]^\alpha \tag{3.31}$$

$$\times \left[\frac{H_1(\xi, t) - H_1(\xi, x)\left(\frac{a^4b - t^4}{a^4b - x^4}\right)^\alpha\left(\frac{x}{t}\right)^{4-4\alpha}}{t^4 - x^4} - a^4\frac{H_1(\xi, t)}{a^8 - x^4t^4}\right] t^3 dt,$$

$$F_2(x) = \frac{4\lambda}{1 + \lambda^2\pi^2} \int_0^{1-2\beta} \left[\frac{((1-2\beta)^4 - x^4)(a^4b - x^4)}{((1-2\beta)^4 - t^4)(a^4b - t^4)}\right]^\alpha$$

$$\times \left[\frac{F(t) - F(x)\left(\frac{a^4b - t^4}{a^4b - x^4}\right)^\alpha\left(\frac{x}{t}\right)^{4-4\alpha}}{t^4 - x^4} - \frac{a^4F(t)}{a^8 - x^4t^4}\right] t^3\,dt \tag{3.32}$$

$$+ \frac{C_2}{\left(1 - \left(\frac{x}{1-2\beta}\right)^4\right)^{1-\alpha}\left(b^2 - \left(\frac{x}{1-2\beta}\right)^4\right)^{1-\alpha}},$$

$$\lambda_1 = \frac{\lambda^2}{1 + \lambda^2\pi^2}.$$

Since we are interested in solutions of (3.30) which are continuous for $x = 1 - 2\beta$, we must set $C_2 = 0$ in (3.32). Fredholm theory may now be applied to (3.30), and so the uniqueness theorem for the Frankl′ problem implies that (3.30) is solvable. Having found $\tau(x)$ from (3.30), we now determine $\nu(x)$ from (3.11) and then, using (3.4) and (3.7), obtain the solution of the Frankl′ problem in the domains D^- and D^+, respectively.

REMARK. If $\Gamma + \Gamma^*$ is the normal curve

$$x^2 + \frac{4}{(m+2)^2}y^{m+2} = \frac{4}{(m+2)^2}$$

and the points B and C coincide, then $H_1(x, \xi) \equiv 0$, and the kernel of equation (3.30) is simplified, with $b = 1$. The boundary-value problem for equation (3.1) is replaced by

$$\frac{\partial u(x, 0)}{\partial y} = \psi_2(x), \qquad a_1 < x < a.$$

CHAPTER V

THE MODIFIED TRICOMI PROBLEM FOR AN EQUATION OF MIXED TYPE

§1. Statement of the problem

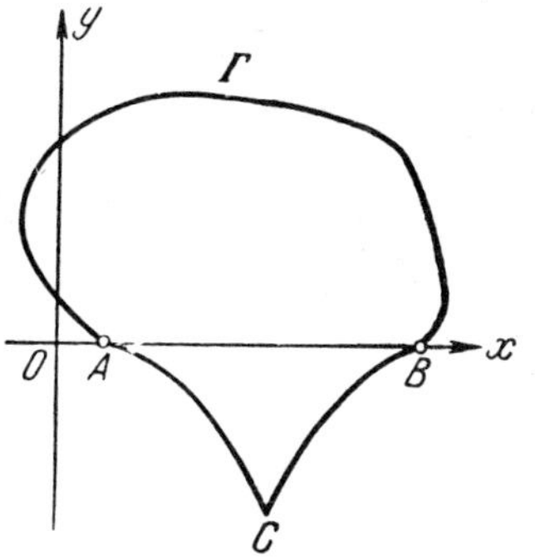

FIGURE 10

Frankl′ [22j] reduced the problem of flow in a plane-parallel symmetric Laval nozzle of given shape (the direct problem of Laval nozzle theory) to a new boundary-value problem for the equation

$$\operatorname{sgn} y |y|^m \frac{\partial^2 u}{\partial x^2} + \frac{\partial^2 u}{\partial y^2} = 0 \qquad (-1 < m < 0) \tag{1.1}$$

with exponent $m = -\frac{1}{2}$. To be more specific, considering a domain D in the hodograph plane as shown in Figure 10, where AC and BC are characteristic arcs of equation (1.1) and Γ is a rectifiable Jordan curve, Frankl′ showed that the boundary conditions

$$u|_\Gamma = \varphi(s), \qquad u|_{AC} = \psi(x), \tag{1.2}$$

are no longer sufficient to ensure the existence and uniqueness of a solution $u(x, y)$ of equation (1.1) in D when $-1 < m < 0$. One must add the following assumption (the Frankl′ discontinuity condition): on the parabolic line AB, instead of the usual continuity condition $u_y(x, +0) = u_y(x, -0)$, the solution must satisfy the condition

$$\frac{\partial u(x,\ +0)}{\partial y} = -\frac{\partial u(x,\ -0)}{\partial y}. \tag{1.3}$$

§2. Uniqueness theorem[1]

We shall prove the uniqueness theorem in the class of functions satisfying the following conditions:

1) $u(x, y) \in C^{(2)}(D^+ \cup D^-)$.

2) The integrals

$$\int_A^B u(x,\ 0)\, u_y(x,\ 0)\, dx, \qquad \iint_{D^+} (y^m u_x^2 + u_y^2)\, dx\, dy,$$

$$\iint_{D^-} \left[(-y)^m u_x^2 - u_y^2\right] dx\, dy$$

exist, and Green's formula is applicable to the integrals

$$\iint_{D^+} u(y^m u_{xx} + u_{yy})\, dx\, dy, \qquad \iint_{D^-} u[(-y)^m u_{xx} - u_{yy}]\, dx\, dy.$$

3) $(-y)^{m/2} u^2(B) = 0$.

Set

$$\frac{\partial u(x,\ +0)}{\partial y} = -\frac{\partial u(x,\ y)}{\partial y} = \nu(x).$$

We shall first show that if $u(x, y)$ is a solution of (1.1), vanishing on the characteristic AC, then

$$\int_0^1 \tau(x)\, \nu(x)\, dx \geqslant 0. \tag{2.1}$$

Indeed, in the domain D^- in the lower half-plane,

$$0 = \iint_{D^-} u[(-y)^m u_{xx} - u_{yy}]\, dx\, dy$$

$$= \iint_{ACBA} u[(-y)^m u_x\, dy + u_y\, dx] + \iint_{D^-} [u_y^2 - (-y)^m u_x^2]\, dx\, dy.$$

Hence, since $u|_{AC} = 0$, it follows that

$$\int_0^1 \tau(x)\, \nu(x)\, dx$$

$$= -\iint_{D^-} [u_y^2 - (-y)^m u_x^2]\, dx\, dy - \int_C^B u[(-y)^m u_x\, dy + u_y\, dx]. \tag{2.2}$$

---

[1] See [22j].

On the characteristic CB we have $dx = (-y)^{m/2}dy$. Then

$$\int_C^B u\left[(-y)^m u_x\,dy + u_y\,dx\right] = \int_C^B (-y)^{m/2} u\,(u_x\,dx + y_y\,dy)$$
$$= \int_C^B (-y)^{\frac{m}{2}} u\,du = \frac{1}{2}(-y)^{\frac{m}{2}} u^2(B) + \frac{m}{4}\int_C^B (-y)^{\frac{m-2}{2}} u^2\,dy$$

or, by condition 3),

$$\int_C^B u\left[(-y)^m u_x\,dy + u_y\,dx\right] = \frac{m}{4}\int_C^B (-y)^{\frac{m-2}{2}} u^2\,dy \leqslant 0. \tag{2.3}$$

We claim that the first integral on the right of (2.2) is nonpositive. To prove this, we transform to characteristic coordinates:

$$\xi = x - \frac{2}{m+2}(-y)^{\frac{m+2}{2}}, \qquad \eta = x + \frac{2}{m+2}(-y)^{\frac{m+2}{2}}. \tag{2.4}$$

This gives

$$\iint_{D^-} \left[u_y^2 - (-y)^m u_x^2\right] dx\,dy = -2 \iint_{\triangle AB'C'} (-y)^{\frac{m}{2}} u_\xi u_\eta\,d\xi\,d\eta. \tag{2.5}$$

In the hyperbolic half-plane equation (1.1) becomes

$$\frac{\partial^2 u}{\partial\xi\,\partial\eta} - \frac{\beta}{\eta-\xi}\left(\frac{\partial u}{\partial\eta} - \frac{\partial u}{\partial\xi}\right) = 0, \tag{2.6}$$

where $\beta = m/2(m+2)$. From (2.6) follows

$$u_\xi u_\eta = u_\eta^2 - \frac{\eta-\xi}{\beta} u_\eta u_{\xi\eta}. \tag{2.7}$$

Substituting (2.7) into (2.6), we get

$$\iint_{D^-} \left[u_y^2 - (-y)^m u_x^2\right] dx\,dy$$
$$= -2\left(\frac{m+2}{4}\right)^{\frac{m}{m+2}} \left[\iint_{\triangle AB'C'} (\eta-\xi)^{\frac{m}{m+2}} u_\eta^2\,d\xi\,d\eta - \frac{1}{\beta}\iint_{\triangle AB'C'} (\eta-\xi)^{\frac{2(m+1)}{m+2}} u_\eta u_{\xi\eta}\,d\xi\,d\eta\right]$$

or, integrating the last integral by parts,

$$\iint\limits_{D^-} [u_y^2 - (-y)^m u_x^2]\, dx\, dy$$
$$= \frac{2(m+2)}{m}\left(\frac{m+2}{4}\right)^{\frac{m}{m+2}} \iint\limits_{\triangle AB'C'} (\eta - \xi)^{\frac{m}{m+2}} u_\eta^2\, d\xi\, d\eta \leqslant 0. \tag{2.8}$$

In view of (2.3) and (2.8), inequality (2.1) now follows from (2.2).

It is now easy to prove the uniqueness theorem. Let $u(x, y)$ be a solution of (1.1), vanishing on Γ and on the characteristic AC. Then

$$0 = \iint\limits_{D^+} u\,(y^m u_{xx} + u_{yy})\, dx\, dy = -\iint\limits_{D^+} (y^m u_x^2 + u_y^2)\, dx\, dy$$
$$- \int_0^1 \tau(x)\, \nu(x)\, dx - \int\limits_{\Gamma} u\,[y^m u_x \cos(nx) + u_y \cos(ny)]\, ds,$$

and since $u|_\Gamma = 0$, it follows that

$$\int_0^1 \tau(x)\, \nu(x)\, dx + \iint\limits_{D^+} (y^m u_x^2 + u_y^2)\, dx\, dy = 0. \tag{2.9}$$

Since $u = 0$ on the characteristic AC, inequality (2.1) is true, and then it follows easily from (2.9) that $u \equiv 0$ in D^+. By the uniqueness of the solution to the Cauchy problem for equation (1.1) (see [66g]), we see that $u \equiv 0$ in D^-.

Thus $u = 0$ in D. Q.E.D.

§3. Some results from the theory of derivatives and integrals of fractional order for a real function of one variable

Let $f(x)$ be an integrable function on the interval (a, b). The *integral of fractional order* $(-\alpha)$ $(\alpha < 0)$ of $f(x)$ is defined as

$$D_a^\alpha f(x) = \frac{1}{\Gamma(-\alpha)} \int_a^x (x-t)^{-(\alpha+1)} f(t)\, dt \qquad (a \leqslant x \leqslant b). \tag{3.1}$$

The function $D_a^\alpha f(x)$ exists for almost all x and is integrable; if $\alpha \leqslant -1$, it is continuous.

If $\alpha > 0$, $n < \alpha < n + 1$ (where n is an integer), and $f^{(n+1)}(t)$ exists and is integrable on (a, b), then the *derivative of order* α of $f(x)$ is defined as

$$D_a^\alpha f(x) = \sum_{k=0}^{n} \frac{f^{(k)}(a)\,(x-a)^{k-\alpha}}{\Gamma(k-\alpha+1)} + D_a^{\alpha-(n+1)} f^{(n+1)}(x); \tag{3.2}$$

since $\alpha - (n + 1) < 0$, the last term in (3.2) is defined by (3.1). The derivative of order α of $f(x)$ can also be written as

$$D_a^\alpha f(x) = \frac{d^{n+1}}{dx^{n+1}} D_a^{\alpha-(n+1)} f(x) \qquad (n < \alpha < n+1). \tag{3.3}$$

This definition is also meaningful for integers $\alpha = n$: $D_a^\alpha f(x) = f^{(n)}(x)$.

It is readily seen that if $\alpha < 0$ and $\beta < 0$, then

$$D_a^\alpha D_{al}^\beta f(x) = D_a^{\alpha+\beta} f(x). \tag{3.4}$$

If

$$f(a) = f'(a) = \ldots = f^{(n-1)}(a) = 0, \tag{3.5}$$

then, applying the operator $D_a^{-\alpha}$ $(n < \alpha < n + 1)$ to the function $D_a^\alpha f(x)$ defined by (3.2), we get $D_a^{-\alpha} D_a^\alpha f(x) = f(x)$.

Let $f(x) = D_a^{-\alpha}\varphi(x)$, $n < \alpha < n + 1$; then we have (3.5) and

$$D_{al}^\alpha f(x) = D_a^\alpha D_a^{-\alpha} \varphi(x) = \varphi(x). \tag{3.6}$$

§4. Investigation of equation (1.1) in the hyperbolic half-plane

1. *Cauchy problem. Properties of generalized solutions.* In the half-plane $y < 0$, equation (1.1) has the form

$$(-y)^m u_{xx} - u_{yy} = 0. \tag{4.1}$$

In characteristic coordinates

$$\xi = x - \frac{2}{m+2}(-y)^{\frac{m+2}{2}}, \qquad \eta = x + \frac{2}{m+2}(-y)^{\frac{m+2}{2}} \tag{4.2}$$

equation (4.1) is an Euler-Darboux equation:

$$\frac{\partial^2 u}{\partial \xi\, \partial \eta} - \frac{\beta}{\eta - \xi}\left(\frac{\partial u}{\partial \eta} - \frac{\partial u}{\partial \xi}\right) = 0, \tag{4.3}$$

where $\beta = m/2(m + 2)$ and $-\frac{1}{2} < \beta < 0$ for $-1 < m < 0$.

It is well known that the solution of equation (4.3) with initial data (see [40a])

$$u(x, (0)) = \tau(x), \qquad 0 \leqslant x \leqslant 1, \tag{4.4}$$

$$\lim_{y \to 0} \frac{\partial u}{\partial y} = [2(1 - 2\beta)]^{-2\beta} \lim_{\eta - \xi \to 0} (\eta - \xi)^{2\beta} \left(\frac{\partial u}{\partial \xi} - \frac{\partial u}{\partial \eta}\right) = \nu(x) \tag{4.5}$$

is

$$\begin{aligned} u(\xi, \eta) = {} & \varkappa_1 \int_0^1 \tau(t)\, \zeta^\beta (1 - \zeta)^\beta\, d\zeta \\ & + \frac{\varkappa_1}{2(1 + 2\beta)} (\eta - \xi) \int_0^1 \tau'(t)\, \zeta^\beta (1 - \zeta)^\beta (2\zeta - 1)\, d\zeta \\ & - \varkappa_2 (\eta - \xi)^{1-2\beta} \int_0^1 \nu(t)\, \zeta^{-\beta} (1 - \zeta)^{-\beta}\, d\zeta, \end{aligned} \tag{4.6}$$

where

$$t = \xi + (\eta - \xi)\zeta,$$

$$\varkappa_1 = \frac{\Gamma(2+2\beta)}{\Gamma^2(1+\beta)}, \qquad \varkappa_2 = [2(1-2\beta)]^{2\beta-1}\frac{\Gamma(2-2\beta)}{\Gamma^2(1-\beta)}.$$

Returning to the old variables (x, y), we get the solution of the Cauchy problem for equation (4.1) in the half-plane $y < 0$ with initial data (4.4), (4.5):

$$\begin{aligned} u(x, y) = {} & \varkappa_1 \int_0^1 \tau(t)\zeta^\beta(1-\zeta)^\beta\, d\zeta \\ & + \frac{\varkappa_1}{m+2}(-y)^{\frac{m+2}{2}} \int_0^1 \tau'(t)\zeta^\beta(1-\zeta)^\beta(2\zeta-1)\, d\zeta \\ & + [2(1-2\beta)]^{1-2\beta}\varkappa_2 y \int_0^1 \nu(t)\zeta^{-\beta}(1-\zeta)^{-\beta}\, d\zeta, \end{aligned} \tag{4.7}$$

where

$$t = x + \frac{2}{m+2}(-y)^{\frac{m+2}{2}}(2\zeta - 1).$$

If $\tau(x) \subset C^3[0, 1]$ and $\nu(x) \in C^2[0, 1]$, then the function $u(x, y)$ defined by (4.6) or (4.7) is a twice continuously differentiable solution of the Cauchy problem for equation (4.1) with initial data (4.4), (4.5), in the domain bounded by the characteristics

$$\xi = x - \frac{2}{m+2}(-y)^{\frac{m+2}{2}} = 0, \qquad \eta = x + \frac{2}{m+2}(-y)^{\frac{m+2}{2}} = 1$$

and by the segment $[0, 1]$ of the x-axis.

An expression of the form (4.6) or (4.7) is called a *generalized solution* of equation (4.1) in the domain D^-: $0 < \xi < \eta < 1$, if $\tau(t)$ and $\nu(t)$ are continuous for $0 < t < 1$. The degree of smoothness of a generalized solution depends on that of the functions $\tau(t)$ and $\nu(t)$.

Consider the following class of generalized solutions of the Cauchy problem (4.1), (4.4), (4.5).

Class R_2 (see [40a]). A generalized solution (4.6) or (4.7) belongs to *class* R_2 if $\nu(x)$ is continuous and integrable on $(0, 1)$, and $\tau(x)$ is the integral of fractional order $-(2\beta - 1)$ of a continuous integrable function $T(x)$ defined on $(0, 1)$, i.e.

$$\tau(x) = \tau(0) + \int_0^x (x-t)^{-2\beta}\, T(t)\, dt \qquad (-1 < \beta < 0). \tag{4.8}$$

It follows easily from (4.8) that $\tau(x) \in C[0, 1]$ and that $\tau(x)$ has a derivative $\tau'(x) \in C(0, 1)$. Without loss of generality, we may assume that $\tau(0) = 0$.

LEMMA 1. *A generalized solution $u \in R_2$ is continuous in the closed domain $\overline{D}^-$.*

Indeed, setting $t = \xi + (\eta - \xi)\zeta$, we express $u(\xi, \eta)$ as

$$u(\xi, \eta) = \varkappa_1 (\eta - \xi)^{-1-2\beta} \int_\xi^\eta \tau(t)(\eta - t)^\beta (t - \xi)^\beta \, dt$$

$$+ \frac{\varkappa_1}{2(1+2\beta)} (\eta - \xi)^{-1-2\beta} \int_\xi^\eta \tau'(t) \left[(\eta - t)^{1+\beta} (t - \xi)^\beta \right. \tag{4.9}$$

$$\left. - (\eta - t)^\beta (t - \xi)^{1+\beta} \right] dt - \varkappa_2 \int_\xi^\eta \nu(t)(\eta - t)^{-\beta}(t - \xi)^{-\beta} \, dt.$$

It is easily checked that the finctions $\tau'(t)t^\beta$ and $\tau'(t)(1 - t)^\beta$ are integrable on $[0, 1]$, so that $u(\xi, \eta)$ is continuous for $\xi \longrightarrow 0$, $0 \leqslant \eta \leqslant 1$, and for $\eta \longrightarrow 1$, $0 \leqslant \xi \leqslant 1$. It follows easily from (4.6) that

$$\lim_{\eta - \xi \to 0} u(\xi, \eta) = \tau(x).$$

Q.E.D.

LEMMA 2. *For any generalized solution $u \in R_2$, the derivatives u_x and u_y are continuous in D^-, u_y is continuous right up to the parabolic line, and*

$$\lim_{y \to 0} \frac{\partial u}{\partial y} = \nu(x) \qquad (0 < x < 1). \tag{4.10}$$

PROOF. Substituting (4.8) into (4.9), we get

$$u(\xi, \eta) = \frac{\varkappa_1}{2(1+2\beta)} (\eta - \xi)^{-1-2\beta}$$

$$\times \int_\xi^\eta \left\{ \int_0^t T(\zeta) \left[2(1+2\beta)(\eta - t)^\beta (t - \xi)^\beta (t - \zeta)^{-2\beta} \right. \right.$$

$$+ 2\beta (\eta - t)^{1+\beta} (t - \xi)^\beta (t - \zeta)^{-1-2\beta} \tag{4.11}$$

$$\left. \left. - 2\beta (\eta - t)^\beta (t - \xi)^{1+\beta} (t - \zeta)^{-1-2\beta} \right] d\zeta \right\} dt$$

$$- \varkappa_2 \int_\xi^\eta \nu(t)(\eta - t)^{-\beta}(t - \xi)^{-\beta} \, dt = \frac{\varkappa_1}{2(1+2\beta)} (\eta - \xi)^{-1-2\beta} J_1 - J_2.$$

In the integral J_1, divide the integration interval on the ζ-axis into two parts: $(0, \xi)$ and (ξ, t). Changing the order of integration, we get

$$J_1 = \int_0^{\xi} I_1(\xi, \eta; \zeta)\, T(\zeta)\, d\zeta + \int_{\xi}^{\eta} I_2(\xi, \eta; \zeta)\, T(\zeta)\, d\zeta, \qquad (4.12)$$

where

$$I_1(\xi, \eta; \zeta) = 2(1+2\beta) \int_{\xi}^{\eta} (\eta - t)^{\beta} (t-\xi)^{\beta} (t-\zeta)^{-2\beta}\, dt$$
$$+ 2\beta \int_{\xi}^{\eta} (\eta - t)^{1+\beta} (t-\xi)^{\beta} (t-\zeta)^{-1-2\beta}\, dt$$
$$- 2\beta \int_{\xi}^{\eta} (\eta - t)^{\beta} (t-\xi)^{1+\beta} (t-\zeta)^{-1-2\beta}\, dt,$$

$$I_2(\xi, \eta; \zeta) = 2(1+2\beta) \int_{\zeta}^{\eta} (\eta - t)^{\beta} (t-\zeta)^{-2\beta} (t-\xi)^{\beta}\, dt$$
$$+ 2\beta \int_{\zeta}^{\eta} (\eta - t)^{1+\beta} (t-\zeta)^{-1-2\beta} (t-\xi)^{\beta}\, dt$$
$$- 2\beta \int_{\zeta}^{\eta} (\eta - t)^{\beta} (t-\zeta)^{-1-2\beta} (t-\xi)^{1+\beta}\, dt.$$

To evaluate these expressions, we set

$$\Phi(k, l, m; \xi, \eta, \zeta) = \int_{\xi}^{\eta} (\eta - t)^{k} (t-\xi)^{l} (t-\zeta)^{m}\, dt.$$

Making the substitution $t = \eta - (\eta - \xi)\mu$ and using the integral representation of hypergeometric functions, we get

$$\Phi(k, l, m; \xi, \eta, \zeta)$$
$$= \frac{\Gamma(k+1)\,\Gamma(l+1)}{\Gamma(k+l+2)} (\eta - \xi)^{k+l+1} (\eta - \zeta)^{m}$$
$$\times F\left(k+1,\ -m,\ k+l+2;\ \frac{\eta-\xi}{\eta-\zeta}\right).$$

Evaluating I_1, we have

$$I_1(\xi, \eta; \zeta)$$
$$= \frac{2\Gamma^2(1+\beta)}{\Gamma(1+2\beta)} (\eta - \xi)^{1+2\beta} (\eta - \zeta)^{-2\beta} \Big[F(1+\beta,\ 2\beta,\ 2+2\beta;\ z)$$
$$+ \frac{\beta}{2(1+2\beta)} z F(2+\beta,\ 1+2\beta,\ 3+2\beta;\ z)$$
$$- \frac{\beta}{2(1+2\beta)} z F(1+\beta,\ 1+2\beta,\ 3+2\beta;\ z) \Big],$$

where $z = (\eta - \xi)/(n - \zeta)$. Since the bracketed expression equals $(1-z)^{-\beta}$, it follows that

$$I_1(\xi, \eta; \zeta) = \frac{2\Gamma^2(1+\beta)}{\Gamma(1+2\beta)} (\eta - \xi)^{1+2\beta} (\eta - \zeta)^{-\beta} (\xi - \zeta)^{-\beta}. \tag{4.13}$$

Similarly, we obtain

$$I_2(\xi, \eta; \zeta) = \frac{\Gamma(1+\beta)\Gamma(1-2\beta)}{\Gamma(1-\beta)} (\eta - \xi)^{1+2\beta} (\eta - \zeta)^{-\beta} (\zeta - \xi)^{-\beta}. \tag{4.14}$$

Substituting (4.13) and (4.14) into (4.12) and inserting the resulting expression for J_1 into (4.11), we get

$$u(\xi, \eta) = \int_0^\xi (\eta - \zeta)^{-\beta} (\xi - \zeta)^{-\beta} T(\zeta)\, d\zeta + \int_\xi^\eta (\eta - \zeta)^{-\beta} (\zeta - \xi)^{-\beta} N(\zeta)\, d\zeta, \tag{4.15}$$

where

$$N(\zeta) = \frac{1}{2\cos\pi\beta} T(\zeta) - \varkappa_2 \nu(\zeta). \tag{4.16}$$

By the definition of the class R_2, it follows from (4.16) that $N(\zeta) \in C(0, 1)$ and $N(\zeta)$ is integrable on $(0, 1)$. Since $-\beta > 0$, it follows easily from (4.15) that $\partial u/\partial \xi$ and $\partial u/\partial \eta$ exist and are continuous in D^-.

We now prove (4.10). We have from (4.15) that

$$\begin{aligned}\frac{\partial u}{\partial y} &= [2(1-2\beta)]^{-2\beta} (\eta - \xi)^{2\beta} \left(\frac{\partial u}{\partial \xi} - \frac{\partial u}{\partial \eta} \right) = -\beta [2(1-2\beta)]^{-2\beta} \\ &\times \left\{ (\eta - \xi)^{1+2\beta} \int_0^\xi (\eta - \zeta)^{-1-\beta} (\xi - \zeta)^{-1-\beta} T(\zeta)\, d\zeta \right. \\ &\quad \left. - (\eta - \xi)^{1+2\beta} \int_\xi^\eta (\eta - \zeta)^{-1-\beta} (\zeta - \xi)^{-1-\beta} N(\zeta)\, d\zeta \right\} \\ &= -\beta [2(1-2\beta)]^{-2\beta} (I_3 - I_4).\end{aligned} \tag{4.17}$$

The substitution $\zeta = \xi + (\eta - \xi)t$ transforms the integral I_4 into

$$I_4 = \int_0^1 N[\xi + (\eta - \xi)t]\, t^{-1-\beta} (1-t)^{-1-\beta}\, dt.$$

The limit passage gives

$$\lim_{\eta-\xi\to 0} I_4 = N(x)\int_0^1 t^{-1-\beta}(1-t)^{-1-\beta}\,dt = \frac{\Gamma^2(-\beta)}{\Gamma(-2\beta)}\,N(x). \qquad (4.18)$$

Let δ be a constant such that $0<\delta<x<1$, and split the integral I_3 into two: $\int_0^\xi = \int_0^\delta + \int_\delta^\xi$. Making the substitution $\zeta = \xi - (\eta-\xi)\mu$ in the second integral, we easily find that

$$I_3 = (\eta-\xi)^{1+2\beta}\int_0^\delta (\eta-\zeta)^{-1-\beta}(\xi-\zeta)^{-1-\beta}\,T(\zeta)\,d\zeta$$
$$+\int_0^{\frac{\xi-\delta}{\eta-\xi}} T[\xi-(\eta-\xi)\mu]\,\mu^{-1-\beta}(1+\mu)^{-1-\beta}\,d\mu.$$

By the choice of δ, the first integral tends to zero as $\eta,\xi\to x$, $0<x<1$. Therefore,

$$\lim_{\eta-\xi\to 0} I_3 = T(x)\int_0^\infty \mu^{-1-\beta}(1+\mu)^{-1-\beta}\,d\mu = T(x)\,\frac{\Gamma(-\beta)\,\Gamma(1+2\beta)}{\Gamma(1+\beta)}. \qquad (4.19)$$

Letting $y\to 0$ in (4.17), we infer, via (4.18), (4.19) and (4.16), that

$$\lim_{y\to 0}\frac{\partial u}{\partial y} = -\beta\,[2(1-2\beta)]^{-2\beta}\,\frac{\Gamma^2(-\beta)}{\Gamma(-2\beta)}\,\varkappa_2 \nu(x) = \nu(x).$$

Q.E.D.

2. *Functional relation between* $\tau(x)$ *and* $\nu(x)$. Using the explicit integral representation (4.15) for the solution of the Cauchy problem (4.1), (4.4), (4.5) in D^-, we shall find the functional relation between $\tau(x)$ and $\nu(x)$, and also the prescribed values $\psi(\eta)$ of the solution $u(\xi,\eta)$ on the characteristic AC ($\xi=0$, $0\leqslant\eta\leqslant 1$). Let $u\in R_2$. Then, since u is continuous in D^-, it follows that

$$\psi(0)=0. \qquad (4.20)$$

Set $\xi=0$ in (4.15); then

$$\psi(\eta) = \int_0^\eta N(\zeta)\,\zeta^{-\beta}(\eta-\zeta)^{-\beta}\,d\zeta.$$

Let us solve this equation for $T(\zeta)$. Set

$$\Phi(\zeta) = N(\zeta)\,\zeta^{-\beta}. \qquad (4.21)$$

Then

$$\int_0^\eta (\eta-\zeta)^{-\beta}\,\Phi(\zeta)\,d\zeta = \psi(\eta). \qquad (4.22)$$

We shall assume that $\psi(\eta)$ has a second derivative $\psi''(\eta)$ which is integrable on (0, 1). Suppose there exists a solution of (4.22). Write it in the form (see (3.1))

$$D_0^{\beta-1}\Phi(\eta) = \frac{1}{\Gamma(1-\beta)}\psi(\eta). \tag{4.23}$$

Applying the operator $D_0^{1-\beta}$ to both sides of (4.23), we see that, by virtue of (3.6),

$$\Phi(\eta) = \frac{1}{\Gamma(1-\beta)} D_0^{1-\beta}\psi(\eta). \tag{4.24}$$

Thus, if a solution of (4.22) exists, it necessarily admits a representation (4.24). This implies the uniqueness of the solution of (4.22). A direct check readily shows that the function (4.24) is indeed a solution of (4.22).

Now, combining (4.16), (4.21) and (4.24), we get

$$T(\zeta) = \varkappa_3 \nu(\zeta) + \frac{2\cos\pi\beta}{\Gamma(1-\beta)}\zeta^\beta D_0^{1-\beta}\psi(\zeta), \tag{4.25}$$

where

$$\varkappa_3 = 2\varkappa_2 \cos\pi\beta. \tag{4.26}$$

Substituting (4.25) into (4.8), we find, finally, that

$$\tau(x) = \varkappa_3 \int_0^x (x-t)^{-2\beta}\nu(t)\,dt + \Phi_1(x), \tag{4.27}$$

where

$$\Phi_1(x) = \frac{2\Gamma(1+\beta)}{\Gamma(1+2\beta)} D_0^{2\beta-1}\left[x^\beta D_0^{1-\beta}\psi(x)\right]. \tag{4.28}$$

§5. Investigation of equation (1.1) in the elliptic half-plane

In the half-plane $y > 0$, equation (1.1) has the form

$$y^m u_{xx} + u_{yy} = 0 \qquad (-1 < m < 0). \tag{5.1}$$

PROBLEM *N*. *Find a solution in* D^+ *of equation* (5.1), *continuous in the closed domain* $\overline{D}^+$, *which satisfies the boundary conditions*

$$u|_\Gamma = \varphi(s), \qquad (0 \leqslant s \leqslant l), \tag{5.2}$$

$$\lim_{y\to 0}\frac{\partial u}{\partial y} = \nu(x), \qquad 0 < x < 1, \tag{5.3}$$

where $\varphi(s) \in C[0, l]$ *and* $\nu(x) \in C(0, 1)$.

Concerning the curve Γ, we assume that 1) the functions $x(s)$ and $y(s)$ defining Γ parametrically have second derivatives satisfying a Hölder condition in the interval $0 \leqslant s \leqslant 1$, and 2) in neighborhood of the endpoints of Γ,

$$\left|\frac{dx}{ds}\right| \leqslant C_1 y^{m+1}(s). \tag{5.4}$$

Equation (5.1) has a fundamental solution

$$q_1(x, y; x_0, y_0) = k_1 r_1^{-2\beta} F(\beta, \beta, 2\beta, 1-\sigma), \tag{5.5}$$

where

$$\left.\begin{matrix} r^2 \\ r_1^2 \end{matrix}\right\} = (x-x_0)^2 + \frac{4}{(m+2)^2}\left(y^{\frac{m+2}{2}} \mp y_0^{\frac{m+2}{2}}\right)^2,$$

$$\sigma = \frac{r^2}{r_1^2}, \quad \beta = \frac{m}{2(m+2)} < 0, \quad k_1 = \frac{1}{4\pi}\left(\frac{4}{m+2}\right)^{2\beta} \frac{\Gamma^2(\beta)}{\Gamma(2\beta)}.$$

It is readily seen that the potential theory constructed for equation (5.1) with $m > 0$ (see [66g]) is also valid for $-1 < m < 0$.

DEFINITION. A function $G(x, y; x_0, y_0)$, is called a *Green's function of problem N for equation* (5.1) if it satisfies the following conditions:

1) In the interior of D^+, the function is a regular solution of (5.1) everywhere except at the point (x_0, y_0).

2) It satisfies the boundary conditions

$$G(x, y; x_0, y_0)|_\Gamma = 0, \qquad \left.\frac{\partial G}{\partial y}\right|_{y=0} = 0 \qquad (x_0, y_0) \in D^+. \tag{5.6}$$

3) It can be represented as

$$G(x, y; x_0, y_0) = q_1(x, y; x_0, y_0) + v(x, y; x_0, y_0), \tag{5.7}$$

where $v(x, y; x_0, y_0)$ is a regular solution of (5.1) inside D^+.

If D_0^+ is the domain bounded by the segment $[0, 1]$ of the x-axis and the normal curve $C_{1/2}$:

$$\left(x - \frac{1}{2}\right)^2 + \frac{4}{(m+2)^2} y^{m+2} = \frac{1}{4}, \tag{5.8}$$

the Green's function of Problem N can be written in explicit form:

$$\begin{aligned} &G_0(x, y; x_0, y_0) \\ &= q_1(x, y; x_0, y_0) - (4R_0^2)^{-\beta} q_1\left(x - \frac{1}{2}, y; \bar{x}_0, \bar{y}_0\right), \end{aligned} \tag{5.9}$$

where

$$\bar{x}_0 = \frac{x_0 - \frac{1}{2}}{4R_0^2}, \qquad \bar{y}_0^{\frac{m+2}{2}} = \frac{y_0^{\frac{m+2}{2}}}{4R_0^2},$$

$$R_0^2 = \left(x_0 - \frac{1}{2}\right)^2 + \frac{4}{(m+2)^2} y_0^{m+2}.$$

If Γ is not the normal curve $C_{1/2}$, the Green's function of Problem N can be represented in the form [66g]

$$G(x, y; x_0, y_0) = G_0(x, y; x_0, y_0) + H(x, y; x_0, y_0). \quad (5.10)$$

Here

$$H(x, y; x_0, y_0) = \int_0^l \rho(t; x, y)\, G_0(\xi, \eta; x_0, y_0)\, dt, \quad (5.11)$$

and $\rho(t; x, y)$ is a solution of the integral equation

$$\rho(s, x, y) - 2\int_0^l K(t, s)\rho(t, x, y)\, dt = 2A_s[q_1(\xi, \eta; x, y)],$$

where

$$A_s[q_1] = \eta^m \frac{d\eta}{ds}\frac{\partial q_1}{\partial \xi} - \frac{d\xi}{ds}\frac{\partial q_1}{\partial \eta}, \quad (\xi(s), \eta(s)) \in \Gamma,$$
$$K(t, s) = A_s[q_1(\xi(t), \eta(t); x(s), y(s))].$$

We have the following assertion (see [66g]):

THEOREM. *The function*

$$u(x_0, y_0) = -\int_0^1 \nu(x)\, G(x, 0; x_0, y_0)\, dx - \int_0^l \varphi(s)\, A_s[G(\xi, \eta, x_0, y_0)]\, ds \quad (5.12)$$

is a solution of Problem N.

For the normal domain D_0^+, the solution of Problem N can be written as

$$u(x_0, y_0) = -k_1 \int_0^1 \nu(x)\left\{\left[(x - x_0)^2 + \frac{4}{(m+2)^2} y_0^{m+2}\right]^{-\beta} - \left[(x + x_0 - 2xx_0)^2 + \frac{16\left(x - \frac{1}{2}\right)^2 y_0^{m+2}}{(m+2)^2}\right]^{-\beta}\right\} dx$$
$$- \frac{k_1 m}{2}\left(\frac{1}{4} - R_0^2\right)\int_0^l \eta^{-1}\varphi(s)(r_1^2)^{-\beta-1} F(\beta, \beta+1, 2\beta; 1-\sigma)\frac{d\xi}{ds}\, ds. \quad (5.13)$$

Letting $y_0 \to +0$ in (5.13), we get

$$\tau(x_0) = -k_1 \int_0^1 \nu(x)\left[|x - x_0|^{-2\beta} - (x + x_0 - 2xx_0)^{-2\beta}\right] dx$$
$$- \frac{k_1 m}{2} x_0(1 - x_0)\int_0^l \eta^{-1}\varphi(s) \times \left[(\xi - x_0)^2 + \frac{4}{(m+2)^2}\eta^{m+2}\right]^{-\beta-1}\frac{d\xi}{ds}\, ds.$$

Using the equation of the normal curve $C_{1/2}$,

$$y = (1-2\beta)^{2\beta-1}[x(1-x)]^{\frac{1}{2}-\beta}$$

and setting $\varphi(s) = \varphi_1(x)$, we get

$$\tau(x_0) = -k_1 \int_0^1 \nu(x)\left[|x-x_0|^{-2\beta} - (x+x_0-2xx_0)^{-2\beta}\right]dx + \Phi_2(x_0), \quad (5.14)$$

where

$$\Phi_2(x_0) = 2k_1\beta(1-2\beta)^{-\beta}x_0(1-x_0)\int_0^1 \varphi_1(x)\frac{[x(1-x)]^{\beta-\frac{1}{2}}}{\left[x_0^2+(1-2x_0)x\right]^{\beta+1}}dx. \quad (5.15)$$

(5.14) gives the second functional relation between $\tau(x)$ and $\nu(x)$; it follows from the assumption that the solution $u(x, y)$ of (5.1) in D^+ takes the values $\varphi(s)$ on the normal curve $C_{1/2}$.

§6. Existence theorem for the modified Tricomi problem

In this section we prove the existence of a solution to the modified Tricomi problem when Γ is the normal curve

$$C_{1/2}\colon \left(x-\frac{1}{2}\right)^2 + \frac{4}{(m+2)^2}y^{m+2} = \frac{1}{4}.$$

1. *Reduction of the modified Tricomi problem to a singular equation.* By virtue of the results of §§4 and 5 and condition (1.3), the existence problem for the modified Tricomi problem for equation (1.1) is equivalent to solvability of the following system:

$$\tau(x) = -\varkappa_3\int_0^x (x-t)^{-2\beta}\nu(t)\,dt + \Phi_1(x), \quad (6.1)$$

$$\tau(x) = -k_1\int_0^1 \left[|t-x|^{-2\beta} - (t+x-2xt)^{-2\beta}\right]\nu(t)\,dt + \Phi_2(x), \quad (6.2)$$

where $\Phi_1(x)$ and $\Phi_2(x)$ are defined by (4.28) and (5.15), respectively.

Eliminating the function $\tau(x)$ from (6.1) and (6.2) and noting that $\kappa_3 = 2k_1 \sin \pi\beta$, we get

$$2 \sin \pi\beta \int_0^x (x-t)^{-2\beta} \nu(t)\, dt$$

$$- \int_0^1 \left[|t-x|^{-2\beta} - (t+x-2xt)^{-2\beta} \right] \nu(t)\, dt = \frac{1}{k_1} f(x), \tag{6.3}$$

where

$$f(x) = \Phi_1(x) - \Phi_2(x). \tag{6.4}$$

Thus, if (6.3) has a solution, the system (6.1), (6.2) also has a solution, i.e. the modified Tricomi problem is solvable.

Following Tricomi, we reduce (6.3) to a singular integral equation, by inverting the Volterra term in (6.3). We shall consider only those solutions $\nu(x)$ of (6.3) which satisfy a Hölder condition in the interval (0, 1) and are in $L(0, 1)$.

By (3.1), the first term on the left of (6.3) can be written as

$$2 \sin \pi\beta \Gamma(1-2\beta) D_0^{2\beta-1} \nu(x).$$

Applying the operator $\Gamma(1 + 2\beta) D_0^{1-2\beta}$ to both sides of (6.3) and using (3.6) and (3.3), we get

$$\frac{2\pi\beta}{\cos \pi\beta} \nu(x) - \frac{d^2}{dx^2} \int_0^x (x-\zeta)^{2\beta}\, d\zeta \int_0^1 \left[|t-\zeta|^{-2\beta} - (t+\zeta-2\zeta t)^{-2\beta} \right] \nu(t)\, dt$$
$$= \frac{\Gamma(1+2\beta)}{k_1} D_0^{1-2\beta} f(x).$$

Then, changing the order of integration and setting

$$I_1(x, t) = \int_0^x (x-\zeta)^{2\beta} |t-\zeta|^{-2\beta}\, d\zeta, \tag{6.5}$$

$$I_2(x, t) = \int_0^x (x-\zeta)^{2\beta} (t+\zeta-2\zeta t)^{-2\beta}\, d\zeta, \tag{6.6}$$

we get

$$\frac{2\pi}{\cos \pi\beta} \nu(x)$$
$$- \frac{d^2}{dx^2} \left[\int_0^1 I_1(x, t) \nu(t)\, dt - \int_0^1 I_2(x, t) \nu(t)\, dt \right] \tag{6.7}$$
$$= \frac{\Gamma(1+2\beta)}{k_1} D_0^{1-2\beta} f(x).$$

To justify differentiating under the integral sign in (6.7), we investigate the derivatives of $I_1(x, t)$ and $I_2(x, t)$.

First, we note that the evaluation of I_2 is reducible to that of I_1 via the easily-checked formula

$$I_2(x, t) = |1 - 2t|^{-2\beta} I_1\left(x, \frac{t}{2t-1}\right), \tag{6.8}$$

It suffices, therefore, to investigate $I_1(x, t)$. By (6.5),

$$I_1(x, t) = \int_0^x \left(\frac{x-\zeta}{t-\zeta}\right)^{2\beta} d\zeta \quad \text{for} \quad t > x \tag{6.9}$$

and

$$\begin{aligned} I_1(x, t) &= \int_0^t \left(\frac{x-\zeta}{t-\zeta}\right)^{2\beta} d\zeta + \int_t^x \left(\frac{x-\zeta}{\zeta-t}\right)^{2\beta} d\zeta \\ &= \frac{2\pi\beta}{\sin 2\pi\beta}(x-t) + \int_0^t \left(\frac{x-\zeta}{t-\zeta}\right)^{2\beta} d\zeta \quad \text{for} \quad t < x. \end{aligned} \tag{6.10}$$

Making the substitution $\mu = (t - \zeta)/(x - \zeta)$, we find that

$$I_1(x, t) = (t-x) \int_{t/x}^{\infty} \frac{\mu^{-2\beta}\, d\mu}{(1-\mu)^2} \quad \text{for} \quad t > x,$$

$$I_1(x, t) = (x-t)\left[\frac{2\pi\beta}{\sin 2\pi\beta} + \int_0^{t/x} \frac{\mu^{-2\beta}\, d\mu}{(1-\mu)^2}\right] \quad \text{for} \quad t < x$$

or, integrating by parts,

$$I_1(x, t) = x^{1+2\beta} t^{-2\beta} + 2\beta(t-x) \int_{t/x}^{\infty} \frac{\mu^{-1-2\beta}}{1-\mu}\, d\mu \quad (t > x),$$

$$I_1(x, t) = x^{1+2\beta} t^{-2\beta} + (x-t)\left[\frac{2\pi\beta}{\sin 2\pi\beta} + 2\beta \int_0^{t/x} \frac{\mu^{-1-2\beta}}{1-\mu}\, d\mu\right] \quad (t < x).$$

Differentiation with respect to x gives

$$\frac{\partial I_1(x, t)}{\partial x} = \left(\frac{t}{x}\right)^{-2\beta} - 2\beta \int_{t/x}^{\infty} \frac{\mu^{-1-2\beta}\, d\mu}{1-\mu} \quad (t > x), \tag{6.11}$$

$$\frac{\partial I_1(x, t)}{\partial x} = \frac{2\pi\beta}{\sin 2\pi\beta} + \left(\frac{t}{x}\right)^{-2\beta} + 2\beta \int_0^{t/x} \frac{\mu^{-1-2\beta}\, d\mu}{1-\mu} \quad (t < x). \tag{6.12}$$

Consider the behavior of $\partial I_1(x, t)/\partial x$ as $t \to x$. We shall find an estimate for the integral in (6.11) as $t \to x$ $(t > x)$ and for the integral in (6.12) as $t \to x$ $(t < x)$.

For the first of these integrals, assuming $t < 2x$, we have

$$\int_{t/x}^{\infty} \frac{\mu^{-1-2\beta}\,d\mu}{\mu-1} \leqslant \int_{t/x}^{2} \frac{d\mu}{\mu-1} + \int_{2}^{\infty} \frac{\mu^{-1-2\beta}\,d\mu}{\mu-1} = O(\ln|t-x|) + O(1),$$

and for the second, if $t > x/2$,

$$\int_{0}^{t/x} \frac{\mu^{-1-2\beta}\,d\mu}{1-\mu} \leqslant \int_{0}^{1/2} \frac{\mu^{-1-2\beta}\,d\mu}{1-\mu} + 2^{1+2\beta} \int_{1/2}^{t/x} \frac{d\mu}{1-\mu} = O(1) + O(\ln|t-x|).$$

Therefore, from (6.11) and (6.12) we obtain an estimate valid for $0 < x < 1$:

$$\frac{\partial I_1(x,t)}{\partial x} = O(\ln|t-x|). \tag{6.13}$$

Hence, in order to differentiate the expression

$$H(x) = \int_0^1 I_1(x,t)\,\nu(t)\,dt \tag{6.14}$$

we can differentiate once under the integral sign:

$$\frac{dH(x)}{dx} = \int_0^1 \frac{\partial I_1(x,t)}{\partial x}\,\nu(t)\,dt.$$

Differentiating (6.11) and (6.12) with respect to x, we get

$$\frac{\partial^2 I_1(x,t)}{\partial x^2} = \left(\frac{t}{x}\right)^{1-2\beta} \frac{2\beta}{t-x}. \tag{6.15}$$

Hence it is clear that we cannot differentiate a second time under the integral sign in (6.14).

Consider the expression

$$j(x,\varepsilon) = \int_0^{x-\varepsilon} \frac{\partial I_1(x,t)}{\partial x}\,\nu(t)\,dt + \int_{x+\varepsilon}^{1} \frac{\partial I_1(x,t)}{\partial x}\,\nu(t)\,dt. \tag{6.16}$$

Obviously,

$$\lim_{\varepsilon\to 0} j(x,\varepsilon) = \frac{dH}{dx}.$$

Differentiating (6.16) with respect to x, and using (6.15), we get

$$\frac{dj(x, \varepsilon)}{dx} = 2\beta \left[\int_0^{x-\varepsilon} + \int_{x+\varepsilon}^1 \right] \left(\frac{t}{x}\right)^{1-2\beta} \frac{\nu(t)\,dt}{t-x}$$
$$+ \nu(x-\varepsilon) \left.\frac{\partial I_1(x, t)}{\partial x}\right|_{t=x-\varepsilon} - \nu(x+\varepsilon) \left.\frac{\partial I_1(x, t)}{\partial x}\right|_{t=x+\varepsilon} \tag{6.17}$$

Let us evaluate the limit of the difference

$$I = \left.\frac{\partial I_1(x, t)}{\partial x}\right|_{t=x-\varepsilon} - \left.\frac{\partial I_1(x, t)}{\partial x}\right|_{t=x+\varepsilon} \quad \text{as} \quad \varepsilon \to 0.$$

By (6.11) and (6.12), we have

$$\lim_{\varepsilon\to 0} I = \frac{2\pi\beta}{\sin 2\pi\beta} + \lim_{\varepsilon\to 0} \left[\left(\frac{x-\varepsilon}{x}\right)^{-2\beta} - \left(\frac{x+\varepsilon}{x}\right)^{-2\beta} \right.$$
$$\left. + 2\beta \left(\int_0^{\frac{x-\varepsilon}{x}} \frac{\mu^{-1-2\beta}\,d\mu}{1-\mu} + \int_{\frac{x+\varepsilon}{x}}^{\infty} \frac{\mu^{-1-2\beta}\,d\mu}{1-\mu} \right) \right] \tag{6.18}$$
$$= \frac{2\pi\beta}{\sin 2\pi\beta} + 2\beta \int_0^\infty \frac{\mu^{-1-2\beta}\,d\mu}{1-\mu}.$$

Obviously, the convergence as $\epsilon \longrightarrow 0$ is uniform in x, $0 < x < 1$. The integral on the right side is interpreted in the sense of the Cauchy principal value.

It is readily shown that

$$\int_0^\infty \frac{\mu^{-1-2\beta}\,d\mu}{1-\mu} = -\pi \cot 2\pi\beta.$$

Therefore, from (6.18),

$$\lim_{\varepsilon\to 0} \left[\left.\frac{\partial I_1(x, t)}{\partial x}\right|_{t=x-\varepsilon} - \left.\frac{\partial I_1(x, t)}{\partial x}\right|_{t=x+\varepsilon} \right] = 2\pi\beta \tan \pi\beta. \tag{6.19}$$

Now, add and subtract $\nu(x-\epsilon)\partial I_1(x, t)/\partial x|_{t=x+\epsilon}$ on the right of (6.17). Then

$$\frac{dj(x, \varepsilon)}{dx} = 2\beta \left[\int_0^{x-\varepsilon} + \int_{x+\varepsilon}^1 \right] \left(\frac{t}{x}\right)^{1-2\beta} \frac{\nu(t)\,dt}{t-x}$$
$$+ \left[\left.\frac{\partial I_1(x, t)}{\partial x}\right|_{t=x-\varepsilon} - \left.\frac{\partial I_1(x, t)}{\partial x}\right|_{t=x+\varepsilon} \right] \nu(x-\varepsilon)$$
$$+ \left.\frac{\partial I_1(x, t)}{\partial x}\right|_{t=x+\varepsilon} [\nu(x-\varepsilon) - \nu(x+\varepsilon)].$$

Hence, by (6.19) and (6.13), the limit

$$\lim_{\varepsilon\to 0}\frac{dj(x,\varepsilon)}{dx}=2\beta\int_0^1\left(\frac{t}{x}\right)^{1-2\beta}\frac{\nu(t)\,dt}{t-x}+2\pi\beta\tan\beta\nu(x), \tag{6.20}$$

exists, and the convergence as $\epsilon\longrightarrow 0$ is uniform in x $(0<x<1)$. Then, by a standard theorem of classical analysis, the derivative d^2H/dx^2 exists and

$$\lim_{\varepsilon\to 0}\frac{dj(x,\varepsilon)}{dx}=\frac{d^2H}{dx^2}\qquad (0<x<1). \tag{6.21}$$

Obviously, by (6.11) and (6.15),

$$\frac{\partial^2 I_2(x,t)}{\partial x^2}=\left(\frac{t}{x}\right)^{1-2\beta}\frac{2\beta}{t+x-2tx}. \tag{6.22}$$

This derivative is continuous for $0\leqslant t\leqslant 1$ and $x\neq 0, 1$. Consequently, the second integral in (6.7) can be differentiated in the same way.

Thus, in view of (6.14) and (6.20)–(6.22), equation (6.7) can be written as

$$\nu(x)-\lambda\int_0^1\left(\frac{t}{x}\right)^{1-2\beta}\left(\frac{1}{t-x}-\frac{1}{t+x-2xt}\right)\nu(t)\,dt=f_1(x), \tag{6.23}$$

where

$$\lambda=\frac{\cos\pi\beta}{\pi(1-\sin\pi\beta)}, \tag{6.24}$$

$$f_1(x)=\frac{\Gamma(1+2\beta)\lambda}{2\beta k_1}D_0^{1-2\beta}f(x)=\frac{\lambda}{2\beta k_1}\frac{d^2}{dx^2}\int_0^x(x-t)^{2\beta}f(t)\,dt. \tag{6.25}$$

2. *Investigation of the function* $f_1(x)$. By (6.25) and (6.4) we have

$$f_1(x)=\frac{\lambda\Gamma(1+2\beta)}{2\beta k_1}\left[D_0^{1-2\beta}\Phi_1(x)-D_0^{1-2\beta}\Phi_2(x)\right]. \tag{6.26}$$

We shall assume that the set of boundary values of problem (2.1)–(2.3) on the characteristic $\xi=0$, i.e. the function $\psi(\eta)$, has a second derivative $\psi''(\eta)$ which is bounded and integrable on (0, 1). Then, by (4.28) and the equality $\psi(0)=0$, it follows that

$$\frac{\lambda\Gamma(1+2\beta)}{2\beta k_1} D_0^{1-2\beta}\Phi_1(x) = \frac{\lambda\Gamma(1+2\beta)}{k_1\beta} x^\beta D_0^{1-\beta}\psi(x)$$

$$= \frac{\lambda}{k_1\beta} x^\beta \frac{d^2}{dx^2}\int_0^x (x-t)^\beta \psi(t)\,dt \tag{6.27}$$

$$= \frac{\lambda}{k_1\beta}\left[\psi'(0)\,x^{2\beta} + x^\beta \int_0^x (x-t)^\beta \psi''(t)\,dt\right].$$

We now consider the second term in (6.26). To investigate it, we need only study the derivatives of $\Phi_2(x)$. Let us assume that the boundary values $\varphi_1(x)$ of the solution on the normal curve $C_{1/2}$ can be expressed as

$$\varphi_1(x) = y^{1+\gamma}\bar{\varphi}(x), \tag{6.28}$$

where $\gamma > 0$ and $\bar{\varphi}(x) \in C[0, 1]$.

If these boundary values are prescribed as a continuous function $\varphi^*(s)$ of arc-length on the curve $C_{1/2}$, a representation of type (6.28) is possible, provided the following conditions hold:

1) $\varphi^*(s)$ has a continuous derivative $\varphi^{*\prime}(s)$ in neighborhoods of the end-points A and B, including these points; without loss of generality, we may assume that

$$\varphi^*(0) = \varphi^*(l) = \varphi^{*\prime}(0) = \varphi^{*\prime}(l) = 0;$$

2) In neighborhoods of A and B, we have finite limits

$$\lim_{s\to 0}\frac{\varphi^{*\prime}(s)}{s^\gamma} \quad (\text{at } B) \qquad \text{and} \qquad \lim_{s\to 0}\frac{\varphi^{*\prime}(s)}{(l-s)^\gamma} \quad (\text{at } A).$$

Substituting (6.28) into (5.15) and using the equation of the normal curve $C_{1/2}$, we get

$$\Phi_2(x) = k_2 x(1-x)\int_0^1 \frac{[t(1-t)]^{\gamma\left(\frac{1}{2}-\beta\right)}\bar{\varphi}(t)\,dt}{[x^2+(1-2x)t]^{1+\beta}}, \tag{6.29}$$

where

$$k_2 = 2\beta k_1(1-2\beta)^{-1+(2\beta-1)\gamma}.$$

Hence, $\Phi_2(x)$ has derivatives of any order in the interior of the interval $(0, 1)$. Let us investigate the behavior of $\Phi_2(x)$ and its derivatives $\Phi_2'(x)$ and $\Phi_2''(x)$ as $x \to 0$ and $x \to 1$. Note that if we replace x by $(1-x)$ and t by $(1-t)$ on the right of (6.29), the expression remains unchanged, except that the argument of $\bar{\varphi}$ is not t but $(1-t)$. Thus $\Phi_2(x)$ displays the same behavior near the points $x = 0$ and $x = 1$, so we can confine ourselves to the limits of $\Phi_2(x)$ and its derivatives as $x \to 0$.

After this remark, we differentiate (6.29) twice with respect to x, and for brevity set

$$H_{kl} = k_2 \int_0^1 \bar{\varphi}(t)\, t^{k+\gamma\left(\frac{1}{2}-\beta\right)} (1-t)^{\gamma\left(\frac{1}{2}-\beta\right)} [x^2 + (1-2x)\,t]^{-(l+1+\beta)}\, dt. \tag{6.30}$$

Then we can write (6.29) and the derivatives in the form

$$\begin{aligned}
\Phi_2(x) &= x(1-x) H_{00},\\
\Phi_2'(x) &= (1-2x) H_{00} - 2(1+\beta) x^2 (1-x) H_{01}\\
&\qquad + 2(1+\beta)\, x(1-x) H_{11},\\
\Phi_2''(x) &= -2H_{00} - 2(1+\beta)(3-5x) H_{01}\\
&+ 4(1+\beta)(1-2x) H_{11} + 4(1+\beta)(2+\beta)\, x^3 (1-x) H_{02}\\
&- 8(1+\beta)(2+\beta)\, x^2 (1-x) H_{12} + 4(1+\beta)(2+\beta)\, x(1-x) H_{22}.
\end{aligned} \tag{6.31}$$

The kernel of the integrand in (6.30) is positive in the integration interval. By the mean-value theorem,

$$H_{kl} = k_2 \bar{\varphi}(\xi_{kl})\, x^{-2(l+1+\beta)} \int_0^1 t^{k+\gamma\left(\frac{1}{2}-\beta\right)} (1-t)^{\gamma\left(\frac{1}{2}-\beta\right)} \left(1 - \frac{2x-1}{x^2}\, t\right)^{-(l+1+\beta)} dt,$$

where $0 < \xi_{kl} < 1$.

But the integral here is a hypergeometric integral, and so

$$H_{kl} = k_2 \bar{\varphi}(\xi_{kl}) x^{-2(l+\beta+1)} \frac{\Gamma\left(k+\gamma\left(\frac{1}{2}-\beta\right)+1\right)\Gamma\left(\gamma\left(\frac{1}{2}-\beta\right)+1\right)}{\Gamma(k+\gamma(1-2\beta)+2)}$$

$$\times F\left(l+1+\beta,\ k+\gamma\left(\frac{1}{2}-\beta\right)+1,\ k+\gamma(1-2\beta)+2;\ \frac{2x-1}{x^2}\right).$$

Next, using the functional relation between hypergeometric functions with arguments z and $1/(1-z)$ (see [49], §9.5), we get

$$\begin{aligned}
H_{kl} = k_2 \bar{\varphi}(\xi_{kl}) &\left[\frac{\Gamma\left(\gamma\left(\frac{1}{2}-\beta\right)+1\right)\Gamma\left(k-l-(1+\gamma)\beta+\frac{\gamma}{2}\right)}{\Gamma(k-l+1+\gamma-(1+2\gamma)\beta)} (1-x)^{-2(l+1+\beta)}\right.\\
&\times F\left(l+1+\beta,\ \left(\frac{1}{2}-\beta\right)\gamma+1,\ l-k+1+(1+\gamma)\beta-\frac{\gamma}{2};\ \left(\frac{x}{1-x}\right)^2\right)\\
&+ \frac{\Gamma\left(k+\gamma\left(\frac{1}{2}-\beta\right)+1\right)\Gamma\left(l-k+(1+\gamma)\beta-\frac{\gamma}{2}\right)}{\Gamma(l+1+\beta)}\\
&\times x^{2(k-l)-2\beta(1+\gamma)+\gamma} (1-x)^{-2\left(k+\gamma\left(\frac{1}{2}-\beta\right)+1\right)}\\
&\times F\left(k-l+\gamma+1-\beta(1+2\gamma),\ k+\gamma\left(\frac{1}{2}-\beta\right)+1,\ k-l+1+\frac{\gamma}{2}-(1+\gamma)\beta;\ \left.\left(\frac{x}{1-x}\right)^2\right)\right].
\end{aligned}$$

Hence, for small values of x, we have

$$H_{00} = O(1), \quad H_{11} = O(1), \quad H_{22} = O(1)$$

(since $\gamma - 2\beta(1 + \gamma) > 0$ for $\gamma > 0$), and

$$H_{01} = O(x^{\gamma-2-2\beta(1+\gamma)}), \qquad H_{02} = O(x^{\gamma-4-2\beta(1+\gamma)}),$$
$$H_{12} = O(x^{\gamma-2-2\beta(1+\gamma)}).$$

Substituting these estimates into (6.31), we get

$$\Phi_2(x) = O(x), \quad \Phi_2'(x) = O(1), \quad \Phi_2''(x) = O(x^{\gamma-1-2\beta(1+\gamma)}). \tag{6.23}$$

We can now determine γ from the condition

$$\gamma - 1 - 2\beta(1+\gamma) \geqslant 0, \tag{6.33}$$

i.e.

$$\gamma \geqslant \frac{1+2\beta}{1-2\beta} = 1 + m \qquad (-1 < m < 0),$$

thus obtaining an estimate

$$\Phi_2''(x) = O(1) \tag{6.34}$$

as $x \longrightarrow 0$.

Thus, $\Phi_2'(x)$ and $\Phi_2''(x)$ have finite limits as $x \longrightarrow 0$ and $x \longrightarrow 1$, provided the boundary values on the normal curve $C_{1/2}$ satisfy (6.28) and (6.33).

It follows from (6.32) that $\Phi_2(0) = 0$. Thus (6.26) and (6.27) imply the following formula for the free term $f_1(x)$:

$$f_1(x) = \frac{\lambda}{k_1\beta}\left\{\left[\psi'(0) - \frac{1}{2}\Phi_2'(0)\right]x^{2\beta} + x^{\beta}\int_0^x (x-t)^{\beta}\psi''(t)\,dt - \frac{1}{2}\int_0^x (x-t)^{2\beta}\Phi_2''(t)\,dt\right\}. \tag{6.35}$$

Note that in view of the restrictions imposed on $\psi(x)$, and by virtue of (6.34), we have the following estimate for the integrals in (6.35):

$$x^{\beta}\int_0^x (x-t)^{\beta}\psi''(t)\,dt - \frac{1}{2}\int_0^x (x-t)^{2\beta}\Phi_2''(t)\,dt = O(x^{1+2\beta}).$$

Let the boundary values on $C_{1/2}$ and on the characteristic AC be such that

$$\psi'(0) - \frac{1}{2}\Phi_2'(0) = \psi'(0) - \frac{k_2}{2}\int_0^1 \varphi(t)\,t^{\gamma\left(\frac{1}{2}-\beta\right)-(1+\beta)}(1-t)^{\gamma\left(\frac{1}{2}-\beta\right)}\,dt = 0. \tag{6.36}$$

Then $f_1(x) = O(x^{1+2\beta})$; that is, $f_1(0) = 0$, and it follows from (6.35) that $f_1(x) \in C[0, 1]$. Moreover, by the Hardy-Littlewood theorem (see [30]), $f_1(x)$ satisfies a Hölder condition with exponent $1 + 2\beta$ on $[0, 1]$. If (6.36) fails to hold, the same theorem shows that, because of (6.35), $f_1(x)$ satisfies a Hölder condition on $(0, 1]$ with exponent $1 + 2\beta$, and

$$f_1(x) = O(x^{2\beta}). \tag{6.37}$$

3. *Solution of the integral equation* (6.23). Setting

$$x^{1-2\beta}\nu(x) = \rho(x), \qquad x^{1-2\beta} f_1(x) = g(x), \tag{6.38}$$

we reduce (6.23) to the form

$$\rho(x) - \lambda \int_0^1 \left(\frac{1}{t-x} - \frac{1}{t+x-2xt}\right) \rho(t)\, dt = g(x), \tag{6.39}$$

where

$$g(x) = \frac{\lambda}{k_1\beta}\left\{\left[\psi'(0) - \frac{1}{2}\Phi_2'(0)\right]x + x^{1-\beta}\int_0^x (x-t)^\beta \psi''(t)\, dt - \frac{x^{1-2\beta}}{2}\int_0^x (x-t)^{2\beta}\Phi_2''(t)\, dt\right\}, \tag{6.40}$$

and $g(x)$ satisfies a Hölder condition with exponent $1 + 2\beta$ for $0 \leqslant x \leqslant 1$.

We seek a solution $\rho(x)$ of equation (6.39) in the class of functions satisfying the following conditions:

a) $\rho(x)$ satisfies a Hölder condition on any interval $a \leqslant x \leqslant b$, $0 < a < b < 1$.

b) The products $\rho(x)\ln x$ and $\rho(x)\ln(1-x)$ are summable on $0 \leqslant x \leqslant 1$.

By a result of Mihlin [55a], the solution of (6.39) is given by

$$\rho(x) = \frac{1}{1+\lambda^2\pi^2}\left\{g(x) + \lambda\int_0^1 \left[\frac{(1-x)\,t}{x\,(1-t)}\right]^{2\theta}\left(\frac{1}{t-x} - \frac{1}{t+x-2xt}\right) g(t)\, dt\right\} + \frac{A}{x^{1+2\theta}(1-x)^{-2\theta}}, \tag{6.41}$$

where $\theta = (1/\pi)\arctan \lambda\pi$, $-\tfrac{1}{2} < \theta < \tfrac{1}{2}$, and A is an arbitrary constant. Replacing λ by its value (6.24), we get

$$\theta = \frac{1}{2}\left(\beta + \frac{1}{2}\right). \tag{6.42}$$

Hence $0 < 2\theta < \tfrac{1}{2}$ for $-\tfrac{1}{2} < \beta < 0$.

Substituting (6.42) into (6.41) and returning to the functions $\nu(x)$ and $f_1(x)$, we get the solution of (6.23):

$$\nu(x)=\frac{1}{2}(1-\sin\pi\beta)\Big[f_1(x)$$
$$+\lambda\int_0^1\left(\frac{t}{x}\right)^{\frac{3}{2}-\beta}\left(\frac{1-x}{1-t}\right)^{\beta+\frac{1}{2}}\left(\frac{1}{t-x}-\frac{1}{t+x-2xt}\right)f_1(t)\,dt\Big] \qquad (6.43)$$
$$+Ax^{\beta-\frac{5}{2}}(1-x)^{\beta+\frac{1}{2}}.$$

Let us study the behavior of the function $\nu(x)$ defined by (6.43) as $x\to 0$. Set

$$\int_0^1\left(\frac{t}{x}\right)^{\frac{3}{2}-\beta}\left(\frac{1-x}{1-t}\right)^{\beta+\frac{1}{2}}\left(\frac{1}{t-x}-\frac{1}{t+x-2xt}\right)f_1(t)\,dt=x^{2\beta}I(x), \qquad (6.44)$$

where

$$I(x)=\int_0^1\left(\frac{t}{x}\right)^{\frac{3}{2}+\beta}\left(\frac{1-x}{1-t}\right)^{\beta+\frac{1}{2}}\left(\frac{1}{t-x}-\frac{1}{t+x-2xt}\right)f_2(t)\,dt, \qquad (6.45)$$
$$f_2(x)=x^{-2\beta}f_1(x).$$

From (6.35) it follows that $f_2(x)\in C[0,1]$. Substituting

$$\frac{x(1-t)}{t(1-x)}=\mu,\qquad t=\frac{x}{x+(1-x)\mu},$$

we get

$$I(x)=\int_0^\infty f_2\left(\frac{x}{x+(1-x)\mu}\right)\frac{\mu^{-\left(\frac{1}{2}+\beta\right)}2\mu\,d\mu}{[x+(1-x)\mu]^2(1-\mu^2)}.$$

Using the expansion of $2\mu/[x+(1-x)\mu]^2(1-\mu^2)$ in partial fractions as a function of μ, we have

$$I(x)=\int_0^\infty f_2\left(\frac{x}{x+(1-x)\mu}\right)\frac{\mu^{-\left(\frac{1}{2}+\beta\right)}d\mu}{1-\mu}-\frac{1}{(2x-1)^2}\int_0^\infty f_2(t)\frac{\mu^{-\left(\frac{1}{2}+\beta\right)}d\mu}{1+\mu}$$
$$+\frac{2(1-x)(2x^2-2x+1)}{(2x-1)^2}\int_0^\infty f_2(t)\frac{\mu^{-\left(\frac{1}{2}+\beta\right)}d\mu}{x+(1-x)\mu}$$
$$+\frac{2x(1-x)}{2x-1}\int_0^\infty f_2(t)\frac{\mu^{-\left(\frac{1}{2}+\beta\right)}d\mu}{[x+(1-x)\mu]^2}.$$

In the last two integrals, we now return to the old variable t. Then

$$I(x)=\int_0^\infty f_2(t)\frac{\mu^{-\left(\frac{1}{2}+\beta\right)}d\mu}{1-\mu}-\frac{1}{(2x-1)^2}\int_0^\infty f_2(t)\frac{\mu^{-\left(\frac{1}{2}+\beta\right)}d\mu}{1+\mu}$$

$$+\frac{2(2x^2-2x+1)}{(2x-1)^2}\left(\frac{x}{1-x}\right)^{-\left(\frac{1}{2}+\beta\right)}\int_0^1\frac{f_2(t)\,dt}{t^{\frac{1}{2}-\beta}(1-t)^{\frac{1}{2}+\beta}} \tag{6.46}$$

$$+\frac{2x^{-\left(\frac{1}{2}+\beta\right)}(1-x)^{\frac{1}{2}+\beta}}{2x-1}\int_0^1\frac{f_2(t)\,t^{\frac{1}{2}+\beta}\,dt}{(1-t)^{\frac{1}{2}+\beta}}.$$

Since $-\frac{1}{2}<\beta<0$, all the integrals in (6.46) exist, while the first integral is interpreted in the sense of the Cauchy principal value.

From (6.46) we obtain the obvious estimate $I(x)=O(x^{-(\frac{1}{2}+\beta)})$ as $x\to 0$, and by virtue of (6.44) the integral in (6.34) is $O(x^{\beta-\frac{1}{2}})$ as $x\to 0$. Hence, in view of (6.37), we must set $A=0$ in (6.43). Then $\nu(x)$ is $O(x^{2\beta})$ as $x\to 0$, Thus, finally,

$$\nu(x)=\frac{1-\sin\pi\beta}{2}\left[f_1(x)+\lambda\int_0^1\left(\frac{t}{x}\right)^{\frac{3}{2}-\beta}\left(\frac{1-x}{1-t}\right)^{\beta+\frac{1}{2}}\right.$$

$$\left.\times\left(\frac{1}{t-x}-\frac{1}{t+x-2xt}\right)f_1(t)\,dt\right]. \tag{6.47}$$

It follows easily from (6.47), (6.46), (6.44) and (6.35) that $\nu(x)=O(1)$ as $x\to 1$. Thus, $\nu(x)$ is continuous and integrable in the interval (0, 1).

Substituting (6.47) into (6.1), we find $\tau(x)$. It follows from (4.27) and (4.28) that $\tau(x)$ may be expressed in the form (4.8), where

$$T(t)=\varkappa_3\nu(t)+\frac{2\cos\pi\beta}{\Gamma(1-\beta)}t^\beta D_0^{1-\beta}\psi(t).$$

It is easy to see that the function $T(t)$ is continuous and integrable on (0, 1). Knowing $\tau(x)$ and $\nu(x)$, we use (5.13) and (4.7) to determine the solution $u(x, y)$ in the domains D^+ and D^-, respectively; while in D^- the function $u(x, y)$ is a generalized solution in class R_2. We have thus proved the existence of a solution to the modified Tricomi problem for equation (1.1), in the case of a normal domain. The uniqueness of this solution follows immediately from the uniqueness of the solution of the singular equation (6.23) in the above-mentioned class of functions and from the uniqueness theorems for Problem N and the Cauchy problem for equation (1.1).

BIBLIOGRAPHY

1. F. B. ABUTALIEV
   a) *A uniqueness theorem for some problems of gas-dynamics*, Izv. Akad. Nauk UzSSR Ser. Fiz.-Mat. **1961**, no. 1, 12–21. (Russian) MR **24** #A2767.
   b) *On the numerical solution of equations of mixed type*, Dokl. Akad. Nauk UzSSR **1961**, no. 7, 3–5. (Russian) RŽ Mat. **1963** #6B16.
2. SHMUEL AGMON
   a) *The fundamental solution and Tricomi's problem for a class of equations of mixed type*, Proc. Internat. Congr. Math. (Amsterdam, 1954), vol. II, Noordhoff, Groningen; North-Holland, Amsterdam, 1954, p. 77.
   b) *Boundary value problems for equations of mixed type*, Convegno Internaz. Equazioni Lineari alle Derivate Parziali (Trieste, 1954), Edizioni Cremonese, Rome, 1955, pp. 54–68. MR **17**, 859.
3. S. AGMON, L. NIRENBERG AND M. H. PROTTER
   A maximum principle for a class of hyperbolic equations and applications to equations of mixed elliptic-hyperbolic type, Comm. Pure Appl. Math. **6** (1953), 455–470. MR **15**, 432.
4. K. I. BABENKO
   a) *On the theory of equations of mixed type*, Doctoral Dissertation, Steklov Inst. Math., Moscow, 1952. (Russian)
   b) *On the theory of equations of mixed type*, Uspehi Mat. Nauk **8** (1953), no. 2 (54), 160. (Russian)
   c) *On conjugate functions*, Dokl. Akad. Nauk SSSR **62** (1948), 157–160. (Russian) MR **10**, 249.
5. N. I. BAKIEVIČ
   a) *Some boundary value problems for equations of mixed type in a strip and in a half-plane*, Dokl. Akad. Nauk SSSR **112** (1957), 793–796. (Russian) MR **19**, 150.
   b) *Some boundary value problems for equations of mixed type arising in the study of infinitesimal bending of surfaces of rotation*, Uspehi Mat. Nauk **15** (1960), no. 1 (91), 171–176. (Russian) MR **22** #8826.
   c) *On a certain equation of mixed type in the theory of infinitesimal bendings of surfaces*, Volž. Mat. Sb. Vyp. 1 (1963), 32–42. (Russian) MR **33** #4791.
   d) *Singular Tricomi problems for the equation* $\eta^\alpha u_{\xi\xi} + u_{\eta\eta} - \mu^2\eta^\alpha u = 0$, Volž. Mat. Sb. Vyp. 1 (1963), 42–52. (Russian) MR **33** #4486.
   e) *Singular Tricomi problems for the equation* $\eta^\alpha u_{\xi\xi} + u_{\eta\eta} - \mu^2\eta^\alpha u = 0$, Izv. Vysš. Učebn. Zaved. Matematika 1964, no. 2(39), 7–13. (Russian) MR **29** #382.
6. R. G. BARANCEV
   Lectures on transonic gas dynamics, Izdat. Leningrad. Univ., Leningrad, 1965. (Russian) MR **34** #7113.

7. Ju. M. BEREZANSKIĬ

a) *Energy inequalities for some classes of equations of mixed type*, Dokl. Akad. Nauk SSSR **132** (1960), 9–12 = Soviet Math. Dokl. 1 (1960), 447–451. MR **22** #11237.

b) *Existence of weak solutions for certain boundary value problems for equations of mixed type*, Ukrain. Mat. Ž. **15** (1963), 347–364. (Russian) MR **28** #370.

c) *Expansion in eigenfunctions of selfadjoint operators*, "Naukova Dumka," Kiev, 1965; English transl., Amer. Math. Soc., Providence, R. I., 1968. MR **36** #5768; **36** #5796.

8. STEFAN BERGMAN

On solutions of linear partial differential equations of mixed type, Amer. J. Math. **74** (1952), 444–474. MR **15**, 229.

9. LIPMAN BERS

Mathematical aspects of subsonic and transonic gas dynamics, Wiley, New York; Chapman & Hall, London, 1958. MR **20** #2960.

10. A. V. BICADZE

a) *On some problems of mixed type*, Dokl. Akad. Nauk SSSR **70** (1950), 561–564. (Russian) MR **11**, 724.

b) *On the uniqueness of solution of a general boundary value problem for an equation of mixed type*, Soobšč. Akad. Nauk Gruzin. SSR **11** (1950), 205–210. (Russian) MR **14**, 281.

c) *On the general problem of mixed type*, Dokl. Akad. Nauk SSSR **78** (1951), 621–624. (Russian) MR **14**, 280.

d) *On the problem of equations of mixed type*, Trudy Mat. Inst. Steklov. **41** (1953); German transl., VEB Deutscher Verlag, Berlin, 1957. MR **16**, 43; **20** #1087.

e) *On a problem of Frankl'*, Dokl. Akad. Nauk SSSR **109** (1956), 1091–1094. (Russian) MR **18**, 743.

f) *On the problem of equations of mixed type in multidimensional domains*, Dokl. Akad. Nauk SSSR 110 (1956), 901–902. (Russian) MR **18**, 657.

g) *On the uniqueness of solution of Frankl''s problem for Čaplygin's equation*, Dokl. Akad. Nauk SSSR **112** (1957), 375–376. (Russian) MR **19**, 861.

h) *Ill-posedness of the Dirichlet problem for equations of mixed type in mixed domains*, Dokl. Akad. Nauk SSSR **122** (1958), 167–170. (Russian) MR **21** #2118.

i) *Equations of mixed type*, Izdat. Akad. Nauk SSSR, Moscow, 1959; English transl., Macmillan, New York, 1964. MR **29** #381.

j) *Equations of mixed type in three-dimensional domains*, Dokl. Akad. Nauk SSSR **143** (1962), 1017–1019 = Soviet Math. Dokl. 3 (1962), 510–512. MR **24** #A3430.

11. TORSTEN CARLEMAN

Sur la résolution de certaines équations intégrales, Ark. Mat. Astr. Fys. **16** (1922), no. 26.

12. L. I. ČIBRIKOVA

a) *On the solution of the Tricomi boundary-value problem for the equation* $\partial^2 u/\partial x^2 + (\operatorname{sgn} y)\partial^2 u/\partial y^2 = 0$, Kazan. Gos. Univ. Učen. Zap. **117** (1957), no. 9, 48–51. (Russian) RŽ Mat. **1959** #2702.

b) *A new method for solving a boundary value problem of mixed type*, Kazan. Gos. Univ. Učen. Zap. **117** (1957), no. 9, 44–47. (Russian) RŽ Mat. **1959** #3827.

13. MARIA CINQUINI-CIBRARIO

a) *Una proprietà degli integrali delle equazioni ellittico-paraboliche del secondo tipo misto*, Atti Accad. Ital. Rend. Cl. Sci. Fis. Mat. Nat. (7) **3** (1942), 502–510. MR **8**, 209.

b) *Alcuni teoremi di esistenza e di unicità per l'equazione* $xz_{xx} + z_{yy} = 0$, Atti Accad. Sci. Torino **68** (1932/33), 35–44.

14. JULIAN D. COLE
Note on the fundamental solution of $wy_{vv} + y_{ww} = 0$, Z. Angew. Math. Phys. **3** (1952), 286–297. MR **14**, 380.

15. R. COURANT
Methods of mathematical physics. Vol. II: *Partial differential equations*, Interscience, New York, 1962. MR **25** #4216.

16. Ju. V. DEVINGTAL'
a) *On the existence of a solution to a problem of Frankl'*, Dokl. Akad. Nauk SSSR **119** (1958), 15–18. (Russian) MR **20** #4703.
b) *The existence and uniqueness of the solution of a problem of F. I. Frankl'*, Izv. Vysš. Učebn. Zaved. Matematika **1958**, no. 2 (3), 39–51. MR **27** #1692.
c) *Existence and uniqueness of the solution of the Frankl' problem*, Uspehi Mat. Nauk **14** (1959), no. 1 (85), 177–182. (Russian) MR **21** #3658.

17. V. P. DIDENKO
a) *Certain systems of differential equations of mixed type*, Dokl. Akad. Nauk SSSR **144** (1962), 709–712 = Soviet Math. Dokl. **3** (1962), 796–799. MR **24** #A3402.
b) *On some systems of differential equations of mixed and mixed-compound types*, Differencial'nye Uravnenija **2** (1966), 33–46 = Differential Equations **2** (1966), 16–22. MR **33** #2971.

18. DING XIA-QI
Differential equations of mixed type, Acta Math. Sinica **5** (1955), 193–204. (Chinese; English summary) MR **17**, 748.

19. DONG GUANG-CHANG
Uniqueness theorem for the Čaplygin problem. I, II, III, Acta. Math. Sinica **6** (1956), 242–247, 250–262; **9** (1959), 365–381; English transl., Amer. Math. Soc. Transl. (2) 86 (1970), 261–267, 269–282; Chinese Math. Acta **9** (1967), 627–642. MR **22** #11221.

20. A. F. FILIPPOV
On the application of the method of finite differences to the solution of the Tricomi problem, Izv. Akad. Nauk SSSR Ser. Mat. **21** (1957), 73–88. (Russian) MR **19**, 773.

21. NORBERT M. FLEISCHER
a) *Boundary value problems for equations of mixed type in the case of unbounded domains*, Rev. Roumaine Math. Pures Appl. **10** (1965), 607–613. (Russian) MR **33** #2972.
b) *On Frankel''s problem for the Lavrent'ev equation in the case of an unbounded domain*, Izv. Vysš. Učebn. Zaved. Matematika **1966**, no. 6 (55), 152–156. (Russian) MR **35** #3273.
c) *The Frankl' problem for the half-plane*, Rev. Roumaine Math. Pures. Appl. **11** (1966), 205–209. (Russian) MR **34** #4688.

22. F. I. FRANKL'
a) *On the problems of Čaplygin for mixed sub- and supersonic flows*, Izv. Akad. Nauk SSSR Ser. Mat. **9** (1945), 121–143; English transl., Tech. Memos Nat. Adv. Comm. Aeronaut. no. 1155 (1947). MR **7**, 496; **8**, 541.
b) *Two gas-dynamical applications of the Lavrent'ev-Bicadze boundary value problem*, Vestnik Leningrad. Univ. Ser. Mat. Meh. Astronom. 6 (1951), no. 11, 3–7. (Russian)
c) *On a new boundary value problem for the equation* $y\partial^2 z/\partial x^2 + \partial^2 z/\partial y^2 = 0$, Moskov. Gos. Univ. Učen. Zap. 152, Meh. 3 (1951), 99–116. (Russian) MR **14**, 654.

d) *On lateral water intake from rapid shallow rivers*, Kirgiz. Gos. Univ. Trudy Fiz.-Mat. Fak. **1953**, no. 2, 33–45. (Russian) RŽ Meh. **1953/54** #4071.

e) *Gas flows past profiles with a local supersonic zone ending in a direct shock wave*, Prikl. Mat. Meh. **20** (1956), 196–202. (Russian) AMR **11** #3160.

f) *A uniqueness theorem for the solution of a boundary value problem for the equation* $u_{xx} + \partial(u_y/y)/\partial y = 0$, Izv. Vysš. Učebn. Zaved. Matematika **1959**, no. 1 (8), 212–217. (Russian) MR **27** #2722.

g) *An existence theorem for a weak solution and first approximation of the direct problem in the theory of the plane-parallel Laval nozzle*, Izv. Vysš. Učebn. Zaved. Matematika **1959**, no. 6 (13), 192–201. (Russian) MR **28** #805.

h) *On the direct problem of the theory of the Laval nozzle*, Kabardino-Balkarsk. Gos. Univ. Učen. Zap. **3** (1959), 35–60. (Russian) RŽ Mat. **1962** #2Б327.

i) *A generalization of the Tricomi problem and its application in solving the direct problem of the theory of the Laval nozzle*, Kabardino-Balkarsk. Gos. Univ. Učen. Zap. **3** (1959), 79–93. (Russian) RŽ Mat. **1962** #2Б329.

j) *A generalization of the Tricomi problem and its application in solving the direct problem of the theory of the Laval nozzle*, Mat. Sb. **54** (96) (1961), 225–236. (Russian) MR **27** #4439.

k) *On the Cauchy problem for partial differential equations of mixed elliptic-hyperbolic type with initial data on the parabolic line*, Izv. Akad. Nauk SSSR Ser. Mat. **8** (1944), 195–224. (Russian) MR **7**, 17.

l) *To the theory of the Laval nozzle*, Izv. Akad. Nauk SSSR Ser. Mat. **9** (1945), 387–422. (Russian) MR **8**, 416.

23. KURT O. FRIEDRICHS

Symmetric positive linear differential equations, Comm. Pure Appl. Math. **11** (1958), 333–418. MR **20** #7147.

24. SVEN GELLERSTEDT

a) *Sur un problème aux limites pour une équation linéaire aux dérivées partielles du second ordre de type mixte*, Thesis, Uppsala, 1935. Jbuch Fortschritte Math. **61**, 1259.

b) *Sur une équation linéaire aux dérivées partielles de type mixte*, Ark. Mat. Astr. Fys. **25A** (1937), no. 29.

c) *Quelques problèmes mixtes pour l'équation* $y^m z_{xx} + z_{yy} = 0$, Ark. Mat. Astr. Fys. **26A** (1938), no. 3.

25. PAUL GERMAIN

a) *Recherche sur une équation du type mixte*, Recherche Aéronautique no. 22 (1951), 7–20. MR **14**, 177.

b) *Remarks on the theory of partial differential equations of mixed type and applications to the study of transonic flow*, Comm. Pure Appl. Math. **7** (1954), 117–143. MR **16**, 485.

c) *Remarques sur les propriétés qualitatives des caractéristiques des équations aux dérivées partielles du type mixte*, Acad. Roy. Belg. Cl. Sci. Mém. **28** (1954), no. 6, 28–36.

d) *New applications of Tricomi solutions to transonic flow*, Proc. Second U. S. Nat. Congr. Appl. Mech. (Ann Arbor, Mich., 1954), Amer. Soc. Mech. Engrs., New York, 1955, pp. 659–666. MR **18**, 90.

e) *Maximum theorems and reflections of simple waves*, NACA Tech. Note #3299 (1955). MR **17**, 100.

f) *Remarks on transforms and boundary value problems*, J. Rational Mech. Anal. **4** (1955), 925–941. MR **17**, 846.

g) *An expression for Green's function for a particular Tricomi problem*, Quart. Appl. Math. **14** (1956), 113–124. MR **18**, 133.

h) *Solutions élémentaires des équations régissant les écoulements des fluides compressibles*, C. R. Acad. Sci. Paris **234** (1952), 1248–1250. MR **14**, 598.

26. PAUL GERMAIN AND ROGER BADER

a) *Application de la solution fondamentale à certains problèmes relatifs à l'équation de Tricomi*, C. R. Acad. Sci. Paris **231** (1950), 1203–1205. MR **14**, 380.

b) *Sur le problème de Tricomi*, C. R. Acad. Sci. Paris **232** (1951), 463–465. MR **14**, 177.

c) *Sur quelques problèmes relatifs à l'équation de type mixte de Tricomi*, O.N.E.R.A.* Publ. No. 54 (1952). MR **14**, 654.

d) *Sur le problème de Tricomi*, Rend. Circ. Mat. Palermo (2) **2** (1953), 53–70. MR **15**, 876.

e) *Problèmes elliptiques et hyperboliques singuliers pour une équation du type mixte*, O.N.E.R.A.* Publ. No. 60 (1953).

f) *Solutions élémentaires de certaines équations aux dérivées partielles du type mixte*, Bull. Soc. Math. France **81** (1953), 145–174. MR **15**, 432.

27. GOTTFRIED GUDERLEY

On the development of solutions of Tricomi's differential equation in the vicinity of the origin, J. Rational Mech. Anal. **5** (1956), 747–790. MR **18**, 314.

28. WOLFGANG HAACK AND GÜNTER HELLWIG

Lineare partielle Differentialgleichungen zweiter Ordnung von gemischtem Typus, Arch. Math. **5** (1954), 60–76. MR **16**, 255.

29. Z. I. HALILOV

a) *Solution of a problem for an equation of mixed type by the method of nets*, Dokl. Akad. Nauk Azerbaĭdžan. SSR **9** (1953), 189–194. (Russian) MR **19**, 178.

b) *Solution of a problem for an equation of mixed type by the method of nets*, Akad. Nauk Azerbaĭdžan. SSR Trudy Inst. Fiz. Mat. **6** (1953), 5–13. (Russian) RŽ Mat. **1960** #10936.

30. G. H. HARDY AND J. E. LITTLEWOOD

Some properties of fractional integrals. I, Math. Z. **27** (1927/28), 565–606.

31. G. H. HARDY, J. E. LITTLEWOOD AND G. PÓLYA

Inequalities, Cambridge Univ. Press, New York, 1934.

32. JU F. HARKEVIČ

Graphical solution of an equation of mixed type, Trudy Irkutsk. Gos. Univ. Ser. Fiz.-Mat. **15** (1957), no. 2, 103–110. (Russian) RŽ Mat. **1958** #8266.

33. GÜNTER HELLWIG

a) *Anfangs- und Randwertprobleme bei partiellen Differentialgleichungen von wechselndem Typus auf den Rändern*, Math. Z. **58** (1953), 337–357. MR **15**, 130.

b) *Über partielle Differentialgleichungen zweiter Ordnung von gemischtem Typus*, Math. Z. **61** (1954), 26–46. MR **16**, 828.

34. KINZŌ HIDA

On some singular solutions of the Tricomi equation relating to transonic flow, J. Phys. Soc. Japan **10** (1955), 869–881. MR **17**, 313.

35. HARRY HOCHSTADT

On a partial differential equation of mixed type with discontinuous coefficients, Arch. Rational Mech. Anal. **10** (1963), 267–272. MR **25** #4256.

36. M. B. KAPILEVIČ

a) *On an equation of mixed elliptic-hyperbolic type*, Mat. Sb. **30** (72) (1952), 11–38. (Russian) MR **13**, 750.

---

*Office National d'Études et de Recherches Aéronautiques, France.

b) *The theory of linear differential equations with two perpendicular lines of parabolicity*, Dokl. Akad. Nauk SSSR **125** (1959), 251–254. (Russian) MR **22** #5804.

c) *Equations of mixed elliptic-hyperbolic type*, Chapter VIII in S. G. Mihlin et al., *The linear equations of mathematical physics*, "Nauka", Moscow, 1964; German transl., Akademie-Verlag, Berlin, 1967. MR **30** #5028; **36** #1851.

37. F. I. KARAMYŠEV

A boundary value problem for a system of partial differential equations of mixed type, Sibirsk. Mat. Ž. **2** (1961), 537–546. (Russian) MR **25** #2332.

38. G. D. KARATOPRAKLIEV

A generalization of the Tricomi problem, Dokl. Akad. Nauk SSSR **158** (1964), 271–274 = Soviet Math. Dokl. **5** (1964), 1220–1223. MR **29** #2551.

39. V. G. KARMANOV

a) *On a boundary problem for an equation of mixed type*, Dokl. Akad. Nauk SSSR **95** (1954), 439–442. (Russian) MR **16**, 369.

b) *On the existence of solutions of certain boundary value problems for equations of mixed type*, Izv. Akad. Nauk SSSR Ser. Mat. **22** (1958), 117–134. (Russian) MR **23** #A2649.

40. I. L. KAROL′

a) *On a boundary value problem for an equation of mixed elliptic-hyperbolic type*, Dokl. Akad. Nauk SSSR **88** (1953), 197–200. (Russian) MR **14**, 757.

b) *On the theory of equations of mixed type*, Dokl. Akad. Nauk SSSR **88** (1953), 397–400. (Russian) MR **14**, 757.

c) *Boundary value problems for an equation of mixed elliptic-hyperbolic type*, Dokl. Akad. Nauk SSSR 101 (1955), 793–796. (Russian) MR **18**, 313.

d) *On boundary value problems for equations of mixed type*, Vestnik Leningrad. Univ. **11** (1956), no. 1, 177–181. (Russian) MR **17**, 858.

e) *On the theory of boundary value problems for an equation of mixed elliptic-hyperbolic type*, Mat. Sb. **38** (80) (1956), 261–282. (Russian) MR **17**, 858.

41. V. V. KLOKOV

Solution of the generalized shock problem of F. I. Frankl′ by the method of integral equations, Trudy Sem. Kraev. Zadačam (Kazan) Vyp. 2 (1964), 42–71. (Russian) RŽ Mat. **1966** #3Б359.

42. M. N. KOGAN

On magnetohydrodynamic flows of mixed type, Prikl. Mat. Mech. **25** (1961), 132–137 = J. Appl. Math. Mech. **25** (1961), 180–188. AMR **16** #3576.

43. L. I. KOVALENKO

a) *The difference method and the uniqueness of the generalized solution of the Tricomi problem*, Dokl. Akad. Nauk SSSR **162** (1965), 751–754 = Soviet Math. Dokl. **6** (1965), 747–751. MR **32** #6078.

b) *Generalized solution of the Tricomi problem*, Dokl. Akad. Nauk SSSR **162** (1965), 988–991 = Soviet Math. Dokl. **6** (1965), 789–793. MR **32** #6079.

44. JU. M. KRIKUNOV

The Tricomi problem with derivatives in the boundary conditions, Kazan. Gos. Univ. Učen. Zap. **123** (1963), kn. 10*, 106–113 (1964). (Russian) MR **33** #6160.

**Editor's note.* In some citations (e.g. RŽ Mat. 1965 #4Б272) this number is given as "9".

45. O. A. LADYŽENSKAJA
On a method of approximate solution of the Lavrent'ev-Bicadze problem, Uspehi Mat. Nauk **9** (1954), no. 4 (62), 187–189. (Russian) MR **16**, 1120.

46. EBERHARD LANCKAU
Eine einheitliche Darstellung der Lösungen der Tricomischen Gleichung, Z. Angew. Math. Mech. **42** (1962), 180–186. MR **25** #336.

47. M. A. LAVRENT'EV AND A. V. BICADZE
On the problem of equations of mixed type, Dokl. Akad. Nauk SSSR **70** (1950), 373–376. (Russian) MR **11**, 724.

48. P. D. LAX AND R. S. PHILLIPS
Local boundary conditions for dissipative symmetric linear differential operators, Comm. Pure Appl. Math. **13** (1960), 427–455. MR **22** #9718.

49. N. N. LEBEDEV
Special functions and their applications, 2nd rev. ed., Fizmatgiz, Moscow, 1963; English transl., Prentice-Hall, Englewood Cliffs, N. J., 1965. MR **30** #4987; 4988.

50. O. M. LEMEŠINSKAJA
Investigation of boundary value problems for the Tricomi equation by the method of the theory of analytic functions, Izv. Vysš. Učebn. Zaved. Matematika **1965**, no. 2 (45), 91–97. (Russian) MR **32** #1458.

51. LIN JIAN-BING
On some problems of Frankl', Vestnik Leningrad. Univ. **16** (1961), no. 13, 28–39. (Russian) MR **25** #3285.

52. I. V. MAĬOROV
a) *On the question of the maximum principle and its consequences for equations of mixed type,* Volž. Mat. Sb. Vyp. 1 (1963), 145–155. (Russian) MR **34** #4705.
b) *The extremum principle for a problem of Frankl',* Sibirsk. Mat. Ž. **7** (1966), 1068–1075 = Siberian Math. J. **7** (1966), 850–855. MR **34** 3106.

53. B. V. MELENT'EV
a) *Uniqueness theorems for the solution of certain boundary value problems for equations of mixed type,* Dokl. Akad. Nauk SSSR **143** (1962), 38–41 = Soviet Math. Dokl. **3** (1962), 338–342. MR **26** #4062.
b) *An existence theorem for the solution of a boundary value problem for an equation of mixed type,* Dokl. Akad. Nauk SSSR **142** (1962), 1251–1254 = Soviet Math. Dokl. **3** (1962), 292–295. MR **24** #A3431.
c) *On a boundary value problem for equations of mixed type,* Dokl. Akad. Nauk SSSR **154** (1964), 1262–1265 = Soviet Math. Dokl. **5** (1964), 270–273. MR **28** #3264.

54. V. P. MIHAĬLOV
On the generalized Tricomi problem, Dokl. Akad. Nauk SSSR **175** (1967), 1012–1014 = Soviet Math. Dokl. **8** (1967), 968–970. MR **36** #2984.

55. S. G. MIHLIN
a) *On the integral equation of F. Tricomi,* Dokl. Akad. Nauk SSSR **59** (1948), 1053–1056. (Russian) MR **9**, 443.
b) *Singular integral equations,* Uspehi Mat. Nauk **3** (1948), no. 3 (25), 29–112; English transl., Amer. Math. Soc. Transl. (1) **10** (1962), 84–198. MR **10**, 305.

56. CATHLEEN S. MORAWETZ
a) *A uniqueness theorem for Frankl's problem,* Comm. Pure Appl. Math. **7** (1954), 697–703. MR **16**, 484.

b) *Note on a maximum principle and a uniqueness theorem for an elliptic-hyperbolic equation*, Proc. Roy. Soc. London Ser. A **236** (1956), 141–144. MR **18**, 133.

c) *Uniqueness for the analogue of the Neumann problem for mixed equations*, Michigan Math. J. **4** (1957), 5–14. MR **19**, 40.

d) *On the non-existence of continuous transonic flows past profiles*. I, II, III, Comm. Pure Appl. Math. **9** (1956), 45–68; **10** (1957), 107–131; **11** (1958), 129–144. MR **17**, 1149; **19**, 490; **20** #2961.

e) *A weak solution of a system of equations of elliptic-hyperbolic type*, Comm. Pure Appl. Math. **11** (1958), 315–331. MR **20** #3375.

57. A. M. NAHUŠEV

An equation of mixed type for the equation $y(y-1)u_{xx} + u_{yy} = 0$, Dokl. Akad. Nauk SSSR **166** (1966), 536–539 = Soviet Math. Dokl. **7** (1966), 113–117. MR **32** #6080.

58. JOHANNES C. C. NITSCHE

On solutions of differential equations of mixed type, Bull. Amer. Math. Soc. **62** (1956), 250.

59. E. I. OBOLAŠVILI

Effective solution of the Riemann-Hilbert problem for a system of equations of mixed type, with application to the theory of shells, Soobšč. Akad. Nauk Gruzin. SSR 36 (1964), 33–39. (Russian)

60. HAJIMU OGAWA

On difference methods for the solution of a Tricomi problem, Trans. Amer. Math. Soc. **100** (1961), 404–424. MR **25** #3272.

61. L. V. OVSJANNIKOV

On Tricomi's problem in a class of generalized solutions of the Euler-Darboux equation, Dokl. Akad. Nauk SSSR **91** (1953), 457–460. (Russian) MR **15**, 432.

62. I. M. PETRUŠKO

The Fredholmicity of certain boundary value problems for the equation $u_{xx} + yu_{yy} + \alpha(x, y)u_y + \beta(x, y)u_x + \gamma(x, y)u = f(x, y)$ *in a mixed domain*, Differencial'nye Uravnenija **4** (1968), 123–135 = Differential Equations **4** (1968), 63–69. MR **37** #618.

63. M. H. PROTTER

a) *A boundary value problem for an equation of mixed type*, Trans. Amer. Math. Soc. **71** (1951), 416–429. MR **14**, 281.

b) *Uniqueness theorems for the Tricomi problem*. I, II, J. Rational Mech. Anal. 2 (1953), 107–114; **4** (1955), 721–732. MR **14**, 654; **17**, 270.

c) *New boundary value problems for the wave equation and equations of mixed type*, J. Rational Mech. Anal. **3** (1954), 435–446. MR **16**, 43.

d) *An existence theorem for the generalized Tricomi problem*, Duke Math J. 21 (1954), 1–7. MR **15**, 628.

e) *On partial differential equations of mixed type*, Proc. Conf. Differential Equations (Dedicated to A. Weinstein), Univ. of Maryland Book Store, College Park, Md., 1956, pp. 91–106. MR **19**, 660.

64. S. P. PUL'KIN

a) *On the question of formulating the Tricomi problem in space*, Kuĭbyšev. Gos. Ped. Inst. Učen. Zap. **14** (1956), 63–77. (Russian) RŽ Mat. **1957** #463.

b) *The Tricomi problem for the generalized Lavrent'ev-Bicadze equation*, Dokl. Akad. Nauk SSSR **118** (1958), 38–41. (Russian) MR **20** #5343.

c) *On the question of solving the Tricomi problem for an equation of Čaplygin type*, Izv. Vysš. Učebn. Zaved. Matematika **1958**, no. 2 (3), 219–226. (Russian)

d) *Some boundary value problems for the equation* $u_{xx} \pm u_{yy} + pu_x/x = 0$, Kuĭbyšev. Gos. Ped. Inst. Učen. Zap **21** (1958), 3–55. (Russian) RŽ Mat. **1959** #11077.

65. B. V. ŠABAT

Examples of solution of the Dirichlet problem for equations of mixed type, Dokl. Akad. Nauk SSSR **112** (1957), 386–389. (Russian) MR **19**, 865.

66. M. M. SMIRNOV

a) *The generalized Tricomi equation*, Belorussk. Gos. Univ. Učen Zap. **12** (1951), 3–9. (Russian)

b) *On a boundary value problem for an equation of mixed type*, Vestnik Leningrad. Univ. **12** (1957), no. 1, 80–96, 209–210. (Russian) MR **19**, 864.

c) *The Dirichlet problem for an equation of mixed type*, Vestnik Leningrad. Univ. **14** (1959), no. 1, 130–133. (Russian) MR **21** #1457.

d) *The uniqueness of the solution of a boundary value problem for an equation of mixed type*, Dokl. Akad. Nauk BSSR **4** (1960), 140–143. (Russian) MR **27** #479.

e) *A mixed boundary-value problem for the equation* $y^m\partial^2u/\partial x^2 + \partial^2u/\partial y^2 = 0$, Sibirsk. Mat. Ž. **4** (1963), 1150–1161. (Russian) MR **28** #3265.

f) *Some boundary-value problems for an equation of mixed composite type*, Sibirsk. Mat. Ž. **5** (1964), 923–928. (Russian) MR **29** #3768.

g) *Degenerating elliptic and hyperbolic equations*, "Nauka", Moscow, 1966. (Russian) MR **36** #1850.

67. V. I. SMIRNOV

A course in higher mathematics, Vol. V, rev. ed., Fizmatgiz, Moscow, 1959; English transl., Pergamon Press, Oxford; Addison-Wesley, Reading, Mass., 1964. MR **29** #5964.

68. N. G. SOROKINA

On the strong solvability of the Tricomi problem, Ukrain. Mat. Ž. 18 (1966), 65–77; English transl., Amer. Math. Soc. Transl. (2) **87** (1970), 29–46. MR **34** #7985.

69. FRANCESCO TRICOMI

a) *Sulle equazioni lineari alle derivate parziali di* 2° *ordine, di tipo misto*, Atti Accad. Naz. Lincei Mem. Cl. Sci. Fis. Mat. Nat. (5) **14** (1923), 133–247 (1924).

b) *Ulteriori richerche sull'equazione* $y\partial^2z/\partial x^2 + \partial^2z/\partial y^2 = 0$, Rend. Circ. Mat. Palermo **52** (1928), 63–90.

c) *Ancora sull'equazioni* $y\partial^2z/\partial x^2 + \partial^2z/\partial y^2 = 0$, Atti Accad. Naz. Lincei Rend. Cl. Sci. Fis. Mat. Nat. (6) **6** (1927), 567–571.

70. SHUN-ICHI TSUGÉ

On a theory of shock waves in a locally supersonic zone and a new boundary-value problem for Tricomi's equation, J. Phys. Soc. Japan **12** (1957), 1412–1419. MR **19**, 800.

71. N. N. VAHANIJA

A special problem for an equation of mixed type, Trudy Vyčisl. Centra Akad. Nauk Gruzin. SSR **3** (1963), 69–80. (Georgian; Russian summary*) MR **32** #2750.

**Editor's note.* The citation in the original is of the summary only.

72. I. N. VEKUA

Generalized analytic functions, Fizmatgiz, Moscow, 1959; English transl., Pergamon Press, Oxford; Addison-Wesley, Reading, Mass., 1962. MR **21** #7288; **27** #321.

73. V. F. VOLKODAVOV

a) *Uniqueness of the solution of problem T for a general Tricomi equation*, Proc. First Sci. Conf. Math. Dept. Ped. Insts. Volga Region (May, 1960), Kuĭbyšev. Gos. Ped. Inst., Kuybyshev, 1961, pp. 45–49. (Russian) MR **34** #1719a.

b) *Solution of problem N for the general Tricomi equation*, Proc. First Sci. Conf. Math. Dept. Ped. Insts. Volga Region (May, 1960), Kuĭbyšev. Gos. Ped. Inst., Kuybyshev, 1961, pp. 49–52. (Russian) MR **34** #1719b.

c) *Solution of the Cauchy-Goursat problem for a Tricomi-type equation of a general form*, Proc. Second Sci. Conf. Math. Dept. Ped. Insts. Volga Region, Vyp. 1, Kuĭbyšev Gos. Ped. Inst., Kuybyshev, 1962, pp. 25–27. (Russian) MR **34** #1718.

d) *On solving problem T for the equation* $y \cdot z_{xx} + z_{yy} = M_0(x, y)$, Volž. Mat. Sb. Vyp. 2 (1964), 25–30. (Russian) MR **34** #1719c.

74. E. A. VOLKOV

On numerical solution of the Lavrent'ev-Bicadze problem, Dokl. Akad. Nauk SSSR **103** (1955), 755–758. (Russian) MR **17**, 539.

75. L. E. VOSTROVA

a) *The mixed boundary value problem for the equation* $\partial^2 u/\partial x^2 + (\operatorname{sgn} y)\partial^2 u/dy^2 - u = 0$, Kuĭbyšev. Gos. Ped. Inst. Učen. Zap. **21** (1958), 219–267. (Russian) RŽ Mat. **1959** #11079.

b) *The mixed boundary value problem for the general Lavrent'ev-Bicadze equation*, Kuĭbyšev. Gos. Ped. Inst. Učen. Zap. **29** (1959), 45–66. (Russian) MR **33** #4489.

c) *Continuity of the solution of problem* T_3 *for the equation* $U_{xx} + (\operatorname{sgn} y)U_{yy} - U = 0$, Proc. First Sci. Conf. Math. Dept. Ped. Insts. Volga Region (May, 1960), Kuĭbyšev. Gos. Ped. Inst., Kuybyshev, 1961, pp. 53–59. MR **33** #1604.

76. WANG GUANG-YIN

a) *Sur l'unicité du problème de Tricomi de l'équation de Chaplygin*, Acta Math. Sinica **5** (1955), 455–461. (Chinese; French summary) MR **17**, 975.

b) *Properties of the solutions of linear partial differential equations with singular coefficients in the neighborhood of a singular line or surface*, Acta Math. Sinica **7** (1957), 590–630. (Chinese; English summary) MR **21** #5816.

77. H. F. WEINBERGER

Sur les solutions fortes du problème de Tricomi, C. R. Acad. Sci. Paris **238** (1954), 1961–1962. MR **16**, 43.

78. WU JI-QIAN

Differential equations of mixed type, J. Sun Yat-sen Univ. Nat. Sci. Ed. **1959**, no. 4, 11–32. (Chinese; English summary) RŽ Mat. **1961** #12Б189.

79. WU XIN-MOU AND DING XIA-QI

Sur l'unicité du problème de Tricomi de l'équation de Chaplygin, Acta Math. Sinica **5** (1955), 393–399. (Chinese; French summary) MR **17**, 748.

80. V. I. ŽEGALOV

a) *A boundary value problem for a fourth-order equation of mixed type*, Izv. Vysš. Učebn. Zaved. Matematika **1960**, no. 4 (17), 73–78. (Russian) MR **27** #6046.

b) *A boundary value problem for a higher-order equation of mixed type*, Dokl. Akad. Nauk SSSR **136** (1961), 274–276 = Soviet Math. Dokl. **2** (1961), 45–47. MR **22** #8228.

c) *A boundary value problem for an equation of mixed type with boundary conditions on both characteristics and with discontinuities on the transition curve*, Kazan. Gos. Univ. Učen. Zap. **122** (1962), kn. 3, 3–16. (Russian) MR **30** #367.

d) *Some boundary-value problems for a system of second-order equations of mixed type*, Kazan. Gos. Univ. Učen. Zap. **122** (1962), kn. 3, 17–29. (Russian) MR **30** #368.

81. A. ZYGMUND

Trigonometrical series, PWN, Warsaw, 1935; reprint, Chelsea, New York, 1952. MR **17**, 844.